METHODS IN MOLECULAR BIOLOGY

Series Editor
John M. Walker
School of Life and Medical Sciences
University of Hertfordshire
Hatfield, Hertfordshire, UK

For further volumes:
http://www.springer.com/series/7651

For over 35 years, biological scientists have come to rely on the research protocols and methodologies in the critically acclaimed *Methods in Molecular Biology* series. The series was the first to introduce the step-by-step protocols approach that has become the standard in all biomedical protocol publishing. Each protocol is provided in readily-reproducible step-by-step fashion, opening with an introductory overview, a list of the materials and reagents needed to complete the experiment, and followed by a detailed procedure that is supported with a helpful notes section offering tips and tricks of the trade as well as troubleshooting advice. These hallmark features were introduced by series editor Dr. John Walker and constitute the key ingredient in each and every volume of the *Methods in Molecular Biology* series. Tested and trusted, comprehensive and reliable, all protocols from the series are indexed in PubMed.

Notch Signaling Research

Methods and Protocols

Edited by

Dongyu Jia

Department of Biology, Georgia Southern University, Statesboro, GA, USA

Humana Press

Editor
Dongyu Jia
Department of Biology
Georgia Southern University
Statesboro, GA, USA

ISSN 1064-3745 ISSN 1940-6029 (electronic)
Methods in Molecular Biology
ISBN 978-1-0716-2200-1 ISBN 978-1-0716-2201-8 (eBook)
https://doi.org/10.1007/978-1-0716-2201-8

Cover illustration: Image provided by the authors of Chapter 10.

This Humana imprint is published by the registered company Springer Science+Business Media, LLC, part of Springer Nature.
The registered company address is: 1 New York Plaza, New York, NY 10004, U.S.A.

Preface

Notch signaling is one of the most conserved signaling pathways across species, which has a wide range of functions in many different tissues in metazoan. The regulation of Notch can occur at different steps of Notch signaling transduction, including ligand and receptor production, ligand-receptor and ligand-ligand interaction at the membrane, endocytic trafficking, and transcriptional regulation. Dysregulation of Notch signaling frequently leads to various diseases, including cancers, highlighting the importance and necessity of Notch research studies.

This book aims to cover a large research area of Notch studies in different model organisms by offering a series of Notch signaling research methods and protocols. The collected research methods and protocols include the dissection of Notch functional sites, Notch regulators, Notch activity reporters and analysis, roles of Notch in development and diseases, wet and dry lab tools and studies, and bioinformatics analysis. This book particularly emphasizes on the exploration of Notch roles in development and diseases and acknowledges the importance of combining classic and novel molecular tools and methods, bioinformatic workflow and applications, and different model organisms to understand Notch.

I would like to thank all our expert authors for their valuable contributions to this volume. I am confident that this book will attract scientists around the world who are working in the Notch research fields. Many of the protocols could potentially be applied to study other signaling pathways, which will further attract broader audience in different research fields.

Statesboro, GA, USA *Dongyu Jia*

Contents

Contributors

SHAAD M. AHMAD • *Department of Biology, Indiana State University, Terre Haute, IN, USA; The Center for Genomic Advocacy, Indiana State University, Terre Haute, IN, USA; Rich and Robin Porter Cancer Research Center, Indiana State University, Terre Haute, IN, USA*

OLEG V. ANDREYENKOV • *Institute of Molecular and Cellular Biology SB RAS, Novosibirsk, Russia*

NATALYA G. ANDREYENKOVA • *Institute of Molecular and Cellular Biology SB RAS, Novosibirsk, Russia*

MERVE ASLAN • *Department of Radiology, Canary Center at Stanford for Cancer Early Detection, Stanford University, Stanford, CA, USA*

ASHLEY AVILA • *Department of Biology, Georgia Southern University, Statesboro, GA, USA*

ALEKSANDRA BABICHEVA • *Section of Physiology, Division of Pulmonary, Critical Care and Sleep Medicine, University of California, San Diego, La Jolla, CA, USA*

VICTORIANO BALADRÓN • *Biochemistry and Molecular Biology Branch, Albacete Medical School/CRIB/Biomedicine Unit, Department of Inorganic and Organic Chemistry and Biochemistry, University of Castilla-LaMancha/CSIC, Albacete, Spain*

MARIANA BATISTA • *Reproduction and Development Laboratory, CIISA - Centro de Investigação Interdisciplinar em Sanidade Animal, Faculdade de Medicina Veterinária, Universidade de Lisboa, Avenida da Universidade Técnica, Lisbon, Portugal; FMV-ULHT – Faculdade de Medicina Veterinária, Universidade Lusófona de Humanidades e Tecnologias, Lisbon, Portugal*

ELENA CANTERO-NAVARRO • *Cellular and Molecular Biology in Renal and Vascular Pathology Laboratory, Fundación Instituto de Investigación Sanitaria-Fundación Jiménez Díaz-Universidad Autónoma Madrid, Madrid, Spain; Red de Investigación Renal (REDINREN), Instituto de Salud Carlos III, Madrid, Spain*

HERBERT CHEN • *Department of Surgery, University of Alabama at Birmingham, Birmingham, AL, USA*

JIYUAN CHEN • *Section of Physiology, Division of Pulmonary, Critical Care and Sleep Medicine, University of California, San Diego, La Jolla, CA, USA*

SARAH L. CRITTENDEN • *Department of Biochemistry, University of Wisconsin-Madison, Madison, WI, USA*

SERGEY A. DEMAKOV • *Institute of Molecular and Cellular Biology SB RAS, Novosibirsk, Russia*

PATRÍCIA DINIZ • *Reproduction and Development Laboratory, CIISA - Centro de Investigação Interdisciplinar em Sanidade Animal, Faculdade de Medicina Veterinária, Universidade de Lisboa, Avenida da Universidade Técnica, Lisbon, Portugal*

FRANCESCA DIOMEDE • *Department of Innovative Technologies in Medicine and Dentistry, University "G. d'Annunzio" Chieti-Pescara, Chieti, Italy*

KOSTIANTYN DREVAL • *The Program in Cellular and Molecular Oncology, University of New Mexico Comprehensive Cancer Center, Albuquerque, NM, USA; Department of Internal Medicine, Division of Molecular Medicine, University of New Mexico Health Science Center, Albuquerque, NM, USA; Department of Molecular Biology and Biochemistry, Simon Fraser University, Burnaby, BC, Canada*

ANTÓNIO DUARTE • *Reproduction and Development Laboratory, CIISA - Centro de Investigação Interdisciplinar em Sanidade Animal, Faculdade de Medicina Veterinária, Universidade de Lisboa, Avenida da Universidade Técnica, Lisbon, Portugal*
JACOB EIDE • *Department of Otolaryngology – Head and Neck Surgery, Northwestern University Feinberg School of Medicine, Chicago, IL, USA*
HUA-YING FAN • *The Program in Cellular and Molecular Oncology, University of New Mexico Comprehensive Cancer Center, Albuquerque, NM, USA; Department of Internal Medicine, Division of Molecular Medicine, University of New Mexico Health Science Center, Albuquerque, NM, USA*
MARIOS GEORGIOU • *School of Life Sciences, University of Nottingham, Nottingham, UK*
RACHAEL GUENTER • *Department of Surgery, University of Alabama at Birmingham, Birmingham, AL, USA*
HIKARU HANDA • *Graduate School of Pharmaceutical Science, Chiba University, Chiba, Japan*
SUSUMU HOSOKAWA • *Section of Physiology, Division of Pulmonary, Critical Care and Sleep Medicine, University of California, San Diego, La Jolla, CA, USA; Department of Pediatrics, Tokyo Medical and Dental University, Tokyo, Japan*
EN-CHI HSU • *Department of Radiology, Canary Center at Stanford for Cancer Early Detection, Stanford University, Stanford, CA, USA*
MOTOYUKI ITOH • *Graduate School of Pharmaceutical Science, Chiba University, Chiba, Japan*
PRITESH P. JAIN • *Section of Physiology, Division of Pulmonary, Critical Care and Sleep Medicine, University of California, San Diego, La Jolla, CA, USA*
RENATA JASKULA-SZTUL • *Department of Surgery, University of Alabama at Birmingham, Birmingham, AL, USA*
DONGYU JIA • *Department of Biology, Georgia Southern University, Statesboro, GA, USA*
JUDITH KIMBLE • *Department of Biochemistry, University of Wisconsin-Madison, Madison, WI, USA*
ANDREW J. KUMP • *Department of Biology, Indiana State University, Terre Haute, IN, USA; The Center for Genomic Advocacy, Indiana State University, Terre Haute, IN, USA; Rich and Robin Porter Cancer Research Center, Indiana State University, Terre Haute, IN, USA*
ROBERT J. LAKE • *The Program in Cellular and Molecular Oncology, University of New Mexico Comprehensive Cancer Center, Albuquerque, NM, USA; Department of Internal Medicine, Division of Molecular Medicine, University of New Mexico Health Science Center, Albuquerque, NM, USA*
CAROLINA LAVOZ • *Division of Nephrology, School of Medicine, Universidad Austral, Valdivia, Chile*
CHANGHWAN LEE • *Department of Biological Sciences, University at Albany, State University of New York, Albany, NY, USA*
DAVID LI-KROEGER • *Jan and Dan Duncan Neurological Research Institute, Texas Children's Hospital, Houston, TX, USA; Department of Neurology, Baylor College of Medicine, Houston, TX, USA*
SHIQIN LIU • *Department of Radiology, Canary Center at Stanford for Cancer Early Detection, Stanford University, Stanford, CA, USA*
LUÍS LOPES-DA-COSTA • *Reproduction and Development Laboratory, CIISA - Centro de Investigação Interdisciplinar em Sanidade Animal, Faculdade de Medicina Veterinária, Universidade de Lisboa, Avenida da Universidade Técnica, Lisbon, Portugal*

TINA LYNCH • *Department of Biochemistry, University of Wisconsin-Madison, Madison, WI, USA*
CHENYU MA • *Department of Molecular Biochemistry, Nagoya University Graduate School of Medicine, Nagoya, Japan*
AYAKO MAKINO • *Division of Endocrinology and Metabolism, Department of Medicine, University of California, San Diego, La Jolla, CA, USA*
GUYA DILETTA MARCONI • *Department of Medical, Oral and Biotechnological Sciences, University "G. d'Annunzio" Chieti-Pescara, Chieti, Italy; Orofacial Development and Regeneration, Institute of Oral Biology, University of Zurich, Zurich, Switzerland*
LAURA MÁRQUEZ-EXPÓSITO • *Cellular and Molecular Biology in Renal and Vascular Pathology Laboratory, Fundación Instituto de Investigación Sanitaria-Fundación Jiménez Díaz-Universidad Autónoma Madrid, Madrid, Spain; Red de Investigación Renal (REDINREN), Instituto de Salud Carlos III, Madrid, Spain*
SERGIO MEZZANO • *Division of Nephrology, School of Medicine, Universidad Austral, Valdivia, Chile*
THIMIOS A. MITSIADIS • *Orofacial Development and Regeneration, Institute of Oral Biology, University of Zurich, Zurich, Switzerland*
TAKAMASA MIZOGUCHI • *Graduate School of Pharmaceutical Science, Chiba University, Chiba, Japan*
DONGQING MO • *Department of Entomology and MOA Key Lab of Pest Monitoring and Green Management, College of Plant Protection, China Agricultural University, Beijing, China*
ASHIM MUKHERJEE • *Department of Molecular and Human Genetics, Institute of Science, Banaras Hindu University, Varanasi, Uttar Pradesh, India*
MOUSUMI MUTSUDDI • *Department of Molecular and Human Genetics, Institute of Science, Banaras Hindu University, Varanasi, Uttar Pradesh, India*
USHA NAGARAJAN • *Department of Biochemistry, School of Interdisciplinary and Applied Sciences, Central University of Haryana, Mahendergarh, India*
MARÍA-LUISA NUEDA • *Biochemistry and Molecular Biology Branch, School of Pharmacy/CRIB/Biomedicine Unit, Department of Inorganic and Organic Chemistry and Biochemistry, University of Castilla-LaMancha/CSIC, Albacete, Spain*
SHUHEI OMARU • *Graduate School of Pharmaceutical Science, Chiba University, Chiba, Japan*
LILY PACULIS • *Department of Biology, Georgia Southern University, Statesboro, GA, USA*
MANOJ PANTA • *Department of Biology, Indiana State University, Terre Haute, IN, USA; The Center for Genomic Advocacy, Indiana State University, Terre Haute, IN, USA*
CRISTINA PORCHERI • *Orofacial Development and Regeneration, Institute of Oral Biology, University of Zurich, Zurich, Switzerland*
RAÚL R. RODRIGUES-DIEZ • *Cellular and Molecular Biology in Renal and Vascular Pathology Laboratory, Fundación Instituto de Investigación Sanitaria-Fundación Jiménez Díaz-Universidad Autónoma Madrid, Madrid, Spain; Red de Investigación Renal (REDINREN), Instituto de Salud Carlos III, Madrid, Spain*
J. BART ROSE • *Department of Surgery, University of Alabama at Birmingham, Birmingham, AL, USA*
MARTA RUIZ-ORTEGA • *Cellular and Molecular Biology in Renal and Vascular Pathology Laboratory, Fundación Instituto de Investigación Sanitaria-Fundación Jiménez Díaz-Universidad Autónoma Madrid, Madrid, Spain; Red de Investigación Renal (REDINREN), Instituto de Salud Carlos III, Madrid, Spain*

NALANI SACHAN • *Department of Cell Biology, NYU Grossman School of Medicine, New York, NY, USA*

JOSE L. SALAZAR • *Department of Molecular and Human Genetics, Baylor College of Medicine, Houston, TX, USA; Jan and Dan Duncan Neurological Research Institute, Texas Children's Hospital, Houston, TX, USA*

CORDULA SCHULZ • *Department of Cellular Biology, University of Georgia, Athens, GA, USA*

KRISTOPHER R. SCHWAB • *Department of Biology, Indiana State University, Terre Haute, IN, USA; The Center for Genomic Advocacy, Indiana State University, Terre Haute, IN, USA; Rich and Robin Porter Cancer Research Center, Indiana State University, Terre Haute, IN, USA*

VARTIKA SHARMA • *Department of Molecular and Human Genetics, Institute of Science, Banaras Hindu University, Varanasi, Uttar Pradesh, India*

JIE SHEN • *Department of Entomology and MOA Key Lab of Pest Monitoring and Green Management, College of Plant Protection, China Agricultural University, Beijing, China*

MICHELLE SHEN • *Department of Radiology, Canary Center at Stanford for Cancer Early Detection, Stanford University, Stanford, CA, USA*

ELISABETE SILVA • *Reproduction and Development Laboratory, CIISA - Centro de Investigação Interdisciplinar em Sanidade Animal, Faculdade de Medicina Veterinária, Universidade de Lisboa, Avenida da Universidade Técnica, Lisbon, Portugal*

TANYA STOYANOVA • *Department of Radiology, Canary Center at Stanford for Cancer Early Detection, Stanford University, Stanford, CA, USA*

HIDEYUKI TAKEUCHI • *Department of Molecular Biochemistry, Nagoya University Graduate School of Medicine, Nagoya, Japan; Institute for Glyco-Core Research (iGCORE), Nagoya University, Nagoya, Japan; Department of Biochemistry, School of Pharmaceutical Sciences, University of Shizuoka, Shizuoka, Japan*

ROXANA GONZALEZ TASCON • *Department of Biology, Georgia Southern University, Statesboro, GA, USA*

PATRICIA A. THISTLETHWAITE • *Division of Cardiothoracic Surgery, Department of Surgery, University of California, San Diego, La Jolla, CA, USA*

ANA TORRES • *Reproduction and Development Laboratory, CIISA - Centro de Investigação Interdisciplinar em Sanidade Animal, Faculdade de Medicina Veterinária, Universidade de Lisboa, Avenida da Universidade Técnica, Lisbon, Portugal*

ALEXANDRE TRINDADE • *Reproduction and Development Laboratory, CIISA - Centro de Investigação Interdisciplinar em Sanidade Animal, Faculdade de Medicina Veterinária, Universidade de Lisboa, Avenida da Universidade Técnica, Lisbon, Portugal; Centro de Investigação Interdisciplinar Egas Moniz (CiiEM), Instituto Universitário Egas Moniz, Egas Moniz - Cooperativa de Ensino Superior, Crl, Quinta da Granja, Monte da Caparica, Portugal*

ORIANA TRUBIANI • *Department of Innovative Technologies in Medicine and Dentistry, University "G. d'Annunzio" Chieti-Pescara, Chieti, Italy*

YOHEI TSUKAMOTO • *Department of Molecular Biochemistry, Nagoya University Graduate School of Medicine, Nagoya, Japan*

ELENA I. VOLKOVA • *Institute of Molecular and Cellular Biology SB RAS, Novosibirsk, Russia*

QIAN XIE • *Morphism Institute, Seattle, WA, USA*

QIUPING XU • *Morphism Institute, Seattle, WA, USA*

SHINYA YAMAMOTO • *Department of Molecular and Human Genetics, Baylor College of Medicine, Houston, TX, USA; Jan and Dan Duncan Neurological Research Institute, Texas Children's Hospital, Houston, TX, USA; Department of Neuroscience, Baylor College of Medicine, Houston, TX, USA; Development, Disease Models and Therapeutics Graduate Program, Baylor College of Medicine, Houston, TX, USA*

SHENG-AN YANG • *Department of Molecular and Human Genetics, Baylor College of Medicine, Houston, TX, USA; Jan and Dan Duncan Neurological Research Institute, Texas Children's Hospital, Houston, TX, USA*

JASON X. -J. YUAN • *Section of Physiology, Division of Pulmonary, Critical Care and Sleep Medicine, University of California, San Diego, La Jolla, CA, USA*

JUNZHENG ZHANG • *Department of Entomology and MOA Key Lab of Pest Monitoring and Green Management, College of Plant Protection, China Agricultural University, Beijing, China*

TENGTENG ZHAO • *Section of Physiology, Division of Pulmonary, Critical Care and Sleep Medicine, University of California, San Diego, La Jolla, CA, USA*

Chapter 1

Using the CRISPR/Cas9 System for Dissection of Functional Sites of the *Notch* Gene in *Drosophila melanogaster*

Oleg V. Andreyenkov, Elena I. Volkova, Natalya G. Andreyenkova, and Sergey A. Demakov

Abstract

The *Notch* gene is a key factor in the signaling cascade that allows communication between neighboring cells in many organisms, from worms and insects to humans. The relative simplicity of the *Notch* pathway in *Drosophila*, combined with a powerful set of molecular and cytogenetic methods, makes this model attractive for studying the fundamental principles of *Notch* regulation and functioning. Here, using the CRISPR/Cas9 system in combination with homologous recombination, for the first time at the level of the whole organism, we obtained a directed deletion of the 5′-regulatory region and the first exon of the *Notch* gene, which were replaced by the *attP* integration site of the ΦC31 phage. Based on this approach, we obtained and characterized new *Notch* mutations. Thus, a new powerful tool is provided for studying the genetic regulation of the *Notch* gene and the organization of chromatin at this locus.

Key words *Notch*, Directed mutagenesis, CRISPR/Cas9, *attP/attB*, Site-specific recombination

1 Introduction

The product of the *Notch* gene is a transmembrane receptor that participates in intercellular interactions and plays a fundamental role in such important processes as differentiation, proliferation, and cell death in many organisms. Homologs of this gene are known in almost all studied animals, from worms and insects to mammals, including humans. Recent studies have shown that disturbances in the *Notch* signaling pathway cause multiple defects in the organism development, and in humans they cause serious diseases. In this regard, the study of the *Notch* gene becomes especially relevant. In *Drosophila*, the pathways of cell signal transmission by the *Notch* receptor have been intensively studied for a long time; however, the regulation of the *Notch* gene expression and its relationship with the chromatin state has almost not been studied. Given the high conservatism of the *Notch* receptor, it can be

Dongyu Jia (ed.), *Notch Signaling Research: Methods and Protocols*,
Methods in Molecular Biology, vol. 2472, https://doi.org/10.1007/978-1-0716-2201-8_1,

expected that the data obtained using the *Drosophila* model will help to understand in more details the mechanisms of *Notch* functioning in more complex organisms, including humans.

The *Notch* (*N*) gene in the *Drosophila* genome is represented by a unique locus on the X chromosome. On the molecular map, the transcribed part of the gene occupies about 37 kb, from which mRNAs with a length of about 10 kb are formed [1]. Currently, in *Drosophila* a huge number of mutations at the *Notch* locus have been generated with X-rays, chemicals, or by transposable element insertion (http://www.flybase.org/reports/FBgn0004647). A significant part of them affects the protein-coding part of the gene and causes developmental disorders in the organs of neuro-ectodermal origin: eyes, wings, bristles, and gonads. A number of mutations in the introns of the gene that affect the development of the fly are also known [1, 2]. Recently, the role of one of the *Notch* introns in the spatial organization and in the regulation of the transcriptional activity of this locus has been demonstrated [3].

It is known that the regulatory region of the *Notch* gene of *Drosophila* has a complex organization and is located in the decompacted region of the X chromosome, which corresponds to the 3C6-C7 interchromomere (interband) on the polytene chromosomes of *Drosophila* salivary glands [4]. However, it should be noted that very few mutations have been characterized affecting the 5′ regulatory region of the gene. One of these mutations, $N^{fa\text{-}swb}$, a recessive mutation affecting eye development, is a deletion localized immediately upstream of the *Notch* promoter region and spans a DNA segment of 880 bp [1]. It was shown that the DNA fragment affected by the fa^{swb} deletion exhibits insulator properties. It protects reporter transgenes from the position effect and blocks enhancer–promoter interactions [5]. Later, several independent experiments on genome-wide chromatin conformation capture (Hi-C) from embryos and cell cultures of *Drosophila* showed that the *Notch* gene body is located within one of the topological associated domains (TADs), and the 5′-regulatory region of the gene matches exactly with the TAD boundary [3, 6–8]. These data indicate that the chromatin of the *Notch* gene promoter zone is in the "open" state and, apparently, contains sequences that perform barrier functions. Thus, the study of the regulatory zone of the *Notch* gene is important in the context of the study of chromatin domains.

The CRISPR/Cas9 system of RNA-guided nucleases has recently been used for the directed editing of the *Drosophila* genome. This effective method is based on the bacterial CRISPR/Cas9 system, adapted for using in eukaryotes. The main elements of the system are chimeric noncoding RNAs (single guide RNAs) and the Cas9 endonuclease. Specific recognition of genomic DNA occurs through a complementary interaction between the directionally edited part of sgRNA and the target sites on the genomic DNA. The resulting complex attracts the Cas9 endonuclease, which causes the formation of a double-stranded break in the target DNA site, which leads to a directed change in the DNA sequence in the studied region during the reparation process [9, 10]. In particular, this system has been successfully applied to obtain some mutations by homologous recombination, that is, replacing the target fragment with the *attP* site, through which the original or mutated fragment of the studied gene is subsequently reintegrated in the site-specific *attP/attB* integration system of the *Φ*C31 bacteriophage. The use of the CRISPR/Cas9 system allowed for increasing the probability of homologous recombination by several orders of magnitude [11]. We applied this approach to obtain targeted mutations and to study the regulatory zone of the *Notch* gene.

In this work, we describe the technique of directed genome editing for dissection of the 5′ region of the *Notch* gene in order to identify the sequences involved in the regulation of this gene activity. At the first stage, we applied the CRISPR/Cas9 system in combination with the homologous recombination method [11]. A founder fly line was created in which a DNA fragment of about 4 kb [**N-4k**] containing the promoter region, the first exon, and a part of the first intron of the *Notch* gene were replaced by the *attP* site. For this purpose, we synthesized the donor plasmid pGX-attP[3′-5′HA-N], which contained the sequences flanking the region of the putative deletion [**N-4k**] (Fig. 1a, b), the *mini-white* reporter gene under the control of the *hsp70* promoter and GMR enhancer surrounded by *loxP* sites for Cre-mediated recombination, and an *attP* site. The transformation was carried out in the line containing the Cas9 nuclease source with a mixture of plasmids encoding gRNAs, and the pGX-attP{5′-3′HA-N} donor plasmid. As a result, 120 flies were obtained out of 198 treated embryos. Among their 17,000 offspring, four red-eyed females were found, and two of them were sterile. Fertile females became the founders of stocks in which the reporter gene *mini-white* [w+] was linked to the X chromosome, and the red-eyed males did not survive. After removing the [w+] reporter flanked by the *loxP* sites using the Cre recombinase, the founder stock dN[w−]/FM0 was obtained, in which about 4 kb from the most distal part of the *Notch* gene, and 2 kb up- and downstream from the gene transcription start site (TSS), were replaced by the *attP* site (Fig. 1c, d).

Then, using the *attP/attB* integration system of the *Φ*C31 phage and DNA of the pGE-attB-GMR [N-4k] vector, we restored the deleted DNA fragment, which led to "rescue"—restoration of the viability of flies containing the founder deletion (Fig. 1e). Thus, as a result of our experiments on the basis of the pGE-attB-GMR [N-4k] vector, there appeared a good opportunity to create the desired mutations affecting possible functional regions of the 5′-regulatory area of the gene and to study the consequences of these disorders.

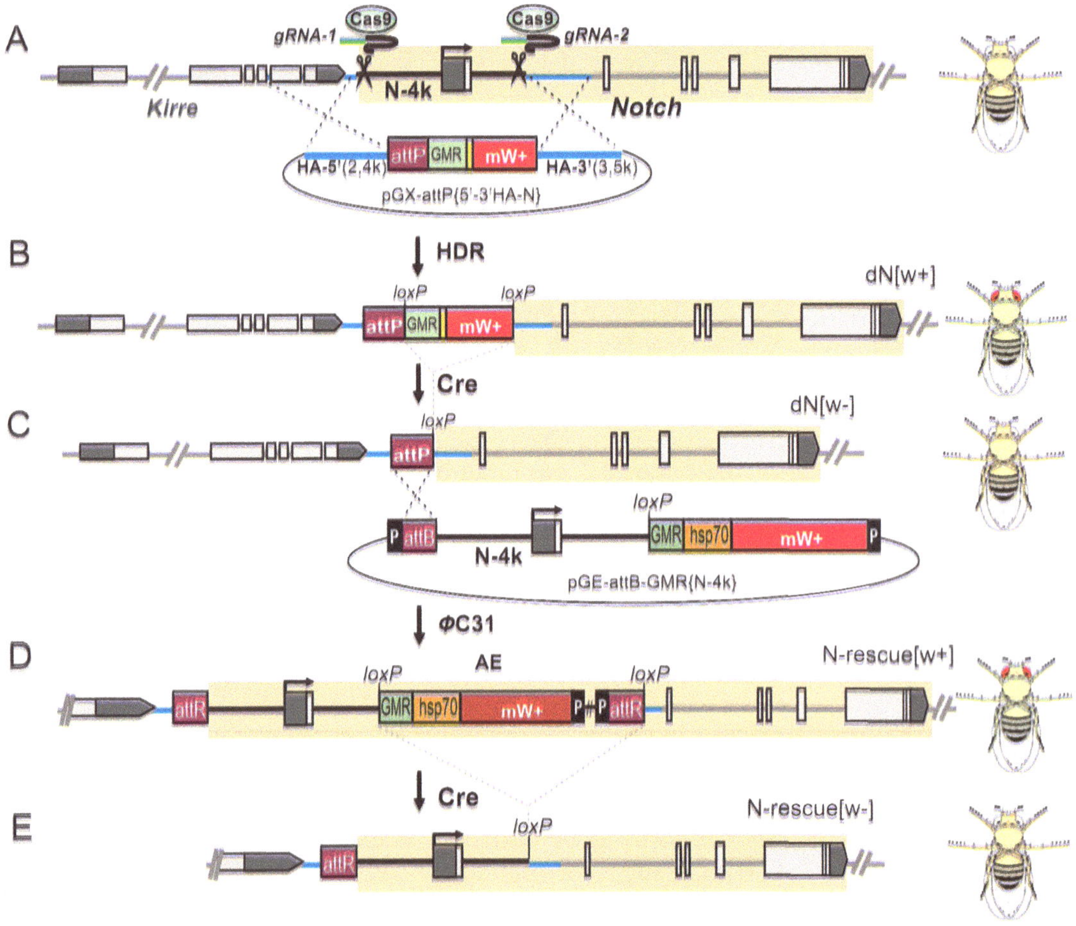

Fig. 1 CRISPR/Cas9-induced homologous recombination (HR) with donor vector pGX-attP {3′–5′HA-N} at the *Notch* locus. (**a**) The scheme of the *Notch* gene organization (shaded) and HR strategy used to replace the 4 kb fragment of the 5′-region of the *Notch* gene (N-4k) to the *attP* insertion site from the *Φ*C31 phage integration system. Target sites for gRNAs/Cas9 binding are marked with "scissors". HA—5′(2,4 k) and HA-3′(3,5k)—homologous DNA regions (arms) of 2.4 kb and 3.5 kb in length, respectively, adjacent to the cleavage sites for Cas9, were cloned into the pGX-attP vector, which contains *attP* site for subsequent integrations of genetically modified variants of the 5′ region of the gene, and the *mini-white* gene under control of the *hsp70* gene promoter and GMR enhancer, which enhances the activity of this gene in the eye, as well as the *loxP* recombination sites flanking this gene necessary for its removing. (**b**) The pGX-attP {5′–3′ HA-N} donor and plasmid DNA encoding gRNAs were injected into vasa-Cas9 embryos to obtain the founder line, dN[w+], resulting from homology directed reparation and recombination. (**c**) After the removal of the *mini-white* gene by Cre recombinase, a founder line dN[w−] carrying only the *attP* site was created. (**d**) The full 5′-region fragment of the *Notch* gene (**N-4k**) was cloned into the pGE-attB-GMR {N-4k} vector and integrated into the deletion region of the founder line at the *attP*/*attB* sites with the formation of the *attR* recombination sites. The integration restores the deleted region to its original genetic state (N-rescue[w+] fly line), but it adds the *mini-white* marker gene and other auxiliary elements (AE) of the vector to the *Notch* gene intron. (**e**) Auxiliary vector DNA sequences along with *mini-white* [w+] were removed by Cre recombinase to obtain a final constructed target region flanked by *attR* and *loxP* sites (N-rescue [w−])

2 Materials

2.1 Chemicals and Reagents

1. Agarose LE.
2. Bacteriological agar for molecular biology.
3. Tryptone bacteriological grade.
4. Yeast extract for microbiology.
5. Glycerol for molecular biology.
6. Sodium chloride (ACS grade).
7. Potassium acetate (ACS grade).
8. Lithium chloride (ACS grade).
9. EDTA (ethylenediaminetetraacetic acid) disodium salt (ACS grade).
10. Tris base (ACS grade).
11. Oil for *Drosophila* embryo transformation. Voltalef® 10S or Halocarbon® 200 Oil.
12. High Fidelity PCR Master Kit, Thermo Fisher Scientific.
13. dNTP mix (10×, 2 mM each) and dATP (2 mM).
14. T4 DNA Ligase with ligase buffer.
15. BpiI (BbsI) restriction endonuclease.
16. FastDigest™ pack (Restriction enzymes), Thermo Fisher Scientific.
17. Plasmid Midi Kit 100, QIAGEN.
18. Sequencing kits and reagents.
19. LB medium (1×) : tryptone (bacteriological grade)—10 g, yeast extract (for microbiology)—5 g, sodium chloride—10 g per liter of distilled water and autoclave. LB plates with 1.5% Bacto—agar and ampicillin (100 μg/mL).

2.2 Supplies

1. Filamented glass capillary tubes: OD = 1.2 mm, $L = 100$ mm, with filaments.
2. Microcon™—30 Centrifugal Filter Devices, Merck.
3. Disposable tissue grinder pestles.
4. Pipette tips, 0.5–10 μL, 200 μL, and 1000 μL.
5. Microcentrifuge tubes 0.6 mL and 1.5 mL.
6. PCR tubes 0.2 mL.
7. Petri dishes.

2.3 Vectors

1. cos163A10 (The 163A10 clone from the cosmid library [Lorist 6] of the *D. melanogaster* genome was kindly provided by Dr. Inga Siden-Kiamos [Greece]).

2. pGX-attP (Drosophila Genomics Resource Center [NIH Grant 2P40OD010949] Stock Number:#1293) [12].
3. pGE-attB-GMR (Drosophila Genomics Resource Center [NIH Grant 2P40OD010949] Stock Number:#1295) [12].
4. pGEM®-T Easy Vector Systems (Promega).
5. pU6-BbsI-chiRNA (Addgene) [9].

2.4 Strain Resources

Bacterial strains

1. XL1-Blue (Genotype: recA1 endA1 gyrA96 thi-1 hsdR17 supE44 relA1 lac [F′ proAB lacI^{q}ZΔM15 Tn10 (Tet^{r})]), Stratagene.

Drosophila strains

1. w^{1118}; PBac{vas-Cas9,U6-tracrRNA}VK00027 (Bloomington Drosophila Stock Center(BDSC), Indiana University, stock #51325).
2. yw; nanos-Cre [13].
3. w^{1118}sov^{ML150}/FM0 (balancer strain, BDSC stock #4591).

2.5 Apparatus

1. Laboratory microscope.
2. Stereo microscope (Zeiss "Stemi 2000" or equivalent instrument).
3. Microinjector (Applied Scientific Instrumentation, MPPI-2, or equivalent instrument).
4. 4-axis micromanipulator (Siskiyou MX130R or equivalent instrument).
5. Micropipet puller.
6. Electroporator (Bio-Rad Gene Pulser XCell or equivalent instrument).
7. Thermal Cycler.
8. Microthermostat for microtubes.
9. Horizontal electrophoretic system with power supply.
10. Gel documentation system (Syngene InGenius 3 or equivalent instrument).
11. Water purification system (MERK Milli-Q or equivalent instrument).
12. Centrifuge (Eppendorf 5424 or an equivalent instrument).
13. Centrifuge/vortex (BioSan microspin FV-2400 or an equivalent instrument).
14. Freezers (−20 °C, −80 °C).
15. Micropipette set (Gilson Pipetman or an equivalent instrument).

3 Methods

3.1 Creation of the Drosophila Founder Line with Replacement of the Notch Promoter Zone at the attP Site

3.1.1 Selection of the Cas9 Source Fly Strain

The line w^{1118}; PBac{vas-Cas9,U6-tracrRNA}VK00027 with the insertion of Cas9 source in the third chromosome is chosen. Substitution of the *Notch* gene sequence with an *attP* site occurred on the X chromosome of this line. The location of the Cas9 source in a linkage group other than X makes it possible to further eliminate Cas9 from the resulting founder line by genetic methods through a series of crosses.

3.1.2 Generation of the Donor Construct

To replace the sequence from the *Notch* gene with *attP* site, the donor construct pGX-attP{5′–3′HA-N} is created (Fig. 1a).

1. The 5′ and 3′ homology arms with lengths of 2381 and 3510 bp (5HA and 3HA, respectively) are generated by PCR using the cos163A10 cosmid vector as a DNA matrix and the following primers (hereinafter, the orientation of primers is 5′–3′):

   5HA_f GCCATTAGTGATGATAGTACATG.

   5HA_r CCGCTGCGAAACTCATCGA.

   3HA_f CATATGGAGCGGTCGTTGTCTA.

   3HA_r GCCCAAAAGCCTGGAAGACA.

2. PCR is performed at the following conditions: *Mastermix*: 10× PCR Buffer—2.5 μL, dNTP mix (10×)—2.5 μL, Forward primer (10 μM)—2.5 μL, Reverse primer (10 μM)—2.5 μL, Enzyme (5 U/μL)—0.5 μL, DNA (cos163A10)—1.0 μL (1 ng), H_2O—14 μL; *Program*: (1) 96 °C, 3 min; (2) 96 °C, 20 s; (3) 62 °C, 20 s; (4) 72 °C, 1 min 30 s; (5) Repeat **steps 2–4** 30 times. (6) 72 °C, 5 min; (7) 4 °C, holding temperature.
3. The amplified DNA fragments are purified on the Microcon columns according to manufacturer's recommendations.
4. Cloning procedure is performed in two stages. At the first stage, PCR with modified primers is carried out using the purified amplicons made in **step 3**. For this purpose, primer pairs with restriction sites NotI and KpnI for 5HA and AscI and StuI for 3HA are used (corresponding restriction sites are underlined):

   5HA(Not)_f GCGGCCGCCATTAGTGATGATAGTACATG.

   5HA(KpnI)_r GGTACCATATGGAGCGGTCGTTGTCTA.

   3HA(Asc)_f GGCGCGCCATATGGAGCGGTCGTTGTC.

   3HA(Stu)_r GCCCAAAGGCCTGGAAGACA.

5. PCR is performed at the following conditions: *PCR Mastermix*: High Fidelity Buffer(10×)—2.5 μL, dNTP mix (2 mM each)—2.5 μL, Forward primer (10 μM)—2.5 μL, Reverse

primer (10 μM)—2.5 μL, High Fidelity Polymerase—0.5 μL, DNA (20 pg/μL)—5 μL, Water—9 μL. *Program*: (1) 96 °C, 3 min; (2) 96 °C, 20 s; (3) 64 °C, 20 s; (4) 72 °C, 1 min 30 s; (5) Repeat **steps 2–4** 30 times; (6) 72 °C, 5 min; (7) 4 °C (holding temperature).

6. PCR fragments are purified similar to **step 3** followed by A-ends extension using Taq polymerase: 10× PCR Buffer—1.5 μL, dATP—1.5 μL, Taq Polymerase—0.5 μL, DNA sample—5.0 μL, Water—6.5 μL, 20 min at 72 °C. Then, the fragments are cloned into the pGEM®-T Easy vector according to the manufacturer's protocol (*see* **Note 1**).
7. The cloned fragments of homology arms are cut out of the vector at the synthesized restriction sites and sequentially integrated into the pGX-attP vector digested at the same restriction sites. As a result, pGX-attP {5′–3′ HA-N} donor construct containing 5′ and 3′ homology arms, the *mini-white* reporter gene under control of *hsp70* gene promoter and GMR enhancer, surrounded by *loxP* sites for Cre-mediated recombination, and an *attP* site, is obtained (Fig. 1).

3.1.3 Generation of the Notch Targeting Guide RNAs (gRNAs)

1. To identify highly specific target CRISPR sites in the *Notch* locus, we used the online resource flyCRISPR Optimal Target Finder tool [11].
2. The following guide RNAs are chosen for the *Notch* gene.

   N-gRNA1_f CTTCGAGCTCGATGAGTTTCGCAG.

   N-gRNA1_r AAACCTGCGAAACTCATCGAGCTC.

   N-gRNA2_f CTTCGGAGAGTTCGCTTCCAAAGC.

   N-gRNA2_r AAACGCTTTGGAAGCGAACTCTCC.

   The oligonucleotides from each pair (sense and antisense) are supplemented with 4-nucleotide motifs (underlined) to form "sticky" ends for ligation into the pU6-BbsI-chiRNA vector (*see* **Note 2**).
3. Annealing of the oligonucleotides (100 μM of each in water) is performed at the following conditions: oligo 1——1 μL, oligo 2——1 μL, T4 ligation buffer (10×)—1 μL, water—7 μL; thermocycler program—95 °C for 5 min, then ramp to 25 °C at a rate of—0.1 °C/s.
4. The pU6-BbsI-chiRNA vector is treated with the restriction endonuclease BbsI (BpiI) and followed inactivation of the enzyme at 65 °C for 20 min.
5. Ligation of the annealed oligonucleotides with the pU6-BbsI-chiRNA vector is performed according to the protocol: Linearized DNA of the pU6-BbsI-chiRNA vector—1 μL (100 ng),

annealed oligonucleotides—1 μL, 10× quick ligase buffer (NEB)—1 μL, T4 DNA ligase (NEB)—1 μL, H_2O—7 μL. The ligase mixture is incubated at room temperature for an hour and purified (*see* **Note 3**).

3.1.4 Bacterial Transformation and Screening

Electroporation is done routinely [14]. A bacterial strain of *E. coli* with a reduced ability for genetic recombination is used. Electroporation is performed with standard instrument settings for prokaryotic cells. Then the cells are spliced on LB agar medium with ampicillin in petri dishes and incubated overnight at 37 °C. Screening of the colonies containing a vector with a target DNA fragment is carried out using PCR with oligonucleotides, one of which is a vector primer (in the case of pGEM®-T Easy Vector, these are T7 or SP6 primers), and the second primer is homologous to the target DNA fragment (*see* **Note 4**). Cells from the target-positive colonies selected according to the PCR results are grown in separate tubes containing 3 mL of LB medium with ampicillin. Isolation of vector DNA from an overnight culture of *E. coli* is carried out by alkaline lysis using a QIAGEN plasmid midi kit.

3.1.5 Genetic Transformation of Drosophila

Since the development of the first successful practical approach to transformation of *Drosophila* [15], many working protocols have been proposed, differing in various details. Differences may relate to the methods of removing the chorionic membrane in embryos (mechanical, chemical) and even the very need for their dechorionization, as well as the use of capillaries for the introduction of a DNA solution under pressure or the combination of glass needles with an emulsion of DNA in oil without using a pressure [16]. Smaller differences may relate to the use of specific brands of oils, capillaries, injectors, and micromanipulators. As a basic protocol, we can recommend the Gompel lab protocol [17], where general recommendations for transformation are well described.

1. For injection into *Drosophila* embryos, a DNA mixture of pGX-attP {5′–3′ HA-N} and both sgRNA source vectors (pU6-N-gRNA1 and pU6-N-gRNA2) in water (mQ) is prepared. The final concentration of sgRNA sources is 100 ng/μL each, and 300 ng/μL for the donor construct (*see* **Note 5**). The line w^{1118}; PBac{vas-Cas9,U6-tracrRNA} VK00027 with an endogenous source of Cas9 is used as a donor.
2. Flies should be kept in large boxes for effective egg-laying (*see* **Note 6**).
3. The embryos are collected from the surface of fly food with a brush and transferred onto a piece of double-sided sticky tape (we recommend using the 3M brand tape), placed on a glass slide.

4. Then, under a binocular microscope, the chorionic sheath is removed from the embryos with dissecting needles, and the embryos are arranged in one line in the same orientation on the surface of the strip of freshly prepared fly food (*see* **Note 7**).
5. When 40–80 embryos are collected, they are transferred to a strip of double-sided tape, glued to the slide. For this it is convenient to use precut squares of microscopic slides.
6. From the moment the dechorionized embryos are transferred to the tape, they began to dry. Drying time depends primarily on the humidity in the room, and at the temperature of 20–23 °C it usually takes from 6 to 10 min (*see* **Note 8**).
7. After drying the embryos, a few drops of oil, which is a polymer of chlorotrifluoroethylene, are added to an adhesive tape with fixed embryos (we have successfully used the oils Halocarbon 200 or Voltalef 10S).
8. The injection of DNA solutions is carried out under a microscope using a micromanipulator and an injector. We used the injectors Applied Scientific Instrumentation MPPI-2. Capillaries must be provided with filaments. We used capillaries with a diameter of 1.0 mm or 1.2 mm produced by the Institute for Biological Instrumentation of the Russian Academy of Sciences (IBI RAS) and a device for micropipette stretching (puller) from the same manufacturer. The device settings should be selected empirically. The resulting micropipette should have a sufficiently pronounced taper to maintain rigidity at the moment of puncturing the embryo and at the same time it should have a sufficiently thin tip so as not to cause too much damage. On the other hand, if the outlet of the capillary is too small, the risk of clogging of the micropipettes increases. For more details on the influence of needle parameters on injection, *see* [18].
9. The slides with injected embryos are placed in a humid chamber until the larvae hatched. The larvae released into the oil are collected, using a thinnest brush, and transferred in the tubes with nutrient medium (*see* **Note 9**).
10. The hatched G0 flies interbred with the flies $w^{1118}sov^{ML150}$/FM0, carrying a balancer X chromosome. Among flies of the next generation (G1), a search is carried out for individuals with red-eye phenotype [w+], corresponding to the presence of the inserted reporter *mini-white* gene. Fertile females became the founders of stocks in which the reporter gene [w +] is linked to the X chromosome, and the red-eyed males did not survive. For further work, we took those fly lines in which the presence of the target deletion and the attP site is confirmed by sequencing. After removing the [*w*+] reporter flanked by the *loxP* sites using Cre recombinase, the founder

stock with a genotype y w dN[w−]/FM0 is obtained, in which 4 kb from the 5′ region of the *Notch* gene are replaced only by the *attP* site.

3.2 Using the Founder Line for Notch Full-Sequence Restoration

The obtained founder line with an *attP* site and a deletion of the 5′ end of the *Notch* gene can be used for analysis of this gene by *attP/attB*—dependent genomic transformation. Some difficulties are caused by the fact that the deletion in this founder line is lethal. As a result, the founder fly line is maintained using the balancer X chromosome. In the founder line, the males lack the insertion of the *attP* site, and the females are heterozygous for *attP*. As a result, the efficiency of the founder line transformation is significantly reduced. Nevertheless, to date, the system remains the most efficient molecular and genetic tool for targeted mutagenesis of the *Notch* gene in *Drosophila*.

As an example of using the obtained *attP* site in the *Notch* gene, consider the restoration of a deleted 4 kb fragment in the founder line to obtain a rescue line of flies. The steps of the procedure for generating PCR fragments and cloning them into the vectors are similar to those described in details in Subheading 3.1.2. So, here we outline only the main stages of the work.

1. The full DNA fragment from the 5′-region of the *Notch* gene, [N-4k], is obtained by PCR on the DNA matrix of the cosmid vector cos163A10 using the following primers:

   N4k_f: CGCAGCGGCAAATTATATC.

   N4k_r: CGACCGCTCCATATGCAAATAC.

   The PCR fragment should be purified using microconcentrators.
2. Using the purified DNA fragment, PCR with the following modified primers is carried out.

   N4k(Not)_f: <u>GCGGCCGC</u>AGCGGCAAATTATATC.

   N4k(Asc)_r: <u>GGCGCGCC</u>GCTCCATATGCAAATAC.
3. The resulting PCR fragment and pGE-attB-GMR vector are digested with NotI and AscI endonucleases and ligated.
4. Transformation of the ligated mixture into bacterial cells and screening of colonies with target construct are carried out similar to those described in Subheading 3.1.4.
5. After isolation and purification of target construct DNA a germline transformation of founder line embryos is performed as described in Subheading 3.1.5.

The integration restores the deleted region to its original genetic state, but it adds the *mini-white* marker gene and other vector auxiliary DNA sequences to the *Notch* gene intron. Auxiliary elements along with *mini-white* are removed by Cre recombinase to obtain a final constructed target region flanked by *attR* and *loxP* sites (Fig. 1d, e). The 4 kb fragment of the construct rescued the lethality of the founder deletion and therefore contained the functional regulatory sequences of *Notch.*

4 Notes

1. Addition of A mononucleotides to the ends of DNA fragments using Taq-polymerase increases the efficiency of their ligation into the pGEM®-T Easy vector.
2. Oligonucleotides should be 5′ phosphorylated. On the physical map the edges of the homology arms should be situated in close proximity to the positions of the N-gRNAs. Annealing of the oligonucleotides and subsequent steps for cloning are carried out by protocol "Generating targeting chiRNAs" (http://flycrispr.molbio.wisc.edu).
3. Before transfer of nucleic acids into bacterial cells by electroporation, it is necessary to purify the ligase mixtures from salts. There are many ways to do this, from simple dilution of the ligase mixture with water or DNA precipitation with ethanol to using commercial purification kits. We used the Microcon™—30 Centrifugal Filters, Ultracel™ YM30 (Merck Millipore LTD). This kit eliminates the risk of sample loss during purification and allows you to use the entire amount of ligated material for transformation, which significantly increases the yield of colonies carrying the target vector.
4. To increase the efficiency of screening, we combined cells from 5 to 10 colonies into one group, and then the material of several such groups is used in PCR with the appropriate pairs of primers. If expected PCR product size corresponding to the target vector is detected in any of the groups, screening of individual colonies from this group is carried out.
5. A DNA mixture with concentration of more than 500 ng/μL increases the viscosity of the solution and complicates the microinjection procedure.
6. We used cylindrical transparent plastic cages with a volume of about 500 mL containing a metal or nylon grid (a cell size of no more than 400 μm) at one of the edges, which provided air access into the cage. At the other edge, removable dishes (e.g., plastic petri dishes of a suitable diameter) with poured substratum are attached. A yeast suspension applied to the substratum in the center of the dish increases the yield of embryos and allows achieving their synchronization. This is achieved by periodically changing the substratum dishes every hour,

starting from the moment 2 h before the work and during the work. It is necessary to achieve egg-laying productivity of about 100–200 per hour.

7. Successful transformation is possible only up to the second stage of early embryonic development [19], since at the third stage the establishment of polar cells has already occurred and the entry of vector DNA into the cells of the embryonic pathway has become impossible. It is necessary to control the stages of development of embryos at all transformation steps. Late embryos should be removed at the first steps of the experiment, since the scope of work directly depends on this.
8. The goal of this step is to reduce the embryo turgor. Without this step the injection will be accompanied by the outflow of syncytial fluid and increased embryo death. Overdrying the embryos also reduces their viability, so you should strive to achieve optimal drying conditions.
9. We recommend to loosen the surface of fly food with a preparation needle in order to facilitate access to food for the larvae, weakened in the experiment.

Acknowledgments

This work was supported by the Russian Science Foundation project #20-14-00074. The authors gratefully acknowledge the resources provided by the "Molecular and Cellular Biology" core facility of the IMCB SB RAS supported by the fundamental scientific research program on the project FWGZ-2021-0014. We thank D.S. Sidorenko at IMCB SB RAS for helpful discussion of this work. We thank the Bloomington Drosophila Stock Center for their kind provision of the fly stocks.

References

1. Ramos RG, Grimwade BG, Wharton KA, Scottgale TN, Artavanis-Tsakonas S (1989) Physical and functional definition of the *Drosophila* Notch locus by P element transformation. Genetics 123:337–348
2. Markopoulou K, Welshons WJ, Artavanis-Tsakonas S (1989) Phenotypic and molecular analysis of the facets, a group of intronic mutations at the Notch locus of *Drosophila melanogaster* which affect postembryonic development. Genetics 122:417–428
3. Arzate-Mejía RG, Cerecedo-Castillo AJ, Guerrero G, Furlan-Magaril M, Recillas-Targa F (2020) In situ dissection of domain boundaries affect genome topology and gene transcription in *Drosophila*. Nat Commun 11: 894
4. Rykowski MC, Parmelee SJ, Agard DA, Sedat JW (1988) Precise determination of the molecular limits of a polytene chromosome band: regulatory sequences for the Notch gene are in the interband. Cell 54:461–472
5. Vazquez J, Schedl P (2000) Deletion of an insulator element by the mutation facet-strawberry in *Drosophila melanogaster*. Genetics 155:1297–1311
6. Hou C, Li L, Qin ZS, Corces VG (2012) Gene density, transcription, and insulators contribute

to the partition of the *Drosophila* genome into physical domains. Mol Cell 48:471–484

7. Sexton T, Yaffe E, Kenigsberg E, Bantignies F, Leblanc B, Hoichman M, Parrinello H, Tanay A, Cavalli G (2012) Three-dimensional folding and functional organization principles of the *Drosophila* genome. Cell 148:458–472
8. Stadler MR, Haines JE, Eisen MB (2017) Convergence of topological domain boundaries, insulators, and polytene interbands revealed by high-resolution mapping of chromatin contacts in the early *Drosophila melanogaster* embryo. eLife 6:e29550
9. Gratz SJ, Cummings AM, Nguyen JN, Hamm DC, Donohue LK, Harrison MM, Wildonger J, O'Connor-Giles KM (2013) Genome engineering of *Drosophila* with the CRISPR RNA-guided Cas9 nuclease. Genetics 194:1029–1035
10. Hsu PD, Lander ES, Zhang F (2014) Development and applications of CRISPR-Cas9 for genome engineering. Cell 157:1262–1278
11. Gratz SJ, Ukken FP, Rubinstein CD, Thiede G, Donohue LK, Cummings AM, O'Connor-Giles KM (2014) Highly specific and efficient CRISPR/Cas9-catalyzed homology-directed repair in *Drosophila*. Genetics 196:961–971
12. Huang J, Zhou W, Watson AM, Jan YN, Hong Y (2008) Efficient ends-out gene targeting in *Drosophila*. Genetics 180:703–707
13. Laktionov PP, White-Cooper H, Maksimov DA, Belyakin SN (2014) Transcription factor COMR acts as a direct activator in the genetic program controlling spermatogenesis in *D. melanogaster*. Mol Biol 48:153–165
14. Dower WJ, Miller JF, Ragsdale CW (1988) High efficiency transformation of *E. coli* by high voltage electroporation. Nucleic Acids Res 16:6127–6145
15. Rubin GM, Spradling AC (1982) Genetic transformation of *Drosophila* with transposable element vectors. Science 218:348–353
16. Shilova IE, Omel'yanchuk LV (2007) A method for transformation of *Drosophila* germline cells with a high-concentration exogenous DNA. Russ J Genet 43:80–83
17. Gompel N, Shröder EA (2015) *Drosophila* germline transformation protocol (www.gompel.org/methods)
18. Miller DFB, Holtzman SL, Kaufman TC (2002) Customized microinjection glass capillary needles for P-element transformations in *Drosophila melanogaster*. Biotechniques 33:366–375
19. Campos-Ortega JA, Hartenstein V (1997) The embryonic development of *Drosophila melanogaster*. Springer, Berlin, p 405

Chapter 2

Screening Mutants by Single Fly Genomic PCR

Usha Nagarajan and Marios Georgiou

Abstract

Imprecise P-element excision or FRT-mediated recombination is routinely performed to mutagenize a gene of interest. It is, however, tedious to maintain all independent and individual excised mutant fly lines before the presence of a mutation is confirmed. Here, we provide a method to detect and confirm the presence of a mutation, as and when mutant flies are generated. By allowing for the maintenance and expansion of only the confirmed mutant lines, this protocol will help to save time, money, and space.

Key words Genetic screening, Mutagenesis, Genomic PCR, Single fly PCR, Mutant, Drosophila

1 Introduction

Drosophila melanogaster is one of the most commonly utilized and popular models to characterize genes of unknown function [1, 2]. Loss-of-function (LoF) alleles are important resources that help to characterize and understand unknown gene function. Loss of gene function can be studied through the generation of null or hypomorphic mutant alleles, or by knocking down gene expression by RNA interference [3]. Many gene functions are still uncharacterized for the want of loss-of-function alleles. Imprecise excision of P-elements or precise recombination of homologous regions in the genome can be employed to generate mutants; however, screening for mutants is a laborious and cumbersome process.

Our laboratory is interested in the study of the ligand-dependent Notch signaling pathway. We carried out a genetic screen to identify the molecular players that interact with and regulate this signaling pathway (unpublished). As many of the novel interactors are uncharacterized, we set out to generate mutants. We set out to train undergraduate students with the aim of saving both time and resources, and we standardized and devised a simple protocol, with this aim in mind. If a large number of mutants are generated, maintaining them is a huge burden (e.g.,

Dongyu Jia (ed.), *Notch Signaling Research: Methods and Protocols*,
Methods in Molecular Biology, vol. 2472, https://doi.org/10.1007/978-1-0716-2201-8_2,

preparing fly food and performing regular fly husbandry). Instead, this procedure gives quick results as and when the mutants are generated, and it helps to reduce the number of stocks that need to be maintained. As this method involves both fly genetics and molecular biology, students are highly motivated when using this method.

2 Materials

2.1 P-Element Excision to Generate Mutant Alleles

1. Control Stocks: W^{1118} (Bloomington Stock #3605/5905/6326)
2. Fly stocks with a P-element inserted in the gene of interest (e.g., Bloomington Stock #30151)
3. Balancer fly stocks.
   (a) Double Balancer—w[1118]; wg[Sp-1]/CyO; MKRS/TM6B, Tb[1] (Bloomington Stock #76357)
   (b) Linked Balancer—T(2;3)SM6a-TM6B, Tb[1] (Bloomington Stock #5687).
4. Transposase stocks:
   (a) w[*]; wg[Sp-1]/CyO; ry[506] Sb[1] P{ry[+t7.2]=Delta2–3}99B/TM6B, Tb[+] (Bloomington Stock #3629)
   (b) y[1] w[1]; Ki[1] P{ry[+t7.2]=Delta2–3}99B (Bloomington Stock #4368).
5. Recombining FRT-stocks flanking the gene of interest [4, 5].

2.2 Single Fly Genomic DNA Extraction and PCR

2.2.1 Materials Required for Method 1

1. Prepare 50 ml of 2× Cracking Buffer by adding 1 ml of 10 N NaOH, 2.5 ml of 10% SDS, and 10 g sucrose in 20 ml of sterile distilled water and make up the volume to 50 ml of sterile distilled water.
2. Optional: Proteinase K 10 mg/ml
3. Bromophenol Blue or 6× BPB dye.

2.2.2 Materials Required for Method 2

1. Prepare 50 ml of Fly Genomic DNA extraction Buffer containing (10 mM Tris–HCl [pH 8.2], 1 mM EDTA, and 25 mM NaCl) by adding 500 µl of 1 M Tris–HCl (pH 8.2), 100 µl of 0.5 M EDTA, and 250 µl of 5 M NaCl in 20 ml of sterile distilled water and make up the volume to 50 ml using sterile distilled water.
2. Proteinase K 10 mg/ml.
3. 4 M or 2.5 M KCl.

2.3 PCR and Agarose Gel Electrophoresis

1. Agarose.
2. Bromophenol Blue (BPB)/6× loading dye.
3. Forward and reverse primers.
4. DNA PCR buffer.
5. Taq polymerase.
6. dNTP mixture,
7. Distilled water.

2.4 Apparatus/Equipment Required

1. Microcentrifuge tubes and PCR tubes.
2. Centrifuge.
3. Water bath.
4. Incubator.
5. PCR machine.
6. Agarose gel electrophoresis apparatus.

3 Methods

3.1 P-Element Excision to Generate Mutant Lines

1. Day 1: (1) Set up a cross between 20 virgin flies containing P-elements inserted within a gene of interest (Bloomington Stock #30151) For example, we have used the stock w^{1118}; P {EP}mgr^{G5308}/TM6C, Sb^1 (Bloomington Stock #30151) inserted in the gene *merry-go-round (mgr*, CG6719) and 10 male flies containing transposase (Bloomington Stock #4368) in a medium-sized bottle. (2) Expand double balancer lines: Sp/Cyo; MKRS/TM6B (Bloomington Stock #76357) or Sp/Cyo-TM6B (this can be generated by combining the linked balancer T(2;3)SM6a-TM6B [Bloomington Stock #76357] together with the double balancer line Sp/Cyo; MKRS/TM6B [Bloomington Stock #76357]) in preparation for **step 3**.
2. Day 3: Transfer the crossed adult flies to a fresh vial/bottle to generate a duplicate of the cross.
3. Day 14: (1) Collect all mosaic eyed males that eclose from the cross. (2) Collect virgins of the double balancer lines.
4. Day 15: In a vial, labeled (1), set up a cross with an individual male fly together with 2–3 virgins from the double balancer line. Repeat this process of single-fly male crosses with double balancer flies in clearly labeled vials (up to 100 crosses). We call these Set-A vials 1–100.
5. Day 18: Move the adult flies to a fresh vial labeled the same as the original cross (Set-B vials 1–100).

6. Day 19: (1) Separate the males to a 1.5 ml microcentrifuge tube labeled same as the original cross and allow it starve for 1-h.

   (2) For control samples, transfer a single male w^{1118} fly and a single male fly from the original P-element line into separate microcentrifuge tubes. Maintain 5 such tubes for both w^{1118} and P-element flies with clear labels. Ensure that all vials and microcentrifuge tubes are labeled correctly.

7. Day 20: Based on the PCR and gel electrophoresis analysis, retain only those vials (both Set A and Set B) that generate mutant stocks and discard all other vials that showed negative results (*see* **Notes 6** and 7).

3.2 Single Fly Genomic DNA Extraction

Day 19: (*see* **Note 1** for advice on whether to use Method 1 or Method 2).

3.2.1 Method 1

Steps to be followed

1. Prepare 50 ml of fresh 2×-cracking buffer (prepare fresh every time).
2. Set a water bath to 70 °C.
3. Leave the labeled microcentrifuge tubes with male flies and control flies on ice for 30 min.
4. Leave the tubes at room temperature for 5 min.
5. Add 100 μl of 2×-cracking buffer.
6. Crush and Grind the fly using a Teflon pestle.
7. Vortex and incubate the tubes in the water bath maintained at 70 °C for 5-min.
8. Allow to cool at room temperature for 5 min.
9. Centrifuge at maximum speed for 3 min.
10. Transfer the supernatant (take only up to 50–80 μl to avoid the fly debris) for further analysis.
11. Add 1.5 μl of 4 M KCl or 2.5 μl of 2.5 M KCl and mix well.
12. Divide the saved supernatant into two aliquots (Aliquot 1 and 2 with 10 μl of the supernatant).
13. Immediately subject the supernatant in Aliquot 1 for gel electrophoresis and Aliquot 2 for PCR analysis.
14. To 5 μl of Aliquot 1 add 0.5 μl of Bromophenol Blue or 6× BPB dye, mix thoroughly, and subject the samples to agarose gel electrophoresis.
15. To 5 μl of Aliquot 2 add PCR master mixture and perform PCR analysis.

3.2.2 Method 2

Steps to be followed.

Day 19

1. Prepare 50 ml of Fly genomic DNA extraction Buffer.
2. Set a water bath to 95 °C and an incubator to 37 °C. Leave the labeled microcentrifuge tubes with male flies and control flies on ice for 30 min.
3. Leave the tubes at room temperature for 5 min.
4. Add 100 μl of Fly Genomic DNA extraction Buffer.
5. And 1 μl of Proteinase K and mix well.
6. Crush and grind the fly using a Teflon pestle (Wheaton).
7. Centrifuge for 1-min at maximum speed.
8. Transfer the supernatant (take only up to 50–80 μl to avoid the fly debris) for further analysis.
9. Incubate the tubes at 37 °C for 30-min.
10. Incubate the tubes at 95 °C for 5-min.
11. Incubate the tubes on ice for 5-min and store at −20 °C.

    Day 20 (or within a week).
12. Take 10 μl of the genomic DNA samples and divide them into three aliquots (Aliquot 1, 2, and 3).
13. To 5 μl Aliquot 1 add 0.5 μl of Bromophenol blue or 6× BPB dye, mix thoroughly, and subject the samples to agarose gel electrophoresis.
14. To 2 μl of Aliquot 2 add PCR master mixture and perform PCR analysis.
15. To 3 μl of Aliquot 3, add 1 μl of primers for sequencing.

3.3 PCR and Agarose Gel Electrophoresis

(*see* **Note 2**)

Day 19

1. Cast either two (up to 4 if there are many samples) 1% agarose gels or one (up to 2) 0.8% agarose gel for running genomic DNA samples and one (up to 2) 1% agarose gel with sufficient number of wells depending on the number of samples [6] (*see* **Note 2**).
2. Set up a PCR reaction (*see* below; 25 μl final volume in each PCR tube).

   If possible two sets of primers, one flanking the P-element and another at a distant location in the genome should be designed for confirmation (*see* **Note 3**). The PCR reaction mixture should be scaled up based on the number of samples to be analyzed.

PCR reaction mix:

10× DNA Buffer: 0.5 μl.

dNTP mixture: 0.2 μl.

Forward Primer: 0.2 μl.

Reverse Primer: 0.2 μl.

Taq polymerase: 0.1 μl.

Distilled Water:22.8 μl.

Template DNA: 1.0 μl.

3. In a 1.5 ml microcentrifuge tube, make up a PCR master mix with all PCR components without template DNA. Volume will depend on the total number of samples to be analyzed (number of mutant samples + two control fly samples (*see* Subheading 3.1) + two negative controls for the PCR reaction; additionally, add two extra volumes of PCR reaction mix to allow for pipetting errors). *See* **Note 4**.
4. Label the PCR tubes clearly.
5. Aliquot 24 μl of the PCR master mix in the labeled PCR tubes.
6. Add 1 μl of corresponding template DNA (1 μl of water for the negative control PCR reactions).
7. Gently spin the reaction mix and carry out PCR using the following settings.

   Step 1—95 °C—2 min.

   Step 2—94 °C—30 s.

   Step 3—60 °C—2 min.

   Step 4—72 °C—2 min (Steps 2–4, 25 cycles).

   Step 5—72 °C—5 min.

   Step 6—4 °C—5 min.
8. In the meantime, run the genomic DNA samples on one of the 1% agarose gels and visualize.
9. Once the PCR is completed, run the samples on another 1% agarose gel. *See* **Note 5**.
10. Based on the analysis of the gels, retain the fly stocks. *See* **Note 6**.

4 Notes

1. If you plan to perform PCR on the same day for further mutant analysis, Method 1 is suggested. If you plan to store the genomic DNA to carry out PCR on another day (but within a

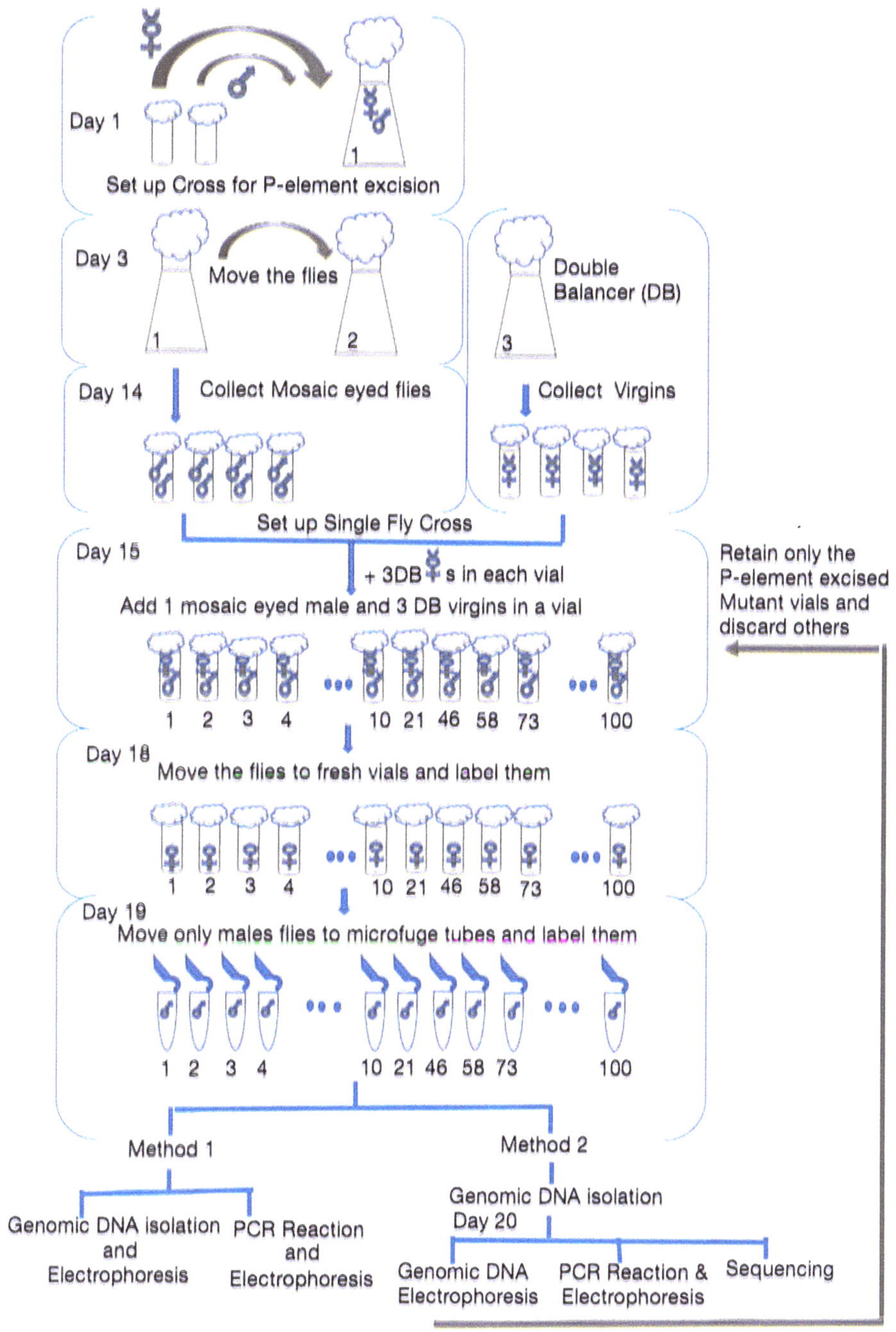

Fig. 1 Scheme for single fly genomic PCR. Genomic PCR and confirmation of mutation. See text for a detailed explanation of all protocol steps (*See* **Note 7**)

week's time) and simultaneously send the sample for sequencing before mutant analysis, Method 2 is suggested. The summary of the protocol and methods followed are shown in Fig. 1.

2. For electrophoresis of genomic DNA, 0.8–1% agarose gels are made. As genomic DNA is usually around 5–10 kb, using 0.8% gels give better separation or resolution. Similarly, 1–2% gels are preferred for smaller sized fragments of 0.2–1 kb (such as PCR products).
3. Primer sequences flanking the P-element regions are designed using standard software, such as Amplify 3.0.
4. Negative controls contain all the ingredients of the PCR mixture except the template. This is to ensure that the amplification of the PCR product is not due to contamination. As the number of samples increases, small volumes of reaction mixture stick to the pipettes and tubes during pipetting, and tend to reduce the total volume. Therefore it is always better to add two extra volumes of reaction mixture to overcome the loss due to pipetting errors.
5. Include a 1 kb DNA marker to run along with the samples. Run the samples in the same order in both the gels for easier comparison of samples between gels.
6. When genomic DNA samples from mutants are subjected to electrophoresis, as mutant flies have shorter DNA than the control WT/P-element containing flies, genomic DNA of mutant flies show faster mobility than the control flies. Figure 2 shows an example experiment. In Fig. 2, genomic DNA from mutants in lanes 6 and 7 (corresponding to mutants 21 and 73) showed faster mobility than control samples.

   For PCR products using primers flanking the P-element, the presence of a DNA fragment of expected size should be seen only in the control samples, but should be absent in mutant samples, and in negative controls; absence of the band would confirm the imprecise excision of the P-element to produce a mutation. At the same time, PCR products of expected size, amplified using the primers away from the P-element, should be seen in both mutants and control samples, other than the negative control.

   In the two PCR product gels in Fig. 2, lane 11 (the negative control) does not show any PCR product, as expected. Lanes 1 and 8 are control samples corresponding to wildtype and P-element flies. In both gels, PCR products of expected sizes of 583 and 185 bp are observed, thereby confirming that the PCR reaction has worked. Only mutant samples 21 and 73 show no band at 583 bp using primers flanking the P-element (upper PCR gel, Fig. 2) but show a band of the expected size (185 bp) using primers away from the P-element (lower PCR gel, Fig. 2). All other mutant samples (lanes 2–7) show bands in both the gels, indicating that the P-elements were not excised in these lines. This indicates that samples

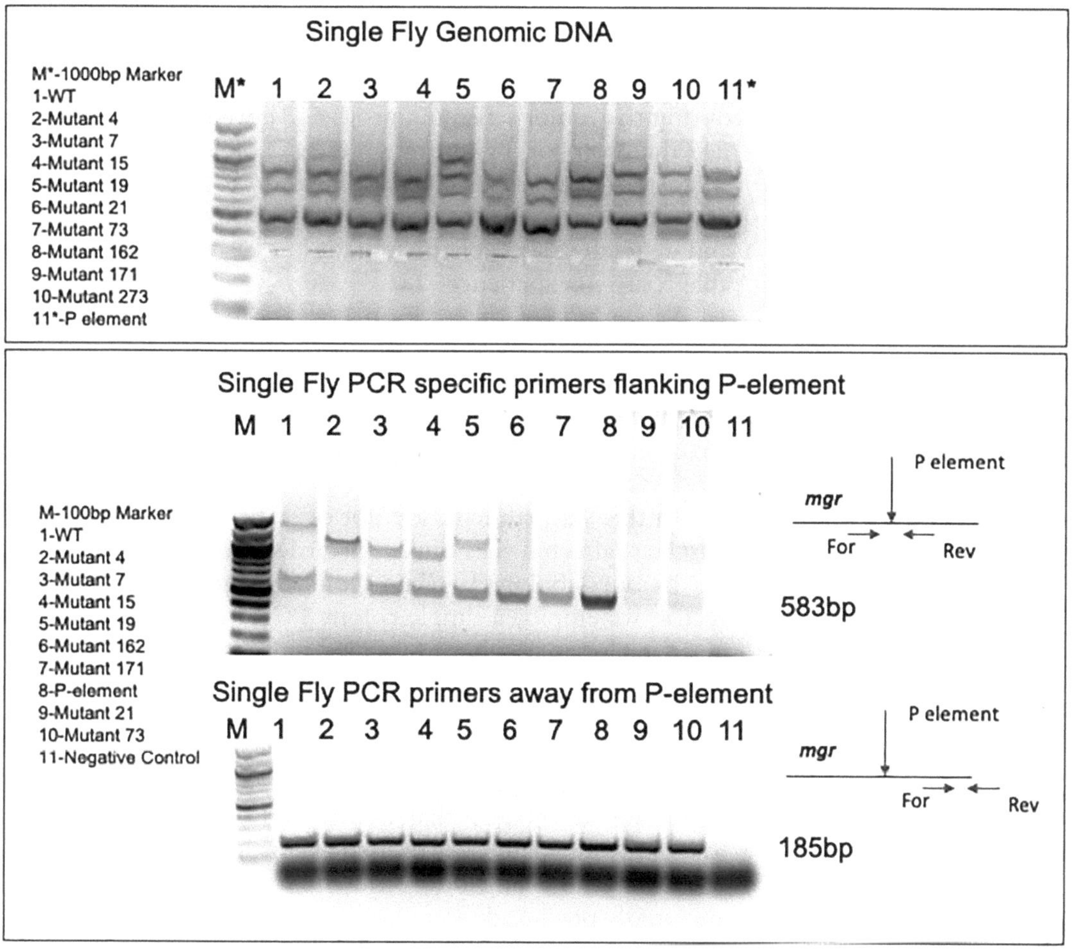

Fig. 2 Agarose gel electrophoresis of Single Fly Genomic DNA and PCR products. Genomic DNA from Mutants Lane 6 and 7 corresponding to Mutants 21 and 73 show faster mobility than other Mutants. Single Fly PCR was set up using the same samples using two sets of primers (one primer set flanking the P-element on gene of interest (mgr) and the other primer set away from the P-element) to confirm the presence of a mutation. In Lane 9 and 10 corresponding to Mutants 21 and 73, a PCR product of 583 bp is absent (upper PCR gel) but a PCR product of 185 bp is present in the lower PCR gel, confirming the excision of the P-element within these lines (*See* **Note 6**)

21 and 73 may be mutants generated due to P-element excision. This can be further confirmed by sequencing the genomic DNA and PCR products. Based on these results, all fly vials other than vials containing mutants 21 and 73 can be discarded and mutants 21 and 73 alone can be expanded for further analysis.

7. The above discussed protocol and method (summarized in Fig. 1) can be performed easily by an experienced researcher. However, for inexperienced researchers, such as undergraduate students, collecting virgins on a large scale and handling

100–200 mosaic-eyed male flies to set up single fly crosses will be difficult. Labeling all samples and duplicating them may be confusing. Instead, based on the experience we have gained during the tumor screening project [3], we revised and simplified the protocol. By setting up crosses on a smaller scale, in a staggered manner, using vials instead of bottles, we could perform a large-scale screen together with undergraduates, and generate novel genetic mutant lines, giving students valuable research experience.

For example, once virgin collection begins, we set up crosses every alternate day, on a small scale in a vial with only 6 females and 3 males. On the third day, the flies were transferred to a fresh vial and labeled correctly. This process of collecting virgins and setting crosses continued for a week to 10 days.

In this way, students are not overburdened and in case of mistakes, affected vials can be discarded.

In our laboratory, when cloning gene fragments into a vector, we employ a simple colony PCR method to screen for positive clones. We have modified and standardized a similar technique to use with fly stocks. This is a simple screening method to identify mutant fly lines. This method is easy to perform, saves time and resources, and can be modified to give undergraduate students valuable research experience. Another important advantage is that, as this method uses individual flies to identify independently excised events, mutant alleles of various strengths falling into different complementation groups are rapidly produced.

Acknowledgments

We thank all undergraduate students who were involved in this process. We thank Science and Engineering Research Board (SERB), Department of Science and Technology (DST), Government of India, for the grants (YSS/2014/000896) and CRG/2018/003725 to UN. MG received funding from Cancer Research UK (Grant Nos. C36430, A12891) and NC3Rs (Grant No. NC/S001417/1).

References

1. Jennings BH (2011) *Drosophila* – a versatile model in biology and medicine. Mater Today 14:190–195
2. Tolwinski NS (2017) Introduction: Drosophila—a model system for developmental biology. J Dev Biol 5:9–10
3. Coutiño BC, Cornhill ZE, Couto A, Mack NA, Rusu AD, Nagarajan U, Fan YN, Hadjicharalambous MR, Uribe MC, Burrows A, Lourdusamy A, Rahman R, May ST, Georgiou M (2020) A genetic analysis of tumor progression in Drosophila identifies the cohesin

complex as a suppressor of individual and collective cell invasion. iScience 23:101237–101280

4. Rong YS, Golic KG (2000) Gene targeting by homologous recombination in Drosophila. Science 288:2013–2018
5. Rong YS, Titen SW, Xie HB, Golic MM, Bastiani M, Bandyopadhyay P, Olivera BM, Brodsky M, Rubin GM, Golic KG (2002) Targeted mutagenesis by homologous recombination in D. melanogaster. Genes Dev 16:1568–1581
6. Ashburner M (1989) Drosophila: a laboratory handbook and manual, 2 vols: xliii + 1331p, 434p

Chapter 3

Generation of Properly Folded Epidermal Growth Factor-Like (EGF) Repeats and Glycosyltransferases Enables In Vitro *O*-Glycosylation

Chenyu Ma, Yohei Tsukamoto, and Hideyuki Takeuchi

Abstract

The epidermal growth factor-like (EGF) domain is an evolutionarily conserved motif found widely distributed among numerous secreted and membrane-anchored proteins, including the Notch receptors. Notch receptors include numerous EGF repeats tandemly connected in the extracellular domain. These EGF repeats must be properly folded in order for them to undergo the three different types of *O*-glycosylation associated with these extracellular proteins: *O*-fucose, *O*-glucose, and *O*-*N*-acetylglucosamine via glycosyltransferases POFUT1, POGLUT1, and EOGT. The *O*-glycosylation of the EGF repeats in the Notch receptors regulates the activation of Notch signaling and mutations in *POFUT1*, *POGLUT1*, and *EOGT* have been linked to specific human diseases. A large amount of EGF repeat and glycosyltransferase protein is required to construct an in vitro *O*-glycosylation system. Here, we describe how we prepared properly folded EGF repeats using two different bacterial expression vectors, generated recombinant glycosyltransferases, and performed in vitro *O*-glycosylation and subsequent product analysis by mass spectrometry. The methods described here are useful for investigating the enzymatic activities of mutated glycosyltransferases, revealing the structural basis of the *O*-glycosylation mechanism by co-crystallization of the glycosyltransferase-EGF repeat complexes, or identifying potential inhibitors of these glycosyltransferases.

Key words Epidermal growth factor-like repeats, *O*-glycosylation, Glycosyltransferases, Protein folding, Reverse phase high-performance liquid chromatography

1 Introduction

Notch signaling is a well-conserved intercellular communication pathway important for building the complex bodies of the metazoans. Notch signaling regulates a multitude of processes, including embryonic development, tissue homeostasis, and the maintenance of stem cells [1]. Notch signaling can be modified in multiple

Chenyu Ma and Yohei Tsukamoto contributed equally to this work.

Dongyu Jia (ed.), *Notch Signaling Research: Methods and Protocols*,
Methods in Molecular Biology, vol. 2472, https://doi.org/10.1007/978-1-0716-2201-8_3,

processes and by diverse components of this pathway, while the *O*-linked glycans on the EGF repeats of the Notch proteins have emerged as one of the major regulators for increasing or decreasing Notch activity via their inhibition or induction of different aspects of the Notch signaling pathway including facilitating the proper conformation of the receptor, receptor trafficking, and recognition by ligands [2].

EGF repeats usually comprise 30–40 amino acid residues, including six conserved cysteine residues that form three disulfide bonds within the endoplasmic reticulum (ER). Glycosylation of EGF repeats occurs within the ER and the Golgi apparatus after translation. EGF repeats that harbor certain consensus sequences can be decorated with *N*-linked glycans and *O*-linked glycans. The specific consensus sequences of different EGF repeats mediate their specific interactions with individual glycosyltransferases which then catalyze the addition of the *O*-glycan onto a serine or threonine residue within the target protein. There are three main types of *O*-glycosylation on the mammalian Notch receptors: *O*-glucose, *O*-fucose [3], and *O*-GlcNAc [4], which can be further extended to a trisaccharide or tetrasaccharide [3]. Protein *O*-glucosyltransferase 1 (POGLUT1) transfers glucose to $C^1X\underline{S}X(P/A)C^2$ from UDP-glucose, and in most EGF repeats glucoside xylosyltransferase 1 and 2 (GXYLT1/2) and xyloside xylosyltransferase 1 (XXYLT1) elongate the *O*-glucose to form a Xyl-Xyl-Glc trisaccharide [5, 6].

Aberrant *O*-glycosylation usually leads to the dysregulation of Notch signaling, which is pertinent in a variety of congenital disorders and some adult-onset diseases, including cancer. Dowling-Degos disease, an autosomal-dominant genodermatosis, has been shown to be the product of several genetic defects, including autosomal dominant mutations in *POGLUT1* leading to the inactivation of POGLUT1. Limb-girdle muscular dystrophy R21, a group of rare inherited disorders, is linked to recessive missense mutations and nonsense variants in *POGLUT1* [7]. Moreover, the overexpression of *POGLUT1* has been found in primary acute myelogenous leukemia (AML) and T-cell acute lymphoblastic leukemia (T-ALL) [8], while the amplification of *XXYLT1* has been reported in squamous cell carcinoma [9].

To further clarify the function of the glycosyltransferases in Notch signaling, it is important to determine whether the above-mentioned disease-related mutations affect enzymatic activity. Our mass spectrometric analysis of NOTCH1 produced in HEK293T cells show that the elongation of *O*-glucose glycans is site-specific [10], and may be the result of differences in the affinity of different EGF repeats for each of the glycosyltransferases. Cocrystallization of EOGT or GXYLT1/2 complexed with (*O*-glucosylated) EGF repeats has not been attained and may be facilitated by the availability of large amounts of high-affinity EGF substrates.

Furthermore, the development of novel Notch-modifying glycosyltransferase inhibitors via high-throughput screening requires the reliable production of a large amount of properly folded EGF repeats to function as active acceptor substrates in enzymatic assays. Here, we describe a set of basic methods for generating properly folded EGF repeats and recombinant glycosyltransferases, as well as establish an enzymatic assay system for the in vitro *O*-glycosylation of EGF repeats.

2 Materials

2.1 Generation of EGF Repeats in *E. coli* Using pET and pMAL

2.1.1 Generation of EGF Repeats by pET20b (+) Vector

1. pET20b (+) expression vector encoding EGF repeats with a terminal 6×His tag.
2. BL21(DE3) *E. coli* (Invitrogen).
3. Ampicillin 100 mg/ml in 50% ethanol.
4. Isopropyl β-D-1-thiogalactopyranoside (IPTG).
5. 200 mM phenylmethanesulfonyl fluoride (PMSF) in ethanol
6. Ni-NTA agarose resin (Wako).
7. Imidazole.
8. Ni-NTA wash buffer: 10 mM Tris–HCl pH 7.4 containing 0.65 M NaCl and 10 mM imidazole.
9. Ni-NTA elution buffer: 250 mM imidazole in TBS pH 7.4.
10. Reverse phase high performance liquid chromatography (HPLC) system (Shimadzu).
11. C18 column (GL Sciences).
12. Acetonitrile.
13. Trifluoroacetic acid.

2.1.2 Generation of EGF Repeats by pMAL-p5X Vector

1. pMAL-p5X vectors encoding EGF repeats (mouse NOTCH1 EGF10, EGF28, and mouse NOTCH2 EGF12, human coagulation factor IX EGF1).
2. XL-10 competent *E. coli* cells.
3. BL21(DE3) *E. coli* (Invitrogen).
4. Glucose.
5. Ampicillin 100 mg/ml in 50% ethanol.
6. Isopropyl β-D-1-thiogalactopyranoside (IPTG).
7. Column buffer: 20 mM Tris–HCl pH 7.4 containing 0.2 M NaCl and 0.1 mM EDTA.
8. pMAL Elution buffer: column buffer with 10 mM maltose.
9. Tris (hydroxymethyl)aminomethane.

10. Amylose resin (New England Biolabs).
11. Maltose.
12. Factor Xa (New England Biolabs).
13. Solvent A: 0.1% TFA in water.
14. Solvent B: 0.1% TFA in acetonitrile.

2.2 Generation of Recombinant Glycosyltransferase

1. HEK293T cells.
2. Expression vector encoding recombinant glycosyltransferases [11].
3. PEI max.
4. DMEM: 4.5 g/l glucose, 10% fetal calf serum, penicillin, and streptomycin.
5. Opti-MEM-I (Thermo Fisher).
6. Glycerol.
7. Tris (hydroxymethyl)aminomethane.

2.3 In Vitro Glycosylation Analysis by Reverse Phase HPLC

1. Single EGF repeats.
2. Recombinant glycosyltransferases.
3. UDP-Glucose (Sigma-Aldrich).
4. 10× Reaction buffer: 500 mM HEPES pH 6.8 containing 100 mM $MnCl_2$
5. Reverse phase HPLC system (Shimadzu).
6. C18 column (GL Sciences).
7. Acetonitrile.
8. Trifluoroacetic acid.

2.4 In Vitro Glycosylation Analysis by Mass Spectrometry

1. Single (glycosylated) EGF repeat.
2. Q-Exactive mass spectrometer (Thermo Fisher).
3. UltiMate3000 RSLCnano LC system (Dionex Co.)
4. nano HPLC capillary column, 150 mm × 75 μm inner diameter (Nikkyo Technos Co.)
5. LC-MS solvent A: 2% acetonitrile and 0.1% formic acid in water.
6. LC-MS solvent B: 95% acetonitrile and 0.1% formic acid in water.
7. Byonic (version 3.5) (Thermo Fisher).
8. Proteome Discoverer (version 2.3) (Thermo Fisher).
9. Xcalibur (version 4.3) (Thermo Fisher).

3 Methods

3.1 Generation of EGF Repeats in E. coli Using pET and pMAL

BL21(DE3) *E. coli* containing a chromosomal copy of the T7 RNA polymerase gene is used to express single EGF repeats. We used the pET-20b (+) expression system to generate single EGF repeats; however, the yields varied significantly among different EGF repeat types. For mouse NOTCH1 EGF10, EGF19, and EGF28, few EGF repeats can be generated using this more traditional method, and this may partly be because these EGF repeats are difficult to fold correctly within the *E. coli* cells. To produce a variety of EGF repeats as potential substrates for our further enzymatic assays, we used the pMAL expression system in which the EGF repeats are expressed as a maltose-binding protein (MBP)-fusion, which generally enhances the solubility of these proteins when expressed in *E. coli*. The MBP-EGF repeats are then cleaved by a protease, Factor Xa, and separated by reverse-phase (RP)-HPLC. The folding status of the EGF repeats was confirmed by in vitro *O*-glycosylation using POGLUT1 and RP-HPLC, taking advantage of the fact that POGLUT1 only uses properly folded EGF repeats as an acceptor substrate. The typical yield of the properly folded single EGF repeats from 1 l of *E. coli* culture was several milligrams. EGF repeats such as human coagulation factor IX EGF1 (hFA9) and mouse NOTCH2 EGF12 (mN2 EGF12) produced a high yield in the pET system, but "difficult" EGF repeats (e.g., mN1 EGF10 and EGF28) were generated using the pMAL system.

3.1.1 Generation of EGF Repeats Using the pET20b (+) Expression System

1. *E. coli* BL21(DE3) cells transformed with the target plasmid are cultured in 4 ml of LB medium supplemented with 100 μg/ml ampicillin at 37 °C overnight.
2. This 4 ml overnight culture is then used to inoculate 200 ml of 2×YT medium until the absorbency at 600 nm reached 0.5.
3. Vector expression is then induced following the addition of 0.4 mM isopropyl-β-thiogalactopyranoside and allowed to grow overnight at 21 °C.
4. The culture is then centrifuged at 4,000 × *g* for 20 min and the supernatant is discarded. The bacterial pellet is then resuspended in 50 mM Tris–HCl pH 8.0 containing 1 mM phenylmethylsulfonyl fluoride.
5. The cell suspension is sonicated ten times for 10 s and then clarified by centrifugation (pellet is resuspended, sonicated, and clarified again).
6. The supernatant is filtered and applied to a Poly-Prep Chromatography Column (Bio-Rad) filled with 1 ml of 50% slurry preequilibrated Ni-NTA agarose.

7. After the supernatant traversed the column using gravity, the agarose column is washed twice with 10 ml of Ni-NTA wash buffer.
8. The protein is then eluted in 1 ml of Ni-NTA elution buffer.
9. The proteins are further purified by reverse-phase HPLC on a 250 × 4.6 mm InertSustainSwift C18 column to separate the properly folded isoform from the misfolded isoform(s). The concentration of the protein products is then determined using a BCA protein assay with a BSA standard.

3.1.2 Generation of EGF Repeats Using the pMAL-p5X Expression System

1. BL21(DE3) *E. coli* cells transformed with the pMAL-p5X vector encoding the desired EGF repeats are cultured in 4 ml of LB medium containing 100 μg/ml ampicillin at 37 °C overnight.
2. This overnight culture (0.8 ml) is inoculated into 80 ml of LB medium containing 2 mg/ml glucose and 100 μg/ml ampicillin and cultured at 37 °C with good aeration. When the cells grew to 2×10^8 cells/ml (~0.5, at 600 nm), 32 μl of 1 M IPTG (final concentration 0.4 mM) is added to the culture, followed by incubation at 37 °C for another 2 h (*see* **Note 1**).
3. The cells are harvested by centrifugation at 4000 × *g* for 10 min (*see* **Note 2**). The supernatant is discarded, and the pellet resuspended in 5 ml of column buffer (*see* **Note 3**).
4. The cell suspension is frozen at −30 °C overnight and then thawed in cold water.
5. The cell suspension is placed in an ice-water bath and sonicated 12 times in short pulses of 10 s, and centrifuged at ~16,000 rpm and 4 °C for 40 min.
6. The supernatant (crude extract) is decanted and diluted at a volume ratio of 1:6 with column buffer.
7. Approximately 2 ml of the amylose resin is added to a Poly-Prep chromatography column, and then equilibrated with 10 ml of column buffer (*see* **Note 4**).
8. The diluted crude extract is loaded onto the column and allowed to flow through the resin by gravity. The resin is then washed with 24 ml of column buffer.
9. The elution buffer is added and allowed to flow through the resin. The fusion protein is eluted in 2.5 ml of elution buffer (*see* **Note 5**).
10. The concentration of the MBP-EGF repeat proteins is then be evaluated using Coomassie brilliant blue staining. If necessary, the elution can be concentrated to more than 1 mg/ml using an Amicon Ultra Centrifugal Filter device.

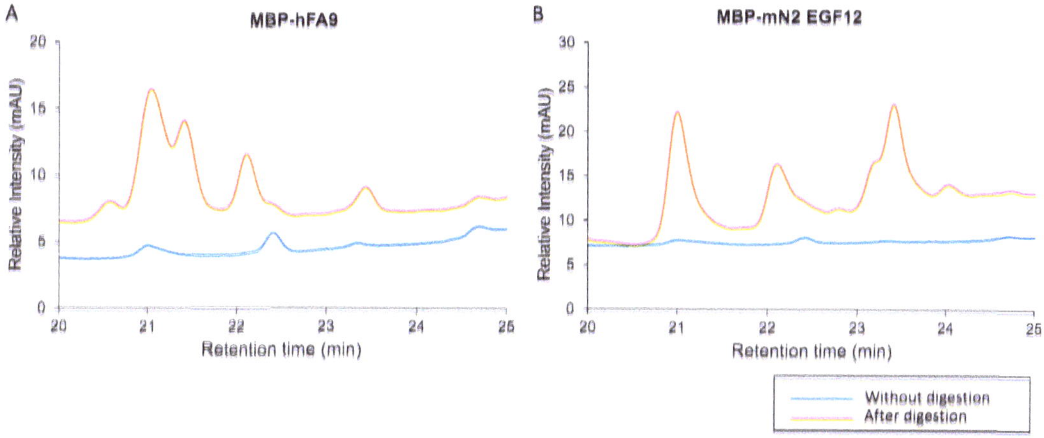

Fig. 1 Separation of the MBP and EGF repeats by RP-HPLC after digesting the MBP-EGF fusion proteins using Factor Xa. The MBP proteins were fused with the EGF repeat from human factor IX (hFA9) (**a**) or with mouse NOTCH2 EGF12 (mN2 EGF12) (**b**) before (blue) and after digestion (orange) by Factor Xa protease and subjected to reverse-phase HPLC. The absorbance at 210 nm was monitored

11. Fusion protein cleavage is then carried out using Factor Xa at a w/w ratio of approximately 1% (the amount of the protease can be adjusted within the range of 0.1%). For example, 100 μl of 4 μg/μl MBP-EGF protein is mixed with 4 μl of Factor Xa (1 mg/ml) and incubated at room temperature for 24 h (*see* **Note 6**).
12. The cleaved EGF repeats are separated from MBP and the protease using reverse-phase HPLC completed on a 250 × 4.6 mm InertSustainSwift C18 column. The column is eluted using a linear gradient from 20 to 38% solvent B in solvent A at a flow rate of 1 ml/min for 30 min. The absorbance of the eluate is monitored at 210 nm. The fractions containing individual peaks are collected, dried, and evaluated using in vitro glycosylation catalyzed by various glycosyltransferases such as POGLUT1 (Fig. 1).
13. In vitro glycosylation and LC-MS analyses are conducted to confirm the folding status and molecular weight of the EGF repeats (described in detail in Subheadings 3.3–4).

3.2 Generation of Recombinant Glycosyltransferases

We generated several recombinant glycosyltransferases, including POGLUT1 [12], POGLUT2/3 [13], POFUT1 [14], and POFUT2 [15], from the culture media of transiently transfected HEK293T cells and then used these recombinant proteins to investigate their enzymatic activities.

1. HEK293T cells are transfected with plasmids encoding different glycosyltransferases using PEI-MAX in regular media containing 10% bovine calf serum for several hours before the media are changed to Opti-MEM I media.

2. Cells are then cultured for about 48 h and the culture medium from the transfected cells is collected and applied to a Ni-NTA agarose column.
3. This is then washed twice with 10 ml of 10 mM Ni-NTA wash buffer and the proteins are then eluted in 1 ml Ni-NTA elution buffer.
4. The proteins are then dialyzed against TBS containing 20% glycerol at 4 °C overnight.
5. The concentration and purity of the recombinant glycosyltransferases are then determined using Coomassie staining and BSA as a standard. Western blot is performed using an anti-Myc primary antibody to confirm the identity of these proteins.

3.3 In Vitro Glycosylation Analysis Using Reverse Phase HPLC

Given the clear need for a robust in vitro glycosylation assay we set out to establish a novel in vitro glycosylation system using recombinant glycosyltransferases, UDP-glucose, and single EGF repeats. These in vitro glycosylation assays are also used to confirm the folding status of the EGF repeats since the glycosyltransferases (e.g., POGLUT1 and POFUT1) utilize only properly folded EGF repeats as an acceptor substrate [11].

1. A total of 10–20 μM of the EGF repeats are incubated with 100–200 μM UDP-Glucose and recombinant POGLUT1 (POGLUT2/3 for mouse NOTCH1 EGF11) in 1x Reaction buffer.
2. The reaction is incubated at 37 °C overnight and stopped by the addition of Milli-Q water.
3. The reaction products are then analyzed by reverse-phase HPLC on a 250 × 4.6 mm InertSustainSwift C18 column using a 30-min linear gradient from 20 to 38% solvent B in solvent A, and the absorbance at 210 nm is monitored.
4. The retention time for certain EGF repeats changes after the addition of glucose, which can be demonstrated by a shift in the HPLC peaks (Fig. 2).
5. The intensities of the peaks are used for quantification. The peak fractions are collected and vacuum-dried for further LC-MS analysis using a Speed-Vacuum-centrifuge as described below.

3.4 Mass Spectrometry Mediated Evaluations of In Vitro Glycosylation

The addition of glycans to EGF repeats can be confirmed in several ways including mass spectrometric analysis [5, 6, 11–13, 16–21]. Here, we briefly describe the method of sample preparation for mass-spectrometric analysis.

1. The EGF repeats are analyzed using LC-MS completed on a Q-Exactive mass spectrometer coupled to an UltiMate3000

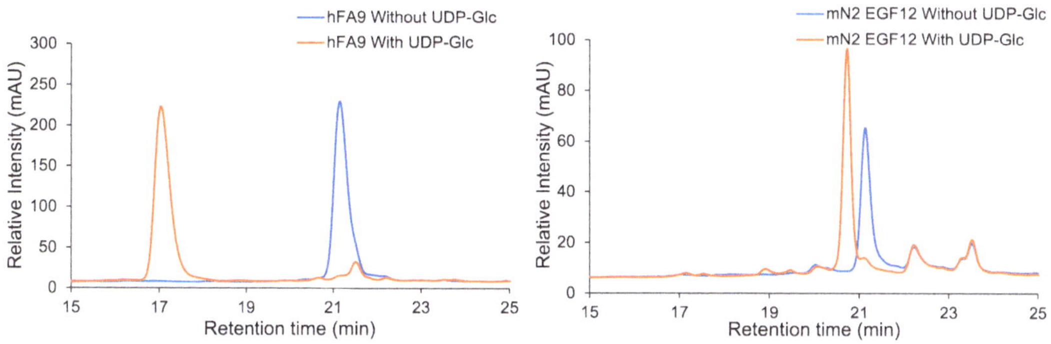

Fig. 2 The addition of *O*-glucose results in a shift in the retention time of single EGF repeats generated using the pMAL system. The in vitro *O*-glycosylation of the EGF repeat from human factor IX (hFA9) (**a**) or mouse NOTCH2 EGF12 (mN2 EGF12) (**b**) in the presence (orange) or absence (blue) of UDP-glucose was evaluated and the reaction products were subjected to reverse-phase HPLC. The absorbance at 210 nm was monitored

RSLCnano LC system using a nano HPLC capillary column with an inner diameter of 150 mm × 75 μm via a nanoelectrospray ion source. Reversed-phase chromatography is performed using a linear gradient (0 min, 5% B; 40 min, 100% B) of LC-MS solvent A and LC-MS solvent B at an estimated flow rate of 3.0 μl/min. A precursor ion scan is carried out at 400–2000 *m/z* prior to MS/MS analysis. High-energy collision-induced dissociation (HCD) at stepped collision energies of 33 ± 13 is employed as the fragmentation method.

2. MS and MS/MS data are analyzed using the analysis software. The *m/z* value should change following the addition of glucose, which is clearly observed in the MS spectra. Qualitative analysis is then performed using these MS/MS spectra.

4 Notes

1. The culture temperature and period should be adjusted to achieve an optimal cultivation.
2. There are two methods of purifying fusion proteins from *E. coli*. You can extract the soluble fusion protein from the (1) total cells or (2) periplasmic extracts. The periplasmic protein contains fewer *E. coli* proteins, but the procedure is more difficult to scale up resulting in a preference for the extraction of fusion proteins from total cell extracts.
3. Adding phenylmethanesulfonylfluoride (PMSF), DTT, or β-mercaptoethanol to the column buffer may help to protect the fusion protein.

4. The volume of the resin should be adjusted based on the amount of fusion protein loaded on the column. Every milliliter of bed volume resin can bind 6–8 mg of fusion protein.
5. We detected the concentration of the fusion proteins in different fractions (fraction size = ½ the resin volume). The MBP-EGF fusion proteins start to elute after the first fraction, and most of the proteins can be eluted in the next four fractions.
6. The cleavage efficiency of Factor Xa is different for different types of MBP-EGF fusion protein. This means that the amount of the protease should be adjusted to produce an acceptable rate of cleavage. For example, in our evaluations, 1% Factor Xa could cleave all MBP-mN2 EGF12 within 3 h at room temperature, but the cleavage of MBP-hFA9 is not complete until 24 h of incubation under the same conditions. In addition, according to our data, the w/w ratio for cleaving MBP-mN2 EGF12 could be reduced to 0.25% (incubated at room temperature for 24 h).
7. The incubation time can be adjusted based on the concentration of the EGF repeats and glycosyltransferases.

Acknowledgments

This work is partly supported by the JSPS KAKENHI grants JP19H03176, JP19KK0195, and 19K22490 (to H.T.). We would like to thank Prof. Tetsuya Okajima at Nagoya University for the critical comments on our research project. Chenyu Ma and Yohei Tsukamoto contributed equally to this work.

References

1. D'Souza B, Miyamoto A, Weinmaster G (2008) The many facets of Notch ligands. Oncogene 27(38):5148–5167. https://doi.org/10.1038/onc.2008.229
2. Pandey A, Niknejad N, Jafar-Nejad H (2020) Multifaceted regulation of Notch signaling by glycosylation. Glycobiology. https://doi.org/10.1093/glycob/cwaa049
3. Moloney D, Shair L, Lu F, Xia J, Locke R, Matta K, Haltiwanger R (2000) Mammalian Notch1 is modified with two unusual forms of O-linked glycosylation found on epidermal growth factor-like modules. J Biol Chem 275(13):9604–9611. https://doi.org/10.1074/jbc.275.13.9604
4. Matsuura A, Ito M, Sakaidani Y, Kondo T, Murakami K, Furukawa K, Nadano D, Matsuda T, Okajima T (2008) O-linked N-acetylglucosamine is present on the extracellular domain of notch receptors. J Biol Chem 283(51):35486–35495. https://doi.org/10.1074/jbc.M806202200
5. Sethi MK, Buettner FF, Krylov VB, Takeuchi H, Nifantiev NE, Haltiwanger RS, Gerardy-Schahn R, Bakker H (2010) Identification of glycosyltransferase 8 family members as xylosyltransferases acting on O-glucosylated notch epidermal growth factor repeats. J Biol Chem 285(3):1582–1586. https://doi.org/10.1074/jbc.C109.065409

6. Sethi MK, Buettner FF, Ashikov A, Krylov VB, Takeuchi H, Nifantiev NE, Haltiwanger RS, Gerardy-Schahn R, Bakker H (2012) Molecular cloning of a xylosyltransferase that transfers the second xylose to O-glucosylated epidermal growth factor repeats of notch. J Biol Chem 287(4):2739–2748. https://doi.org/10.1074/jbc.M111.302406
7. Servián-Morilla E, Cabrera-Serrano M, Johnson K, Pandey A, Ito A, Rivas E, Chamova T, Muelas N, Mongini T, Nafissi S, Claeys KG, Grewal RP, Takeuchi M, Hao H, Bönnemann C, Lopes Abath Neto O, Medne L, Brandsema J, Töpf A, Taneva A, Vilchez JJ, Tournev I, Haltiwanger RS, Takeuchi H, Jafar-Nejad H, Straub V, Paradas C (2020) POGLUT1 biallelic mutations cause myopathy with reduced satellite cells, α-dystroglycan hypoglycosylation and a distinctive radiological pattern. Acta Neuropathol 139(3):565–582. https://doi.org/10.1007/s00401-019-02117-6
8. Wang Y, Chang N, Zhang T, Liu H, Ma W, Chu Q, Lai Q, Liu L, Wang W (2010) Overexpression of human CAP10-like protein 46 KD in T-acute lymphoblastic leukemia and acute myelogenous leukemia. Genet Test Mol Biomarkers 14(1):127–133. https://doi.org/10.1089/gtmb.2009.0145
9. Yu H, Takeuchi M, LeBarron J, Kantharia J, London E, Bakker H, Haltiwanger RS, Li H, Takeuchi H (2015) Notch-modifying xylosyltransferase structures support an SNi-like retaining mechanism. Nat Chem Biol 11(11):847–854. https://doi.org/10.1038/nchembio.1927
10. Urata Y, Saiki W, Tsukamoto Y, Sago H, Hibi H, Okajima T, Takeuchi H (2020) Xylosyl extension of O-glucose glycans on the extracellular domain of NOTCH1 and NOTCH2 regulates Notch cell surface trafficking. Cells 9(5):1220. https://doi.org/10.3390/cells9051220
11. Takeuchi H, Kantharia J, Sethi MK, Bakker H, Haltiwanger RS (2012) Site-specific O-glucosylation of the epidermal growth factor-like (EGF) repeats of notch: efficiency of glycosylation is affected by proper folding and amino acid sequence of individual EGF repeats. J Biol Chem 287(41):33934–33944. https://doi.org/10.1074/jbc.M112.401315
12. Takeuchi H, Fernández-Valdivia RC, Caswell DS, Nita-Lazar A, Rana NA, Garner TP, Weldeghiorghis TK, Macnaughtan MA, Jafar-Nejad H, Haltiwanger RS (2011) Rumi functions as both a protein O-glucosyltransferase and a protein O-xylosyltransferase. Proc Natl Acad Sci U S A 108(40):16600–16605. https://doi.org/10.1073/pnas.1109696108
13. Takeuchi H, Schneider M, Williamson DB, Ito A, Takeuchi M, Handford PA, Haltiwanger RS (2018) Two novel protein *O*-glucosyltransferases that modify sites distinct from POGLUT1 and affect Notch trafficking and signaling. Proc Natl Acad Sci U S A 115(36):E8395–E8402. https://doi.org/10.1073/pnas.1804005115
14. Takeuchi H, Wong D, Schneider M, Freeze HH, Takeuchi M, Berardinelli SJ, Ito A, Lee H, Nelson SF, Haltiwanger RS (2018) Variant in human POFUT1 reduces enzymatic activity and likely causes a recessive microcephaly, global developmental delay with cardiac and vascular features. Glycobiology 28(5):276–283. https://doi.org/10.1093/glycob/cwy014
15. Vasudevan D, Takeuchi H, Johar SS, Majerus E, Haltiwanger RS (2015) Peters plus syndrome mutations disrupt a noncanonical ER quality-control mechanism. Curr Biol 25(3):286–295. https://doi.org/10.1016/j.cub.2014.11.049
16. Ma C, Takeuchi H, Hao H, Yonekawa C, Nakajima K, Nagae M, Okajima T, Haltiwanger RS, Kizuka Y (2020) Differential labeling of glycoproteins with alkynyl fucose analogs. Int J Mol Sci 21(17):6007. https://doi.org/10.3390/ijms21176007
17. Takeuchi H, Yu H, Hao H, Takeuchi M, Ito A, Li H, Haltiwanger RS (2017) *O*-Glycosylation modulates the stability of epidermal growth factor-like repeats and thereby regulates Notch trafficking. J Biol Chem 292(38):15964–15973. https://doi.org/10.1074/jbc.M117.800102
18. Servián-Morilla E, Takeuchi H, Lee TV, Clarimon J, Mavillard F, Area-Gómez E, Rivas E, Nieto-González JL, Rivero MC, Cabrera-Serrano M, Gómez-Sánchez L, Martínez-López JA, Estrada B, Márquez C, Morgado Y, Suárez-Calvet X, Pita G, Bigot A, Gallardo E, Fernández-Chacón R, Hirano M, Haltiwanger RS, Jafar-Nejad H, Paradas C (2016) A POGLUT1 mutation causes a muscular dystrophy with reduced Notch signaling and satellite cell loss. EMBO Mol Med 8(11):1289–1309. https://doi.org/10.15252/emmm.201505815

19. Yu H, Takeuchi H, Takeuchi M, Liu Q, Kantharia J, Haltiwanger RS, Li H (2016) Structural analysis of Notch-regulating Rumi reveals basis for pathogenic mutations. Nat Chem Biol 12(9):735–740. https://doi.org/10.1038/nchembio.2135
20. Takeuchi H, Haltiwanger RS (2013) Enzymatic analysis of the protein O-glycosyltransferase, Rumi, acting toward epidermal growth factor-like (EGF) repeats. Methods Mol Biol 1022:119–128. https://doi.org/10.1007/978-1-62703-465-4_10
21. Acar M, Jafar-Nejad H, Takeuchi H, Rajan A, Ibrani D, Rana NA, Pan H, Haltiwanger RS, Bellen HJ (2008) Rumi is a CAP10 domain glycosyltransferase that modifies Notch and is required for Notch signaling. Cell 132(2):247–258. https://doi.org/10.1016/j.cell.2007.12.016

Chapter 4

Use of FLP/FRT System to Screen for Notch Signaling Regulators in the Drosophila Wing

Dongqing Mo, Jie Shen, and Junzheng Zhang

Abstract

Mutations of genes encoding key components of the Notch signaling pathways often result in lethality at early developmental stages, making it difficult to decipher how they regulate the formation of specific cell types or organs. Mosaic analysis using the FLP/FRT system allows investigating the roles of essential genes during wing development in *Drosophila melanogaster*. This chapter describes the practical methods to isolate Notch signaling regulators by somatic mosaic screen. The fly stocks, cross schemes, and screen parameters are summarized. We also explain how to validate the roles of potential Notch signaling regulators.

Key words FLP/FRT, *Drosophila melanogaster*, Wing, Notch signaling

1 Introduction

Creating mosaic animals enables investigating the roles of essential genes in specific developmental events such as cell fate determination, tissue growth, and pattern formation [1]. The FLP/FRT system remains the premier tool for generating mosaic tissues in *Drosophila melanogaster* ever since they were introduced nearly three decades ago [2, 3]. In mitotically active tissues such as the wing imaginal disc, clones of homozygous mutant cells could be efficiently induced. The detailed mechanism and methods about using FLP/FRT system to generate mutant clones have been thoroughly summarized [1, 4, 5], and therefore are not further discussed in this chapter.

Notch signaling is crucial for various developmental processes, such as cell fate determination, cell proliferation, and cell cycle progression [6]. In fly, the *Notch* gene encodes a trans-membrane receptor that is activated by ligands encoded by Delta or Serrate [7]. The Notch receptor on the signal-receiving cell binds directly to ligands located on the membrane of signal sending cell

Dongyu Jia (ed.), *Notch Signaling Research: Methods and Protocols*,
Methods in Molecular Biology, vol. 2472, https://doi.org/10.1007/978-1-0716-2201-8_4,

[6]. Receptor–ligand engagement triggers the cleavage within the Notch transmembrane domain, and releases the Notch intracellular domain (NICD) from the membrane [6, 7]. The released NICD translocates into the nucleus, where it forms a transcriptional complex with the DNA binding protein CSL (CBF1, Suppressor of Hairless, Lag1), Mastermind (Mam), and transcriptional coactivators to drive the expression of Notch target genes [6]. In the absence of NICD, CSL forms complexes with a variety of corepressors to suppress the transcription of Notch target genes [7].

First discovered in Thomas Hunt Morgan's lab more than a century ago, the "notch" mutant was named after its most obvious phenotype, which is the loss of wing margin tissue at the distal end of the wing [8–11]. Cells located at the dorsal–ventral (D/V) boundary of the wing disc eventually form the margin of the adult wing [12]. Subsequent studies reveal that Notch signaling promote cell proliferation and cell survival in these cells, through activating the expression of target genes such as *vestigial*, *wingless*, and *cut* [13–18]. The Notch signaling activity could also be visualized in the wing disc by an NRE-GFP transgenic reporter [19]. After clonal cells of lethal mutations are induced by the FLP/FRT system, Notch signaling defects could be identified by examination of marginal nicks in the wing and Notch targets expression in the wing disc [20, 21]. Thus, Notch signaling regulators could be rapidly isolated from existing mutant collections.

2 Materials

2.1 Drosophila Stocks

Listed below are used for screening the left arm of the second chromosome, using FRT40A containing lethal mutations from the BruinFly collection [22].

1. The BruinFly stocks. These could be obtained from the KYOTO Stock Center (DGRC, https://kyotofly.kit.jp).
2. The *Ubx-FLP; Ubi-mRFP, FRT40A* stock is used to generate mosaic clones in the developing wing [21].
3. The *hs-Flp; M(2)24F, arm-LacZ, FRT40A/Cyo* stock is used to generate large clones in the *Minute* background [21].
4. The NRE-GFP reporter line could be ordered from the Bloomington Drosophila Stock Center (BDSC, #30728, http://flystocks.bio.indiana.edu).
5. The Deficiency kit contains a predefined set of stocks which are ordered from BDSC.

2.2 Lab Equipment

1. Basic equipment to perform fly works, including stereomicroscope with light source, CO_2 anesthetizer, equipment to prepare and store fly food, 18 °C and 25 °C incubators, −20 °C freezer, and so on.

2. 37 °C water bath.
3. Tabletop rotator for 1.5 mL Eppendorf tubes.
4. Dissecting tools: 9-well glass plate, buffers, and fine forceps.
5. Slides and coverslips.
6. Stereo microscope fluorescence adapter (Nightsea, https://nightsea.com/), for sorting larvae with GFP or RFP marker.
7. Confocal microscope, for acquiring fluorescence images.
8. QImaging QICAM Fast 1394 digital camera, for imaging adult wings.

2.3 Solutions and Reagents

1. Phosphate buffered saline (PBS): PBS (1×, pH 7.4) is prepared from commercial mixtures and used as base for other solutions.
2. 4% paraformaldehyde (PFA): 20.0 g PFA is dissolved in 500.0 mL 1× PBS and the pH is adjusted to 7.2–7.3 (*see* **Note 1**). Aliquot into 1.5 mL Eppendorf tubes and keep frozen at −20 °C. We usually put 250 μL 4% PFA into each tube and avoid repeated freeze-thaw cycles (*see* **Note 2**).
3. PBS with 0.1% Triton (PBST): A 10% Triton X-100 stock solution is prepared with 1× PBS, which is further diluted 100× times with 1× PBS to use as 1× PBST.
4. PBST with 0.2% bovine serum albumin (BSA): A 2% BSA stock solution is prepared with 1× PBS, and then diluted ten times by 1× PBST. The PBST-BSA is used as the blocking solution and for diluting antibodies.
5. 40% glycerol is used as mounting medium for wing discs. Aliquots with a volume of 500 μL are kept at −20 °C, and several tubes can be stored at 4 °C for daily use.
6. Isopropanol: for fixing adult wings before dissection.
7. Euparal mounting medium is purchased from BioQuip products (#6372A, http://bioquip.com) and used for mounting wings.

2.4 Antibodies

1. Primary antibodies: mouse anti-Cut (2B10; 1:200), mouse anti-Wingless (Wg; 4D4; 1:200), mouse anti-Notch intracellular domain (NICD; C17.9C6; 1:200) and mouse anti-Delta (Dl; C594.9B; 1:200). These antibodies are purchased from the Developmental Studies Hybridoma Bank (DSHB, https://dshb.biology.uiowa.edu). The rabbit anti-β-Galactosidase antibody is acquired from MPBio (#085597-CF, https://www.mpbio.com).
2. Alexa 488- and Alexa 568-conjugated secondary antibodies (Thermo Fisher, https://www.thermofisher.com/) are used.

3 Methods

3.1 Primary Screen

1. Culture the *Ubx-FLP;Ubi-mRFP, FRT40A* stock in large bottles at 25 °C and change the food every 3 days. Maintain the bottles at 18 °C when the progenies begin to eclose, so that the time the females remain as virgins could be extended. Collect virgins every 4 h and start to set up crosses when you have around 100 healthy virgins.
2. Each BruinFly mutant strain is crossed with the *Ubx-FLP;Ubi-mRFP, FRT40A* stock individually. Five males of a BruinFly mutant and five virgins of *Ubx-FLP;Ubi-mRFP, FRT40A* stock are put into one vial containing standard food and maintained at 25 °C. The food is changed every 3 days until the progenies start to emerge.
3. For the primary screen, at least 100 F1 progenies of each cross are examined for wing defects. Wing marginal nicks are most likely caused by Notch signaling defects (Fig. 1). Exactly how many progenies are examined, how many of them display wing margin defects, and the characteristic of the defects are recorded.
4. Adults exhibiting wing phenotypes are immersed in isopropanol in a 1.5 mL tube and fixed for at least 24 h.
5. Wings are dissected in isopropanol and mounted in the Euparal mounting medium. Wings from parental stock of the same gender can be used as control. Remove the wings from the

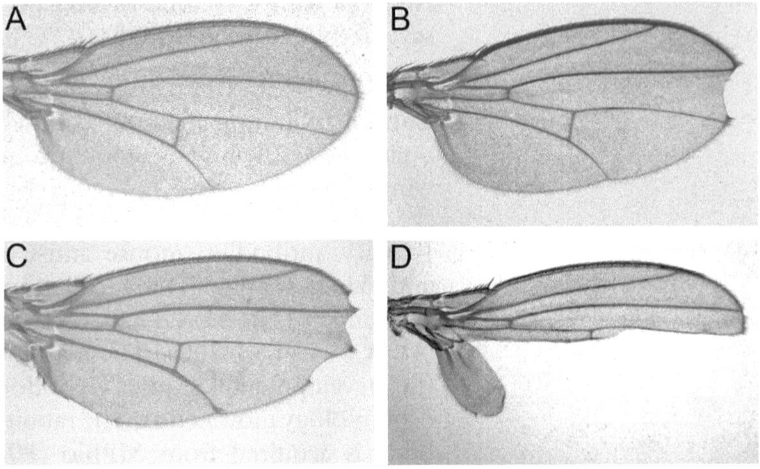

Fig. 1 Wing phenotypes that are potentially related with Notch signaling defects. (**a**) The adult wing of parental *Ubx-FLP;Ubi-mRFP, FRT40A* stock is used as wild-type control. Induction of clones of several BruinFly mutants in the wing blade led to defective wing margin formation, with different degrees of severity (**b** small nicks; **c** medium size notches; **d** large truncation)

trunk by clipping them off at the hinge region. Get rid of any tissues connected with the wings and rinse them with isopropanol.

6. Add a small drop of isopropanol to the middle of a microscope slide. Carefully transfer the wings into the isopropanol drop and use the forceps to separate the wings from each other. Before the isopropanol dries out, add another drop of the Euparal mounting medium onto the wings. Gently lay a coverslip down over the wings, and slightly press the coverslip with forceps to get rid of air bubbles when necessary.
7. Air dry the Euparal mounting medium by leaving the slides on benchtop for at least 24 h. These slides can be stored for several years after images are taken.
8. Analyze the wing phenotypes and group them by the size and penetrance of wing margin nicks.

3.2 Secondary Screen

Mutants that displayed wing margin defects are selected for secondary screen, during which Notch signaling activity was monitored in third-instar larval wing imaginal discs.

1. Males of selected BruinFly mutant strain are crossed with virgins of *Ubx-FLP;Ubi- mRFP, FRT40A*. The crosses are maintained at 25 °C and the food are changed every 3 days. After 6–7 days, larvae in the first vial will be at the third instar stage and ready for dissection.
2. Pick about 10 wondering third instar larvae into a well of the dissection plate and rinse with PBS. Under the dissecting microscope, sort the larvae with fluorescence adapter and discard the ones without RFP marker.
3. Transfer the larvae with RFP marker into a new well filled with ice cold PBS. Dissect the larva by gently removing the tail end and keep the head with approximately half of the larvae. Use the forceps to turn the head inside out and remove excessive fat and guts. You may want to keep the salivary glands so that the dissected heads could be easily observed and handled in following steps.
4. Transfer the dissected heads into 1.5 mL Eppendorf tubes containing 4% PFA and fix the tissues at room temperature for 15 min. Normally the dissected heads can sit in cold PBS for no more than 20 min. After the heads are put into Eppendorf tubes, always keep them in dark to avoid fluorescence quenching. Covering the tubes with foil or small carboard box would be enough.
5. Remove the fixative and wash the heads with 1 mL PBST. Use the tabletop rotator for all washing and incubating steps, and make sure the heads are not sticked to each other. After

washing for 5 min, remove the wash buffer and add 300 μL PBST-BSA. Rotate at room temperature for 1 h and replace with 100 μL primary antibody solution at the working concentration. We use mouse monoclonal antibodies against Cut (DSHB; 2B10) and Wg (DSHB; 4D4) at a dilution of 1:200 for the immunostaining experiments.

6. Incubate overnight at 4 °C with gentle rotation. Carefully transfer the primary antibody solution to a new Eppendorf tube and store at 4 °C, most DSHB antibodies could be reused for two or more times. Wash three times with 1 mL PBST at room temperature, and each wash step should be no less than 5 min.
7. Remove the wash buffer and add 200 μL secondary antibody solution. Incubate at room temperature with gentle rotation for 1 h. Discard the secondary antibody solutions, and wash three times with 1 mL PBST. After the final wash, remove the wash buffer and replace with 1 mL PBS.
8. Transfer the tissues to the 9-well glass plate and dissect in PBS. Gently pull the wing discs away from the rest of the tissue using fine forceps. Never touch the pouch area of the disc and remove the tracheal trunks as much as possible.
9. Add a small drop of 40% glycerol to the middle of a microscope slide. Carefully transfer the wing discs into the glycerol drop by clamping at the notum region with forceps. Use the forceps to separate the wing discs from each other and gently lay a coverslip down over the discs. Remove excess glycerol with paper wipes, and seal the coverslip with clear nail polish.
10. Analyze the discs for generation of mutant clones and Cut/Wg/NRE-GFP expression pattern under a confocal microscope. The mutants that impact the expression of Notch targets expression are considered as Notch signaling modulators, and their effects on Notch and Delta could be further examined (Fig. 2).
11. For mutants that resulting in cell growth defects, generating clones in the *Minute* background will be necessary (*see* **Note 3** and Fig. 3).

3.3 Validation Tests

Mutants that result in wing margin nicks and Cut/Wg expression defects likely disrupt Notch signaling transduction. The majority of BruinFly stocks have been molecularly mapped and the exact gene mutated in a specific stock can be identified. These genes are considered as candidates of Notch signaling regulators. However, before any conclusions are drawn, validation tests should be performed.

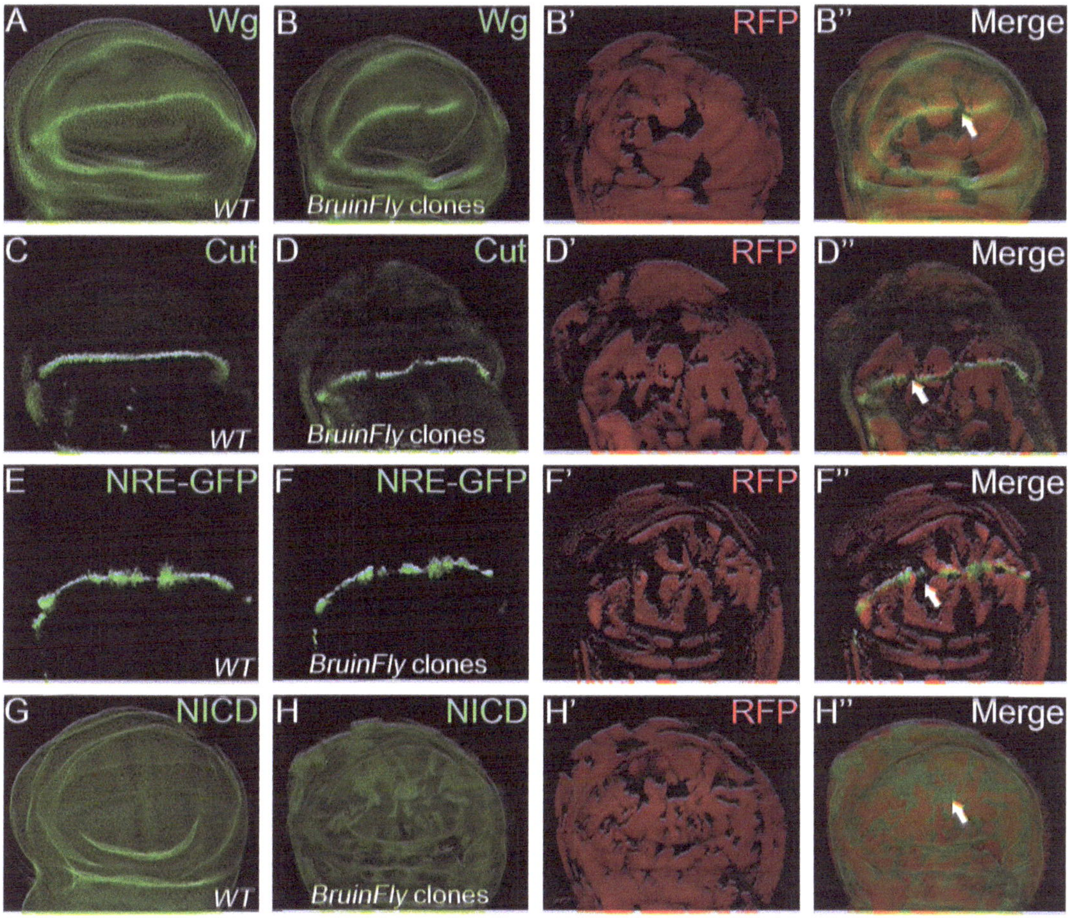

Fig. 2 Example of a mutant that affects Notch signaling activity. (**a**, **b**) Wg are expressed in cells located at the D/V boundary (**a**), and Wg expression are reduced in mutant clones (**b**). (**c**, **d**) Expression of Notch signaling target Cut (**c**) are abolished in mutant clones (**d**). (**e–f**) The expression of Notch activity reporter NRE-GFP (**e**) are disrupted in mutant clones (**f**). (**g–h**) NICD are found to be accumulated in the mutant cells, while the Notch signaling activity are reduced in these mutant cells as shown in (**b**, **d**, and **f**). The clones are marked by absence of RFP, representative clones are indicated by arrows

1. It has been suggested that about 5% of the BruinFly FRT40A stocks might contain second-site mutation alleles [23]. Therefore, a thorough compensation test using the deficiency kit is necessary. Ideally, all but only one deficiency line which covers the genomic region of the candidate gene should be able to compensate the lethality of the mutant stock. If two or more nonoverlapping deficiency lines fail to compensate the lethality of the mutant stock, it is possible that a second mutation allele is flowing in the stock.
2. Search for other mutant alleles of the candidate gene, and examine whether they impact the wing margin and Notch targets expression in a similar fashion.

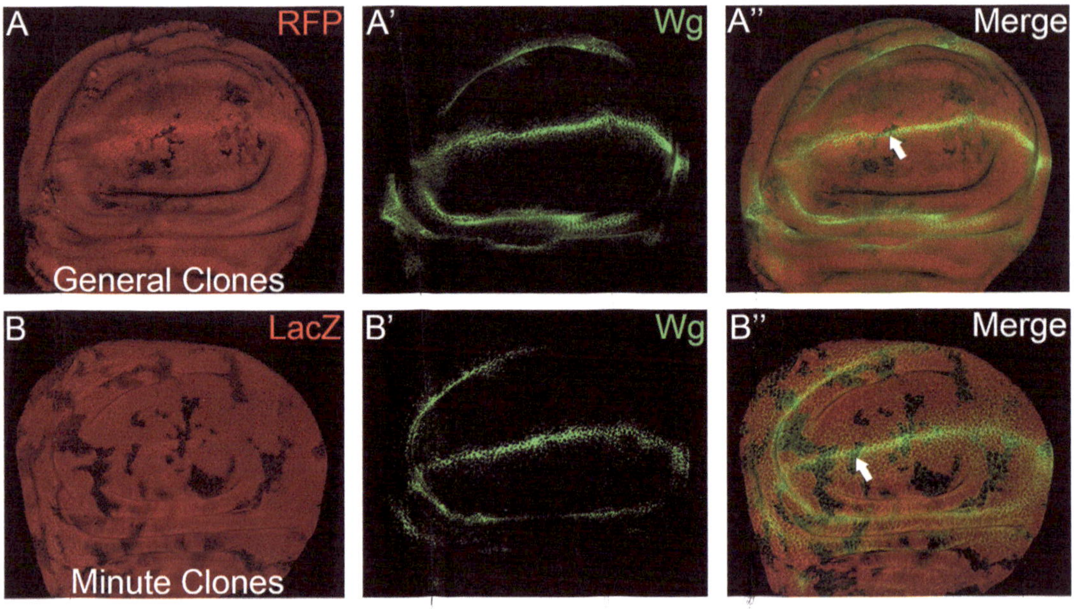

Fig. 3 Increase clone size in the *Minute* background. (**a**) Small clones are obtained in the wild-type background. (**b**) Medium sized clones are generated in the *Minute* background, and reduction of Wg expression is evident in the clones. The clones are marked by absence of RFP (**a**) or LacZ (**b**), representative clones are indicated by arrows

3. Transgenic RNAi lines against the candidate gene are also valuable resources. They can be used to knock down the expression of candidate genes in the wing disc, and the resulting wing and Notch activity defects could be compared with mutant alleles. Transgenic RNAi stocks could be obtained from the Vienna Drosophila Resource Center (VDRC, https://stockcenter.vdrc.at), National Institute of Genetics (NIG-Fly, https://shigen.nig.ac.jp/fly/nigfly/), and BDSC.
4. The traditional PCR method [24] and a modified Splinkerette PCR method [25] could be used to identify the mutation sites of BruinFly stocks if they have not been mapped yet. Alternatively, whole genome sequencing is proven to be affordable and accurate for identifying transposon insertion sites [26].
5. Candidates that pass the validation tests are likely involved in Notch signaling regulation during wing development, experiments could be designed to reveal the underlying mechanism. Any mutant collection could be screened by this scheme as long as an FRT site is included (*see* **Note 4**).

4 Notes

1. To dissolve PFA, heat and stir the mixture on hot plate slowly, and keep the temperature below 60 °C, otherwise PFA will be denatured. PFA dissolves slowly, typically over a period of 20–60 min. Add drops of 1 M NaOH while stirring continuously until all PFA is dissolved. Let the solution cool down to room temperature before adjusting pH.
2. PFA is hazardous and irritating, always wear mask and gloves to avoid direct contact. Weight PFA and prepare the PFA solution in fume hood. All solutions containing PFA should be disposed of properly.
3. Mutants that impairing cell survival and growth give rise to small clones in the wing disc, making it difficult to analyze changes in target gene expression. Generating clones in the *Minute* background will help to increase the size of mutant clones [21, 27]. When males of mutant stock are crossed with the *hs-Flp;M (2)24F, arm-LacZ, FRT40A/Cyo* virgins, medium to large sized clones are induced (Fig. 3). Place 10 males and 10 females in fresh vials of food and change food every day. At day 4, immerse the first vial into 37 °C water bath for 1 h, then return the vial to the 25 °C incubator. Third instar larvae are dissected and the expression of Notch targets are examined by immunostaining. Chances to obtain large enough clones crossing the D/V boundary are quite low, you may need to dissect 20–30 larvae for each experiment.
4. The screen scheme can be applied to other collections of mutations when they are recombined with FRT sites. You can also generate your own mutant library by EMS and P-element mutagenesis [28].

References

1. Bankers CM, Hooper JE (2007) Clonal analysis of Hedgehog signaling in Drosophila somatic tissues. Methods Mol Biol 397:145–160
2. Golic KG, Lindquist S (1989) The FLP recombinase of yeast catalyzes site-specific recombination in the Drosophila genome. Cell 59: 499–509
3. Xu T, Rubin GM (1993) Analysis of genetic mosaics in developing and adult Drosophila tissues. Development 117:1223–1237
4. Rubin T, Huynh JR (2015) Mosaic analysis in the Drosophila melanogaster ovary. Methods Mol Biol 1328:29–55
5. Germani F, Bergantinos C, Johnston LA (2018) Mosaic analysis in Drosophila. Genetics 208:473–490
6. Bray SJ (2006) Notch signalling: a simple pathway becomes complex. Nat Rev Mol Cell Biol 7:678–689
7. Guruharsha KG, Kankel MW, Artavanis-Tsakonas S (2012) The Notch signalling system: recent insights into the complexity of a conserved pathway. Nat Rev Genet 13:654–666
8. Morgan TH, Bridges CB (1916) Sex-linked inheritance in Drosophila. Carnegie Inst Wash Publ 237:1–88

9. Morgan TH (1917) The theory of the gene. Am Nat 51:513–544
10. Dexter JS (1914) The analysis of a case of continuous variation in Drosophila by a study of its linkage relations. Am Nat 48:712–758
11. Mohr OL (1919) Character changes caused by mutation of an entire region of a chromosome in Drosophila. Genetics 4:275–282
12. Blair SS (2007) Wing vein patterning in Drosophila and the analysis of intercellular signaling. Annu Rev Cell Dev Biol 23:293–319
13. de Celis JF, Garcia-Bellido A, Bray SJ (1996) Activation and function of Notch at the dorsal-ventral boundary of the wing imaginal disc. Development 122:359–369
14. Neumann CJ, Cohen SM (1996) A hierarchy of cross-regulation involving Notch, wingless, vestigial and cut organizes the dorsal/ventral axis of the Drosophila wing. Development 122: 3477–3485
15. Micchelli CA, Rulifson EJ, Blair SS (1997) The function and regulation of cut expression on the wing margin of Drosophila: Notch, wingless and a dominant negative role for Delta and Serrate. Development 124:1485–1495
16. Giraldez AJ, Cohen SM (2003) Wingless and Notch signaling provide cell survival cues and control cell proliferation during wing development. Development 130:6533–6543
17. Rafel N, Milán M (2008) Notch signalling coordinates tissue growth and wing fate specification in Drosophila. Development 135: 3995–4001
18. Jia D, Bryant J, Jevitt A, Calvin G, Deng WM (2016) The ecdysone and Notch pathways synergistically regulate cut at the dorsal-ventral boundary in Drosophila wing discs. J Genet Genomics 43:179–186
19. Saj A, Arziman Z, Stempfle D, van Belle W, Sauder U, Horn T, Dürrenberger M, Paro R, Boutros M, Merdes G (2010) A combined ex vivo and in vivo RNAi screen for notch regulators in Drosophila reveals an extensive notch interaction network. Dev Cell 18:862–876
20. Yu Z, Wu H, Chen H, Wang R, Liang X, Liu J, Li C, Deng WM, Jiao R (2013) CAF-1 promotes Notch signaling through epigenetic control of target gene expression during Drosophila development. Development 140: 3635–3644
21. Ren L, Mo D, Li Y, Liu T, Yin H, Jiang N, Zhang J (2018) A genetic mosaic screen identifies genes modulating Notch signaling in Drosophila. PLoS One 13:e0203781
22. Chen J, Call GB, Beyer E, Bui C, Cespedes A, Chan A et al (2005) Discovery-based science education: functional genomic dissection in Drosophila by undergraduate researchers. PLoS Biol 3:e59
23. Roegiers F, Kavaler J, Tolwinski N, Chou YT, Duan H, Bejarano F, Zitserman D, Lai EC (2009) Frequent unanticipated alleles of lethal giant larvae in Drosophila second chromosome stocks. Genetics 182:407–410
24. Spradling AC, Stern D, Beaton A, Rhem EJ, Laverty T, Mozden N, Misra S, Rubin GM (1999) The Berkeley Drosophila Genome Project gene disruption project: single P-element insertions mutating 25% of vital Drosophila genes. Genetics 153:135–177
25. Potter CJ, Luo L (2010) Splinkerette PCR for mapping transposable elements in Drosophila. PLoS One 5:e10168
26. Chang X, Zhang F, Li H, Mo D, Shen J, Zhang J (2021) Characterization of a new mastermind allele identified from somatic mosaic screen. Cells Dev 165:203664
27. Morata G, Ripoll P (1975) Minutes: mutants of Drosophila autonomously affecting cell division rate. Dev Biol 42:211–221
28. Giagtzoglou N (2014) Genetic screens to identify new Notch pathway mutants in Drosophila. Methods Mol Biol 1187:15–28

Chapter 5

High-Throughput Analysis to Identify Activators of Notch Signaling

Rachael Guenter, Jacob Eide, Herbert Chen, J. Bart Rose, and Renata Jaskula-Sztul

Abstract

The Notch pathway regulates many cellular functions in a context-dependent manner. Depending on the cell type, either the activation or inhibition of Notch signaling can influence many processes such as cellular proliferation, specification, differentiation, and survival. The activation of Notch signaling has been shown to have therapeutic advantages in some cancers, thus having a method to identify Notch-activating compounds is needed. In this chapter we outline a method for high-throughput analysis of potential Notch pathway activators in a pancreatic neuroendocrine tumor cell line as an example. We also include the steps for subsequent validation of results and preclinical testing.

Key words High-throughput analysis, Notch pathway, Notch signaling, Notch activation, Cancer therapeutics, Luciferase reporter plasmid

1 Introduction

The Notch pathway is an evolutionarily conserved signaling cascade in mammals responsible for various cellular functions. This pathway comprises of four single-pass transmembrane receptors (Notch1–4) and five ligands (DLL1,3,4, Jagged1,2) [1]. Canonically, Notch signaling operates in a juxtracrine manner, in which the binding of receptors and ligands between adjacent cells activates signal transduction events. The first event after receptor and ligand binding is the S2 cleavage, where a metalloproteinase (ADAM17) cleaves the full-length Notch receptor into a shorter protein called the transmembrane/intracellular domain (TMIC). Subsequently, this TMIC molecule is cleaved again (S3 cleavage) by gamma-secretase to form the Notch intracellular domain (NICD), creating the active form of the Notch protein. The newly formed NICD translocates

Rachael Guenter and Jacob Eide are co-first authors.

Dongyu Jia (ed.), *Notch Signaling Research: Methods and Protocols*,
Methods in Molecular Biology, vol. 2472, https://doi.org/10.1007/978-1-0716-2201-8_5,

to the nucleus of the cell where it binds the CSL/Rasl/CBF repressor complex to consequently regulate various genes. Genes belonging to the Hes (Hairy/enhancer of Split) and the Hey (Hairy/Enhancer of Split related with YRPW motif) families are well categorized as direct targets of Notch signaling.

In human cancer cells, the Notch pathway functions in a context-dependent manner, acting as either a tumor suppressor or an oncogene. In certain cancer types, such as pancreatic cancer, colon cancer, non–small-cell lung cancer, renal cell carcinoma, and several lymphomas, Notch signaling upregulation has been shown to stimulate cell proliferation and prevent apoptosis. Conversely, Notch signaling has a tumor suppressive function in skin cancer [2, 3] and neuroendocrine tumors (NETs), including small-cell lung cancer (SCLC), medullary thyroid cancer (MTC), and both pulmonary and gastrointestinal carcinoids [4–8]. Thus, it could be therapeutically beneficial to identify novel Notch pathway-activating compounds for patients with cancers deficient in Notch signaling.

In order to identify Notch pathway activators, a cell-based high-throughput screening (HTS) assay was developed to screen a library of 7264 compounds. HTS is a drug-discovery process used to evaluate the biochemical activity of a large number of drug-like compounds with the purpose of discovering a potential therapy [9]. One of the most daunting aspects of HTS is determining which compounds to test. Several libraries of drugs were tested for their ability to activate the Notch pathway. In general, libraries of compounds with known safety, bioavailability, and high purity were preferred for HTS. Moreover, a mixture of both untested drugs and currently available compounds were included in the study to increase the number of compounds tested and, therefore, the chance of finding a potent Notch activator.

In our cell-based high-throughput assay, functional Notch activity was measured via a pancreatic NET cell line (BON) expressing a CBF1-luciferase (CBF1-Luc) reporter plasmid. Centromere-binding factor 1 (CBF-1) is a ubiquitous transcription factor and critical downstream protein in the human Notch signaling pathway. Notch activation was therefore detected by an increase in the luciferase activity following treatment with a specific drug. The compounds expressing the largest relative luciferase activity (resveratrol, chrysin, and hesperetin) in HTS screening and Notch pathway activation were chosen for further testing and validation of antitumor properties [10]. Testing included repeating the primary reporter gene assay on the activators ("hits") from the HTS, confirmation of Notch induction at the protein and message levels, examination of downstream effects of Notch activation, reversal of those effects using Notch pathway inhibitors, and in vivo testing. The present protocol provides a description of the process for identifying compounds that induce the Notch pathway for studying the effects of Notch pathway activation.

2 Materials

2.1 Cell Culture

1. The human pancreatic NET cell line (BON-1, RRID: CVCL_3985) was provided by Dr. Mark Hellmich (University of Texas, Galveston, TX). BON-1 cells are grown using Dulbecco's modified Eagle's medium (DMEM)/F12, supplemented with 10% fetal bovine serum (FBS), 100 IU/mL penicillin, and 100 Ag/mL streptomycin. BON-CBF1-Luc cells 1 cells are grown using Dulbecco's modified Eagle's medium (DMEM)/F12, supplemented with 10% fetal bovine serum (FBS), 100 IU/mL penicillin, 100 Ag/mL streptomycin, and puromycin (5 μg/mL).

2.2 Plasmids and Reagents

1. JH23A plasmid (Diane Hayward, Baltimore, MD).
2. pGL4.20[luc2/Puro] vector (Promega, Madison, WI).
3. Kpn1 and hindIII restriction endonucleases (New England Biolabs, Ipswich, MA).
4. T4 DNA Ligase (New England Biolabs, Ipswich, MA).
5. Suberoylanilide hydroxamic acid (SAHA) (Sigma-Aldrich, St. Louis, MO).
6. Valproic acid (VPA) (Sigma-Aldrich, St. Louis, MO).
7. BrightGlo reagent (Promega, Madison, WI).
8. CellTiter 96® nonradioactive cell proliferation assay (MTT) assay (Promega, Madison, WI).

2.3 Specialized Equipment and Resources for High-Throughput Screening

1. Monolight 3010 Luminometer (PharMingen, San Diego, CA).
2. MicroFlo select noncontact reagent dispenser (BioTek, Winooski, VT).
3. Biomek FX liquid handler (Beckman Coulter, Brea, CA).
4. Safire 2 microplate reader (Tecan, Mannedorf, Switzerland) (*see* **Note 1**).
5. Assay database (such as Activitybase, IDBS Inc., Alameda, CA).

2.4 Compound Libraries

1. KBA01 library includes three libraries available commercially. They include compounds that have previously determined bioactivity and safety profiles from Prestwick Chemicals (Washington, DC), drugs that are FDA approved, and natural compounds derived from Microsource Discovery Systems (Gaylordsville, CT), as well as the Library of Pharmacological Active Compounds, which includes well-studied drugs, failed trials, and currently marketed drugs (Sigma, St Louis, MO).
2. NIH Clinical Collection 1 is a National Cancer Institute Developmental Therapeutics Program library that consists of compounds that have been used in human clinical trials.

3 Methods

3.1 Generation of Stable CBF1 Reporter Cell Line

1. Using JH32A plasmid, digest with kpn1 and hindIII restriction endonucleases to extract the 4xCBF1 binding site sequence. Ligate the excised 4xCBF1 segment into pGL4.2 plasmid using T4 DNA Ligase (New England Biolabs, Ipswich, MA).
2. Stably transfect human cells with the previously constructed plasmid (pGL4.2 CBF1) and verify the success of the transfection using puromycin selection. Puromycin-resistant cells are now BON-CBF1-Luc clones (Fig. 1).

3.2 Dose-Dependent Luciferase Induction

1. Validate the functionality of the reporter plasmid. Valproic acid (VPA) or suberoylanilide hydroxamic acid (SAHA) (Sigma-Aldrich, St. Louis, MO)—both known Notch activators in endocrine cancer cell lines—may be used to test for increased luciferase activity as measured by luminometer following treatment of transfected cells compared to a transfected, but untreated, control (*see* **Note 1**). Concentrations of 1 mM for VPA and 3 μM SAHA are appropriate for plasmid testing [11].
2. Significant, dose-dependent increases in luciferase treatment must be observed before BON-CBF1-Luc cells are used for high-throughput screening (Fig. 2).

3.3 High-Throughput Assay

1. Select compound library to test for Notch activation activity.
2. Plates 25,000 BON-CBF1-Luc cells per well in 96 well microtiter dishes using a MicroFlo noncontact reagent dispenser. Allow the cells to adhere overnight. For screens in 384-well format, 5000 cells per well are plated.

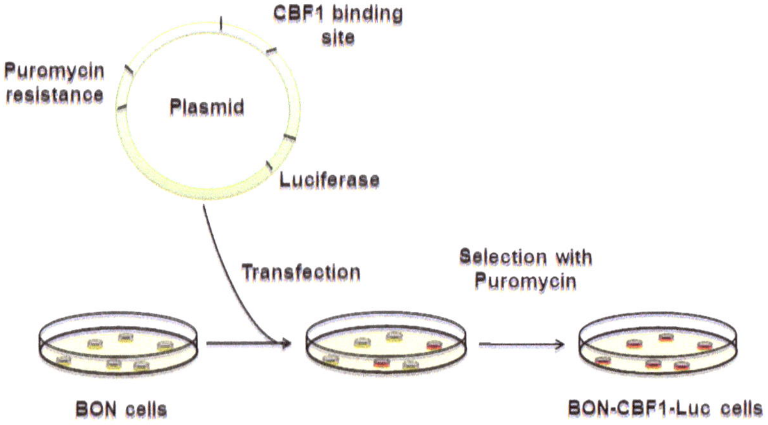

Fig. 1 Creation of BON-CBF1-luc cell line. BON cells are stably transfected with a centromere-binding factor 1 (CBF1)-luciferase reporter vector, which also confers puromycin resistance. Nontransfected cells fail to survive under puromycin selection, and all remaining cells are BON-CBF1-Luc cells

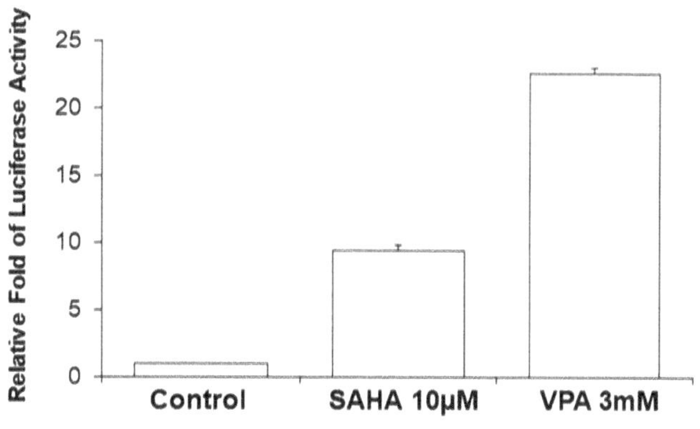

Fig. 2 Dose-dependent induction of luciferase activity in BON-CBF1-Luc cells after treatment with known Notch activators, suberoylanilide hydroxamic acid (SAHA) and valproic acid (VPA). These results confirm that the transfected plasmid is functioning correctly and may be used to test novel compounds in high-throughput screening

3. For 96 well plates, with 100 μL of media. Use a Biomek FX liquid handler to deliver a final volume of 0.5 μL of each library compound to each well. The final DMSO concentration is 0.5%. For screens in 384-well plates, compounds can be added via pin transfer.
4. Concentrations of library compounds may be adjusted depending on the library of compounds being used. The KBA01 library should be tested at 1 μM and the NIH Clinical Collection 1 library should be tested at a final concentration of 10 μM.
5. Following 24-h incubation with compounds, the cells are lysed (*see* **Note 2**). Luciferase activity is detected via BrightGlo reagent (Promega, Madison, WI) using a Safire 2 microplate reader, and a luminescence setting of 1000 ms integration time (*see* **Note 1**).
6. Perform all primary screens in duplicate and compare quality across replicates.
7. Use compounds that elicit a greater than 150% signal increase to calculate the *Z*-factor value and compounds causing more than a 300% increase to calculate a *Z*-prime value. *Z* and *Z*′ values above 0.5 indicate a robust assay fit for HTS [10].
8. To exclude nonspecific activators, query an assay database (e.g., Activitybase) that contains screening data previously acquired in other assays for the compounds being tested as Notch activators.
9. Search for compounds that show >150% in signal for retesting. Perform another round of high-throughput testing on these highly activating compounds.

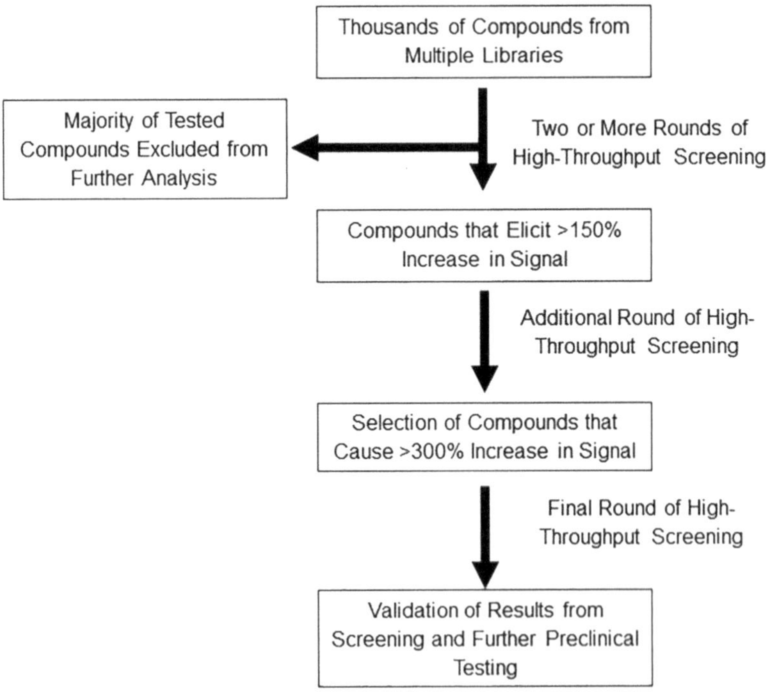

Fig. 3 Flowchart demonstrating how Notch activators were selected using high-throughput screening technology. Many rounds of testing are included to ensure that the compounds identified are specific Notch activators

10. Of the compounds that were retested, select those that consistently show a >300% increase in signal for a final round of high-throughput screening and further validation (Fig. 3).

3.4 Validation of High-Throughput Results in Endocrine Cancer Cell Lines

1. Ascertain that the Notch pathway is being activated. Use either Western-blotting, quantitative real-time PCR, or a combination of both to demonstrate that one of the Notch isoforms shows increased expression on the message or protein level [10–12].
2. Show that expression levels of downstream targets of the Notch pathway are affected. Previous research has demonstrated downregulation of achaete scute complex-like 1 (ASCL1) as well as modulation of levels of members of both the hairy/enhancer of split (HES) and hairy/enhancer of split related with YRPW motif (HEY) families. All should be examined on the protein or message levels [13, 14].
3. Notch pathway inhibitors may be used to reverse the effects upon downstream targets found following treatment, further supporting the designation of a tested compound as a Notch pathway activator. Commonly, a γ-secretase inhibitor (GSI) is added to inhibit cleavage of the Notch protein's intercellular domain—a fragment that travels to the nucleus and affects

transcription of Notch and its downstream targets. If the compound in question is not a Notch activator but is affecting another pathway, γ-secretase inhibitor treatment will not affect levels of Notch or Notch's downstream targets [11, 15, 16] (*see* **Note 3**).

4. The Notch pathway has been found to reduce cell proliferation via cell-cycle and/or apoptotic mechanisms, depending on which Notch isoform is being activated and the identity of the cells in which the activation occurs. Treatment with a Notch activator should reproduce these effects and should be checked by determining the half maximal inhibitory concentration (IC50) to determine the concentrations at which the compound becomes cytotoxic. Once the IC50 is known, a 3-(4,5-dimethylthiazol-2-yl)-2,5-diphenyltetrazolium bromide (MTT) assay (CellTiter 96® nonradioactive cell proliferation assay (MTT) (Promega, Madison, WI)) can be performed using nontoxic concentrations (*see* **Note 4**). Increasing concentration of a Notch activator should reduce cell proliferation in a time- and dose-dependent manner [11, 12, 17].

4 Notes

1. Luciferase detection in HTS. The amount of luciferase signal produced by the BON-CBF1-Luc cells can be low, accordingly, a plate reader with very high sensitivity, such as the PE Enspire reader (PerkinElmer, Waltham, MA) can be used.
2. Complete lysis of BON-CBF1-Luc cells is critical for good luciferase detection. It is recommended that cells are first lysed using Glo lysis buffer (Promega, Madison, WI), and one freeze–thaw cycle (10 min at −80 °C), followed by a 30 min incubation at 37 °C. Once the cells are lysed, add an equal volume of luciferase substrate, shake briefly, and read plates.
3. Examples of γ-secretase inhibitors (GSIs) to use include *N*-[*N*-(3,5-difluorophenacetyl)-L-alanyl]-*S*-phenylglycine *t*-butyl ester (DAPT) and nirogacestat (PF-03084014).
4. To minimize dimethyl sulfoxide (DMSO) concentration during screening, pins can be used for compound addition.

References

1. Takebe N, Nguyen D, Yang SX (2014) Targeting Notch signaling pathway in cancer: clinical development advances and challenges. Pharmacol Ther 141(2):140–149
2. Okuyama R, Ogawa E, Nagoshi H, Yabuki M, Kurihara A, Terui T, Aiba S, Obinata M, Tagami H, Ikawa S (2007) p53 homologue, p51/p63, maintains the immaturity of keratinocyte stem cells by inhibiting Notch1 activity. Oncogene 26(31):4478–4488. https://doi.org/10.1038/sj.onc.1210235

3. Demehri S, Turkoz A, Kopan R (2009) Epidermal Notch1 loss promotes skin tumorigenesis by impacting the stromal microenvironment. Cancer Cell 16(1):55–66. https://doi.org/10.1016/j.ccr.2009.05.016
4. Platta CS, Greenblatt DY, Kunnimalaiyaan M, Chen H (2008) Valproic acid induces Notch1 signaling in small cell lung cancer cells. J Surg Res 148(1):31–37. https://doi.org/10.1016/j.jss.2008.03.008
5. Kunnimalaiyaan M, Traeger K, Chen H (2005) Conservation of the Notch1 signaling pathway in gastrointestinal carcinoid cells. Am J Physiol Gastrointest Liver Physiol 289(4): G636–G642. https://doi.org/10.1152/ajpgi.00146.2005
6. Jang S, Janssen A, Aburjania Z, Robers MB, Harrison A, Dammalapati A, Cheng YQ, Chen H, Jaskula-Sztul R (2017) Histone deacetylase inhibitor thailandepsin—a activates Notch signaling and suppresses neuroendocrine cancer cell growth in vivo. Oncotarget 8(41):70828–70840. https://doi.org/10.18632/oncotarget.19993
7. Mohammed TA, Holen KD, Jaskula-Sztul R, Mulkerin D, Lubner SJ, Schelman WR, Eickhoff J, Chen H, LoConte NK (2011) A pilot phase II study of valproic acid for treatment of low-grade neuroendocrine carcinoma. Oncologist 16(6):835–843. https://doi.org/10.1634/theoncologist.2011-0031
8. Ning L, Jaskula-Sztul R, Kunnimalaiyaan M, Chen H (2008) Suberoyl bishydroxamic acid activates Notch1 signaling and suppresses tumor progression in an animal model of medullary thyroid carcinoma. Ann Surg Oncol 15(9):2600–2605. https://doi.org/10.1245/s10434-008-0006-z
9. High-throughput molecular screening center (2019). In: Scripps Res Inst. https://www.scripps.edu/science-and-medicine/cores-and-services/high-throughput-molecular-screening-center/index.html. Accessed 4 Dec 2021
10. Zhang JH, Chung TDY, Oldenburg KR (1999) A simple statistical parameter for use in evaluation and validation of high throughput screening assays. J Biomol Screen 4(2):67–73. https://doi.org/10.1177/108705719900400206
11. Pinchot SN, Jaskula-Sztul R, Ning L, Peters NR, Cook MR, Kunnimalaiyaan M, Chen H (2011) Identification and validation of Notch pathway activating compounds through a novel high-throughput screening method. Cancer 117(7):1386–1398. https://doi.org/10.1002/cncr.25652
12. Zarebczan B, Pinchot SN, Kunnimalaiyaan M, Chen H (2011) Hesperetin, a potential therapy for carcinoid cancer. Am J Surg 201(3): 329–332. https://doi.org/10.1016/j.amjsurg.2010.08.018
13. Kunnimalaiyaan M, Vaccaro AM, Ndiaye MA, Chen H (2006) Overexpression of the NOTCH1 intracellular domain inhibits cell proliferation and alters the neuroendocrine phenotype of medullary thyroid cancer cells. J Biol Chem 281(52):39819–39830. https://doi.org/10.1074/jbc.M603578200
14. Nakakura EK, Sriuranpong VR, Kunnimalaiyaan M, Hsiao EC, Schuebel KE, Borges MW, Jin N, Collins BJ, Nelkin BD, Chen H, Ball DW (2005) Regulation of neuroendocrine differentiation in gastrointestinal carcinoid tumor cells by Notch signaling. J Clin Endocrinol Metab 90(7):4350–4356. https://doi.org/10.1210/jc.2005-0540
15. Allenspach EJ, Maillard I, Aster JC, Pear WS (2002) Notch signaling in cancer. Cancer Biol Ther 1:466–476
16. Artavanis-Tsakonas S, Rand MD, Lake RJ (1999) Notch signaling: cell fate control and signal integration in development. Science 284(5415):770–776
17. Truong M, Cook MR, Pinchot SN, Kunnimalaiyaan M, Chen H (2011) Resveratrol induces Notch2-mediated apoptosis and suppression of neuroendocrine markers in medullary thyroid cancer. Ann Surg Oncol 18(5):1506–1511. https://doi.org/10.1245/s10434-010-1488-z

Chapter 6

Artificial Notch Signaling Activation Method Using Immobilized Ligand Beads

Takamasa Mizoguchi, Hikaru Handa, Shuhei Omaru, and Motoyuki Itoh

Abstract

Activation of Notch signaling requires physical interaction between ligand- and receptor-expressing cells and pulling force to release the Notch intracellular domain. Therefore, the soluble recombinant ligand protein is not suitable for the activation of Notch signaling in a cell culture system. Here, we describe an efficient method for transient activation of Notch signaling using immobilized ligand beads. Using this method, the timing of Notch signaling can be efficiently controlled.

Key words Notch signaling, Ligand beads, Luciferase reporter assay, Immobilized

1 Introduction

Notch signaling is an important pathway that regulates various cell fates during development and plays critical roles in various fundamental processes in metazoans [1]. Notch signaling is a cell-to-cell communication system in which Notch ligands on neighboring cells activate Notch receptors. Ligand binding to the Notch receptor activates ADAM metallopeptidase 10 (ADAM10) and γ-secretase, cleaving the Notch intracellular domain (NICD) that is then translocated to the nucleus. NICD forms a complex with the DNA-binding protein CBF1/Suppressor of Hairless/LAG-1 (CSL), also known as RBPJ and regulates the transcription of downstream target genes [1, 2].

Four Notch receptor proteins (NOTCH1-NOTCH4), three Delta-like canonical Notch ligands (DLL1, DLL3, and DLL4), and two Jagged ligands (JAG1 and JAG2) have been identified in the Notch signaling pathway in mammals [1, 2]. Moreover, Fringe family proteins, Radical Fringe (RFNG), Manic Fringe (MFNG), and Lunatic Fringe (LFNG) regulate Notch signaling by differentially controlling Notch receptor sensitivity to its ligands by glycosylation [1, 2]. Therefore, the combination of Notch ligands,

Dongyu Jia (ed.), *Notch Signaling Research: Methods and Protocols*,
Methods in Molecular Biology, vol. 2472, https://doi.org/10.1007/978-1-0716-2201-8_6,

receptors and glycosylation status impacts Notch signaling activation.

However, it is challenging to induce artificial Notch activation by simply adding the extracellular domain of the ligands to the culture medium, as is routinely the case with secreted ligands. Although the Notch signaling pathway depends on the interaction between ligand and receptor, their interaction alone is not sufficient to release NICD and activate Notch signaling. Previous studies have shown that ligands need to be activated by intracellular ubiquitination to induce mechanical pulling forces and recycling [3–5].

The coculture system of Notch ligand-expressing cells and Notch receptor-expressing cells has been widely used to investigate the mechanism of Notch signaling activation. However, this approach has drawbacks in performing biochemical analysis because processes in ligand and receptor cells cannot be separated. Alternatively, biomaterials such as self-assembling monolayers [6] or microbeads [7] can be used to immobilize ligand proteins to induce mechanical pulling forces and activate Notch signaling in the cell type of interest.

We previously modified the microbead method and developed an efficient method for transient activation of Notch signaling using immobilized ligand beads [8]. Using this method, the timing and intensity of Notch signaling can be efficiently controlled. Here, we describe the detailed methods of preparing immobilized ligand beads and Notch signaling reporter assays using immobilized ligand beads.

2 Materials

2.1 Equipment

1. CO_2 incubator (37 °C, 5% CO_2).
2. Centrifuges for 15 mL tubes and 1.5 mL tubes.
3. Rotator.
4. Electrophoresis devise for SDS-PAGE.
5. Rocking shaker.
6. UV illuminator and gel imaging device.
7. Software for measuring band intensity.
8. Luminometer (*see* **Note 1**).
9. (optional) Sonicator.

2.2 Reagents and Supplies

1. Plasmids encoding avi-tagged Notch ligands, as shown in Fig. 1.
2. Negative control avi-tagged cytoplasmic beta-galactosidase (cbgal) plasmid, as shown in Fig. 1 (*see* **Note 2**).

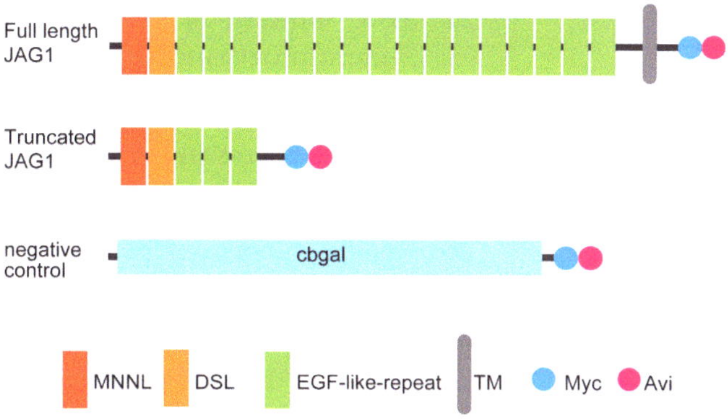

Fig. 1 Ligand plasmid structures. The structure of the C-terminal avi-tagged ligand and cbgal (negative control). Full-length and truncated JAG1 structures are shown as an example. The myc-tag is fused to the C-terminus of the Notch ligand

3. HEK293T cells.
4. 10 cm culture dishes.
5. 1.5 mL tubes.
6. 15 mL tubes.
7. DMEM: Dulbecco's modified Eagle's medium (4.5 g/L glucose) with L Gln, without sodium pyruvate, liquid.
8. DMEM/FBS/PS: DMEM containing 1% stabilized penicillin–streptomycin mixed solution and 10% fetal bovine serum.
9. PEI: 1 mg/mL polyethylenimine in water.
10. 2 mM (+)-biotin in water (sterilized by filtration).
11. Phosphate-buffered saline: PBS: 137 mM NaCl, 8.1 mM Na_2HPO_4, 2.68 mM KCl, and 1.47 mM KH_2PO_4.
12. PBS/BSA/NaN_3: PBS containing 1% bovine serum albumin: BSA and 0.05% NaN_3.
13. Streptavidin MagneSphere® paramagnetic particles (Promega).
14. Dynal MPC™-S (magnetic particle concentrator, Invitrogen).
15. Protease inhibitor cocktail for mammalian cell and tissue extracts.
16. Lysis buffer PI+: 1% protease inhibitor cocktail and 1% Triton X-100 in PBS.
17. Lysis buffer: 1% Triton X-100 in PBS.
18. PBST: 0.1% Tween 20 in PBS.
19. BSA solutions: 10, 20, 50, and 100 ng/μL in distilled water.
20. 7–10% Polyacrylamide gel.
21. 2× Sample buffer: 4% SDS, 125 mM Tris-HCl pH 6.8, 2% glycerol, and 0.05% bromophenol blue in distilled water.

22. SDS buffer: a mixture of 2× sample buffer and 2-mercaptoethanol at a ratio of 9:1.
23. Oriole fluorescent gel stain (Bio-Rad).
24. Notch1-expressing NIH3T3 (N1-3T3) [9] (*see* **Note 3**).
25. Notch reporter plasmid (pGa981–6) [10] (*see* **Note 4**).
26. Renilla luciferase plasmid: internal control for luciferase reporter assay, pRL-TK vector (Promega) (*see* **Note 5**).
27. 1× Passive lysis buffer: made by dilution of passive lysis 5× buffer (Promega) with distilled water (*see* **Note 6**).
28. Luciferase assay buffer: 50 mM Tris-HCl pH 7.8, 10 mM $MgCl_2$, 500 μM CoA, 300 μM ATP, and 200 μg/mL D-luciferin potassium salt in distilled water (*see* **Note 7**).
29. Renilla luciferase assay buffer: 40 μM PTC124, 6.67 μM coelenterazine h, 30 mM EDTA disodium salt dehydrated, 20 mM $Na_4P_2O_7$, and 950 mM NaCl in distilled water (*see* **Note 8**).
30. (optional) Liquid nitrogen.

3 Methods

3.1 Expression of Recombinant Notch Ligands

1. Seed HEK293T cells at 2 × 10^6 in a 10 cm culture dish and cultured with 10 mL of DMEM/FBS/PS at 37 °C and 5% CO_2 for 24 h.
2. Add 30 μL of PEI to 600 μL of DMEM and mix. After centrifugation, incubate the mixture at room temperature for 5 min (*see* **Note 9**).
3. Add 10 μg of Notch ligand expression plasmid to the mixture in Subheading 3.1, **step 2**, mix, centrifuge, and incubate at room temperature for 20 min.
4. Drop 100 μL of 2 mM biotin into the medium of the cultured HEK293T cells.
5. Add the transfection mixture to the medium of the cultured HEK293T cells and culture at 37 °C with 5% CO_2 for 24–48 h.

3.2 Purification of Immobilized Notch Ligand Beads

1. Detach cells from the dish by gentle pipetting and suspending them in culture medium.
2. Transfer suspended cells to a 15 mL tube and centrifuge the cells at 230 × *g* for 2 min at room temperature. After centrifugation, remove the medium without loosening the pellet.
3. Resuspend the pellet by tapping it and adding 500 μL of PBS containing 1% BSA and 0.05% NaN_3. Wash the cells by tapping and inverting the tube gently. Transfer suspended cells to a 1.5 mL tube.

4. Centrifuge the tube at 400 × *g* for 2 min at room temperature and remove the supernatant without loosening the pellet.
5. Resuspend the pellet by tapping and adding 300 μL of PBS/BSA/NaN_3. Wash cells by tapping and inverting the tube gently.
6. Centrifuge at 400 × *g* for 2 min at room temperature and remove the supernatant without loosening the pellet.
7. Repeat **steps 5** and **6** once.
8. Resuspend the pellet by tapping the tube.
9. (optional) For recombinant Notch ligands containing a transmembrane domain (e.g., full length), freeze the cell pellet with liquid nitrogen and wait for the frozen pellet to melt on ice (*see* **Note 10**).
10. Add 1 mL of PBS/BSA/NaN_3 and mix by tapping and inverting the tube gently.
11. Centrifuge at 4 °C at 11,000 × *g* for 10 min.
12. Remove the supernatant without loosening the pellet. Then, resuspend the pellet by tapping the tube.
13. Add 300 μL of lysis buffer PI+ to the tube and mix by inverting the tube.
14. Shake the tube at 4 °C for 30 min–1 h with a rocking shaker set at 15 rpm.
15. Wash the Streptavidin MagneSphere® during **step 14** with lysis buffer. Add 50–100 μL of Streptavidin MagneSphere® slurry into a 1.5 mL tube, and if the slurry is stuck on the tube wall and does not gather at the bottom of the tube, centrifuge it at 11,000 × *g* for a few seconds. Place the tube on a magnetic particle concentrator with the supernatant, remove the supernatant, add 500 μL of lysis buffer, detach the tube from the magnetic particle concentrator, tap, and invert tube for mixing. Then, centrifuge the tube at 11,000 × *g* and 4 °C for 3 min. Place the tube on a magnetic particle concentrator at 4 °C.
16. Centrifuge the cell lysate (obtained in **step 14**) at 11,000 × *g* and 4 °C for 10 min.
17. Remove the washing solution from the tube of Streptavidin MagneSphere® (obtained in **step 15**) and add the supernatant of the cell lysate obtained in **step 16** and mix by tapping and inverting the tube.
18. Rotate the tube at 15–20 rpm and 4 °C for 30 min to overnight (*see* **Note 11**).
19. Centrifuge at 11,000 × *g* and 4 °C for 2 min and then place the tube on a magnetic particle concentrator and remove the supernatant.

20. Add 750 μL of PBST and wash the beads by tapping and inverting the tube.
21. Centrifuge at 11,000 × *g* and 4 °C for 2 min and then place the tube on a magnetic particle concentrator and remove the supernatant.
22. Repeat **steps 20** and **21** five times (six times in total).
23. Add 750 μL of PBS and wash the beads by tapping and inverting the tube.
24. Centrifuge at 11,000 × *g* and 4 °C for 2 min and then place the tube on a magnetic particle concentrator and remove the supernatant.
25. Repeat **steps 23** and **24** once (twice in total).
26. Add 50–100 μL of PBS (the same volume as Streptavidin MagneSphere® used in **step 15**).
27. Store the purified immobilized ligand beads in PBS at 4 °C (*see* **Note 12**).

3.3 Measurement of the Purified Recombinant Protein Concentration by Oriole Staining

1. Add the same volume of SDS buffer as was added to BSA solutions adjusted to 10, 20, 50, and 100 ng/μL to prepare reference BSA samples with final concentrations of 5, 10, 25, and 50 ng/μL, respectively.
2. Boil the samples at 98 °C for 5 min.
3. Let the samples cool at room temperature and then centrifuge at maximum speed for a few seconds at room temperature.
4. The prepared reference BSA samples can be stored at −80 °C for 2–3 months.
5. Resuspend the immobilized ligand beads in PBS by pipetting.
6. Add 3–6 μL of the immobilized ligand beads suspension to the tube, add distilled water up to a total volume of 9 μL, and add 9 μL of SDS buffer.
7. Boil samples at 98 °C for 5 min.
8. Cool the samples at room temperature and vortex the tubes for approximately 5 s.
9. Centrifuge the tube at 11,000 × *g* for a few seconds, place the tubes on a magnetic particle concentrator and transfer the boiled samples to new tubes for SDS-PAGE.
10. Perform SDS-PAGE by adding 15 μL of each boiled ligand sample to a well (*see* **Note 13**).
11. Apply 10 ng (2 μL of 5 ng/μL reference BSA sample), 25 ng (2.5 μL of 10 ng/μL reference BSA sample), 50 ng (2 μL of 25 ng/μL reference BSA sample), 75 ng (3 μL of 25 ng/μL reference BSA sample), 100 ng (2 μL of 50 ng/μL reference BSA sample), and 150 ng (3 μL of 50 ng/μL reference BSA sample) to each well individually.

12. Perform electrophoresis in a stacking gel at 20 mA and in a separation gel at 35 mA.
13. After electrophoresis, wash the gel with ultrapure water for 5–10 min on a rocking shaker set at approximately 20 rpm.
14. Remove the ultrapure water and add 25–30 mL of the Oriole fluorescent gel stain (be sure to cover the entire gel with staining solution) and shake for 90 min by setting the shaker to approximately 20 rpm.
15. Take a fluorescent image of the gel under a UV illuminator and measure the relative intensity of each band for quantification using an appropriate software (e.g., we used CS analyzer ver3.0 [ATTO]; *see* **Note 14**).
16. Estimate the concentration of the ligand on the beads by generating a standard curve on the basis of the band intensity of the BSA (Fig. 2 and Table 1; *see* **Note 15**).

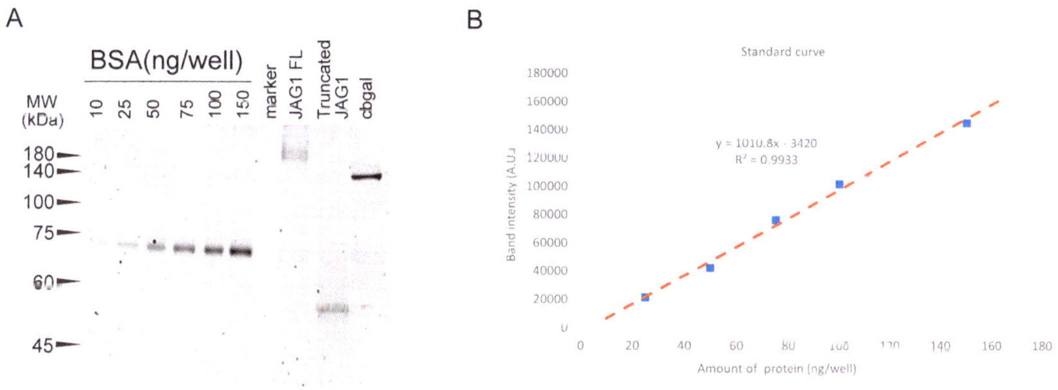

Fig. 2 Representative Oriole-stained polyacrylamide gel sample. (**a**) Representative sample of Oriole-stained polyacrylamide gel after electrophoresis. The reference BSA samples (left 6 lanes), full-length JAG1 (JAG1 FL) beads, truncated JAG1 beads and cbgal beads (right 3 lanes). (**b**) Standard curve generated from measured intensities of standard BSA samples in (**a**). The band intensity value and protein amount showed a high positive correlation ($R^2 = 0.9933$). The immobilized ligand-protein amount could be estimated using their band intensity values and linear approximation formula, calculated by the standard curve. Regarding the estimated protein concentration, please *see* Table 1

Table 1
The estimated protein concentration. The prepared ligand-protein concentration was estimated by the linear approximation formula presented in Fig. 2

Apply volume (μL)	Ligand protein	Band intensity	Amount of protein (ng/well)	Ligand concentration of beads (ng/μL)
2.5	Jag1 FL	54,544	57.3	22.9
5	Truncated JAG1	86,240	88.7	17.7
2.5	cbgal	112,672	114.9	45.9

3.4 Measurement of Notch Signaling Activity by Ligand Immobilized Beads

1. Transfect the Notch reporter plasmid (80 ng/well [96-well plate]) and Renilla luciferase plasmid (1.6 ng/well [96-well plate]) into N1-3T3 cells (Notch receptor-expressing cells) and culture for 24 h in 96-well plates.
2. Prepare 100 μL of DMEM/FBS/PS per well (96-well plate) in 1.5 mL tube (e.g., 400 μL for 4 well). Resuspend the ligand immobilized beads in PBS (prepared in Subheading 3.1) and add beads in an amount equivalent to 1 pmol ligand protein (e.g., 50 ng of truncated JAG1 [MW: 50 kDa]) per well to DMEM/FBS/PS in the tube (*see* **Note 16**).
3. Remove the medium and add the DMEM/FBS/PS containing immobilized ligand beads to each well.
4. After incubation at 37 °C and 5% CO_2 for several hours to several tens of hours, then remove the culture medium, and add 25 μL of 1× passive lysis buffer to each well.
5. Shake the cells for 20 min by rocking the shaker at approximately 20 rpm, and collect the cell lysates.
6. Measure luciferase-dependent luminescence using a luminometer (we used GloMax [Promega], *see* **Note 1**) (Fig. 3). To detect Notch signaling-dependent luciferase activity, add 40 μL of luciferase assay buffer to 10 μL of cell lysate. To detect Renilla luciferase activity as an internal control, add 50 μL of Renilla luciferase assay buffer continuously.
7. Calculate the relative Notch signaling-dependent luciferase activity on the basis of the obtained luciferase activity value and Renilla luciferase activity value (Fig. 3).

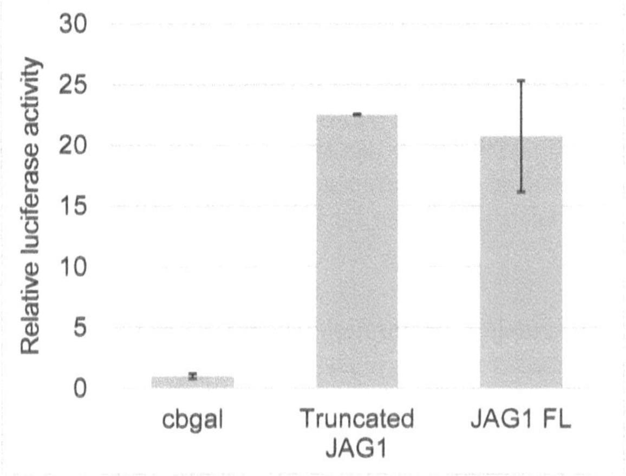

Fig. 3 The relative luciferase activity induced by immobilized ligand beads. The luciferase activity of cbgal-, truncated JAG1- and full-length JAG1-immobilized beads. The graph shows the relative luciferase activity. Error bar indicates the S. D., $n = 2$

4 Notes

1. A luminometer by which a dual reporter assay can be performed is recommended for precise measurement of Notch signaling activity.
2. Any protein that is not involved in the Notch signaling pathway can be used as a negative control.
3. Any Notch-receptor expressing cell line can be used.
4. Any Notch-reporter plasmid can be used.
5. Any internal control plasmid for the luciferase reporter assay can be used.
6. Any lysis buffer suitable for a luciferase reporter assay can be used.
7. Any luminescent reagent that can detect luciferase activity can be used.
8. Any luminescent reagent that can detect Renilla luciferase activity or the activity of another internal control can be used.
9. It is also possible to use appropriate "transfection reagents" other than PEI for HEK293T cells and Notch receptor-expressing cells.
10. This step (Subheading 3.2, **step 9**) is needed to remove immature ligands that are not anchored in the membrane. Each procedure performed with solutions frozen in liquid nitrogen was performed on ice until Subheading 3.2, **step 13**.
11. This step (Subheading 3.2, **step 18**) can be completed using a rocking shaker, which should be set at 15 rpm and used at 4 °C. If the protein beads tend to aggregate, resuspend them by tapping the tube every 10 min.
12. If the purified beads aggregate, sonicate the ligand beads at 4 °C in sonicator with 6 cycles of 10 s on/20 s off at 200 W. To prevent contamination, a sealed sonicator is recommended; we used a Bioruptor UCD-250 (Cosmo Bio).
13. A 15 μL boiled ligand sample contains 2.5–5 μL of ligand beads when the sample is prepared with 3–6 μL of ligand beads. When low-molecular-weight ligands, such as truncated types, are used, more beads should be used for sample preparation to facilitate Oriole fluorescent gel staining of the bands.
14. If no ligand band is detected with Oriole stain, please check the expression of the ligand protein in the HEK293T cells by western blotting using suitable antibodies.
15. Usually, 0.25–1 μg (truncated ligand) or 1–2 μg (full-length ligand) ligand protein can be obtained using 50 μL beads slurry per 10 cm dish.

16. To analyze the effect of the ligand amount and/or incubation time on Notch signaling activation, the amount of Notch ligand and incubation time can be easily changed by modifying Subheadings 3.4, **steps 2** and **3**.

Acknowledgments

This project was supported by JSPS Grants-in-Aid for Scientific Research grant nos. 19 K06454 to T.M. and 18H02568 to M.I.

References

1. Siebel C, Lendahl U (2017) Notch signaling in development, tissue homeostasis, and disease. Physiol Rev 97:1235–1294. https://doi.org/10.1152/physrev.00005.2017
2. Kopan R, Ilagan MXG (2009) The canonical Notch signaling pathway: unfolding the activation mechanism. Cell 137:216–233
3. Le Borgne R, Bardin A, Schweisguth F (2005) The roles of receptor and ligand endocytosis in regulating Notch signaling. Development 132: 1751–1762
4. Musse AA, Meloty-Kapella L, Weinmaster G (2012) Notch ligand endocytosis: mechanistic basis of signaling activity. Semin Cell Dev Biol 23:429–436
5. Itoh M, Kim CH, Palardy G et al (2003) Mind bomb is a ubiquitin ligase that is essential for efficient activation of notch signaling by delta. Dev Cell 4:67–82. https://doi.org/10.1016/S1534-5807(02)00409-4
6. Gonçalves RM, Martins MCL, Almeida-Porada G, Barbosa MA (2009) Induction of notch signaling by immobilization of jagged-1 on self-assembled monolayers. Biomaterials 30: 6879–6887. https://doi.org/10.1016/j.biomaterials.2009.09.010
7. Taqvi S, Dixit L, Roy K (2006) Biomaterial-based notch signaling for the differentiation of hematopoietic stem cells into T cells. J Biomed Mater Res Part A 79A:689–697. https://doi.org/10.1002/jbm.a.30916
8. Liu L, Wada H, Matsubara N et al (2017) Identification of domains for efficient notch signaling activity in immobilized notch ligand proteins. J Cell Biochem 118:785–796. https://doi.org/10.1002/jcb.25744
9. Abe N, Hozumi K, Hirano K et al (2010) Notch ligands transduce different magnitudes of signaling critical for determination of T-cell fate. Eur J Immunol 40:2608–2617. https://doi.org/10.1002/eji.200940006
10. Minoguchi S, Taniguchi Y, Kato H et al (1997) RBP-L, a transcription factor related to RBP-Jkappa. Mol Cell Biol 17:2679–2687. https://doi.org/10.1128/mcb.17.5.2679

Chapter 7

Mammalian NOTCH Receptor Activation and Signaling Protocols

María-Luisa Nueda and Victoriano Baladrón

Abstract

The NOTCH signaling pathway is one of the highly conserved key pathways involved in most cell differentiation and proliferation processes during both developmental and adult stages in most animals. The NOTCH signaling pathway appears to be very simple but the existence of several receptors and ligands, their posttranslational modifications, their activation in the cell surface and its migration to the cell nucleus, as well as their interaction with multiple signaling pathways in the cytoplasm and the nucleus of cells, make the study of its function very complex.

To determine the activation of NOTCH signaling in animal cells, several complementary approaches can be performed. One of them is the analysis of the transcription of NOTCH receptor target genes HES/HEY by qRT-PCR and Western blot. This approach would give us an idea of the global NOTCH activation and signaling. We can also analyze the NOTCH transcriptional activity by luciferase assays to determine the global or specific activation of NOTCH receptors under a given treatment or in response to the modification of gene expression. On the other hand, we can determine the specific activation of each NOTCH receptor by Western blot with antibodies that recognize the active forms of each NOTCH receptor. For this assay will be very important to collect the cells to be analyzed under the appropriate conditions. Finally, we can detect the intracellular domain of each NOTCH receptor into the cell nucleus by confocal microscopy using the appropriate antibodies that recognize the intracellular domain of the receptors.

Key words NOTCH activation, *Hes/Hey* genes and proteins, Luciferase assays, Western blot, NOTCH antibodies, Confocal microscopy, EDTA, DAPT

1 Introduction

The NOTCH signaling is a key pathway involved in many cells' differentiation and proliferation processes during embryogenesis and adult tissues in mammalians [1–13]. In mammalians, four homologous receptors, NOTCH1–4, and 5 canonical ligands homologous to Serrate ligand (Jagged1 and 2) and Delta ligand (DLL1, 3, and 4) have been described [14–18], as well as inhibitory noncanonical ligands such as DNER, CCN3/NOV, MAGP1 y 2 [15, 16] or DLK1 and DLK2 proteins [1, 19–22]. This family of

Dongyu Jia (ed.), *Notch Signaling Research: Methods and Protocols*,
Methods in Molecular Biology, vol. 2472, https://doi.org/10.1007/978-1-0716-2201-8_7,

proteins is involved in interactions with other proteins and modulates cell differentiation through mechanisms of lateral inhibition and induction by a cell contact-dependent manner, cell proliferation and apoptosis.

The NOTCH receptor structure consists of a single-pass transmembrane protein composed of a large extracellular portion, which associates in a calcium-dependent, noncovalent interaction with another region of the NOTCH protein composed of a short extracellular region, a single transmembrane-pass, and an intracellular region [2, 23–26]. NOTCH receptors are synthesized as precursors that are subsequently processed in the Golgi apparatus by a furin protease at the site termed S1, generating the two polypeptides that heterodimerize as transmembrane proteins in the cell surface. The NOTCH receptor extracellular region (NECD) contains EGF repeats, responsible for the interaction with canonical and noncanonical ligands [2, 27]; a negative regulatory region (LNR) that regulates ligand-dependent and ligand-independent signaling; and a heterodimerization domain involved in NOTCH signaling activation upon ligand binding [28]. The intracellular domain (NICD) consists of different important regions, including domains called RAM and ankyrin repeats with high binding affinity for the CSL/RBP-Jκ factor and other coactivating factors that regulate the expression of target genes [2, 24].

Interaction of the amino-terminal DSL domain of canonical ligands with specific EGFs of NOTCH receptors generally activates the signaling pathway [3, 29–34]. This interaction produces a conformational change of the receptor that allows a second proteolysis at an extracellular site close to the plasma membrane called the S2 site. This proteolysis is mediated by the ADAM/TACE disintegrin and metalloprotease, releasing the NECD region, which is endocytosed by the cell expressing the ligand. A membrane-bound intermediate is generated that lacks NECD and it is the substrate for the γ-secretase complex, which performs another intracellular proteolysis, at the S3–S4 site, releasing the NICD region into the cytoplasm. NICD translocates to the cell nucleus and interacts with the transcription factor CSL/RBP-Jκ, through its RAM and ankyrin domains. In the absence of NICD, the CSL factor forms a transcriptional repressor complex together with other corepressors. However, binding of NICD allows it to form a transcriptional activation complex together with other coactivators, such as Mastermind (MAML), which induces the expression of the main NOTCH target genes: the HES and HEY protein families [35, 36]. HES and HEY are transcription factors that regulate the expression of other genes [37, 38].

NOTCH signaling pathway appears to be very simple without any signal amplification through second messengers and each NOTCH receptor molecule present on the plasma membrane can only transmit the signal it picks up once, indicating that the number and availability of NOTCH receptors on the cell surface is critical for modulating signal intensity and duration [2, 23, 24]. In addition, its ligands can activate or inhibit NOTCH signaling under certain conditions. Ligands activate NOTCH receptors present on the neighboring cells (interaction in TRANS), whereas they inhibit receptors present on the same cell (interaction in CIS) [2, 23, 24]. These aspects and other such as their posttranslational modifications, its migration to the cell nucleus, or their interaction with multiple signaling pathways, make the study of its function very complex. We describe here different approaches to analyze the global and specific NOTCH activation and signaling in animal cells.

2 Materials

2.1 Cell Cultures

1. Mouse C3H10T1/2 (ATCC CCL-226, clon 8) and 3T3-L1 (ATCC CL-173) cell lines, and human SK-MEL-2 (ATCC HTB-68) cell line.
2. DMEM (Dulbecco's modified eagle medium, Lonza) supplemented with heat-inactivated fetal bovine serum (10%), L-glutamine (1%), and penicillin–streptomycin (1%).
3. DPBS (Dulbecco's phosphate-buffered saline, Lonza).
4. 0.5 M EDTA (ethylenediaminetetraacetic acid) and trypsin–EDTA solution (Lonza).

2.2 RT-qPCR Assays

Oligonucleotides used to analyze *Hes1* and *Hey1* expression.

Rplp0 Forward: 5′-AAGCGCGTCCTGGCATTGTCT-3′.

Rplp0 Reverse: 5′-CCGCAGGGGCAGCAGTGGT-3′.

Hes1 Forward: 5′-AGCCTATCATGGAGAAGAGGCGAA-3′.

Hes1 Reverse: 5′-TGGAATGCCGGGAGCTATCTTTCT-3′.

Hey1 Forward: 5′-ATGTGGCCTACTTCAGCTCCATGT-3′.

Hey1 Reverse: 5′-TCTCCAGGCAGGTAAACAATGGGA-3′.

2.3 Luciferase Assays

1. pGLucWT plasmid. This plasmid possesses four binding sites for the CSL/RBP-Jκ transcription factor in front of the SV40 promoter that directs the expression of the luciferase gene.
2. pRLTK plasmid. This plasmid contains the gene encoding the renilla protein, another protein capable of generating light under the control of a constitutive promoter. Used to normalize luciferase activity data.

2.4 Western Blot Assays

1. Lysis buffer: 50 mM Tris-HCl pH 7.4, IGEPAL 1%, 150 mM NaCl, 1 mM EDTA, 1 mM PMSF, 1 μg/mL Leupeptin, 1 mM Na_3VO_4, 1 mM NaF, 20 μL Phosphatase Inhibitor Cocktails I and II (Sigma).

2. 4× protein loading buffer: 50 mM Tris-HCl pH 6.8, SDS 2%, 100 mM mercaptoethanol, glycerol 10%, bromophenol blue 0.025%.
3. Electrophoresis buffer pH 8.3: 25 mM Tris-HCl, 192 mM glycine, 0.1% SDS.
4. Transfer buffer: 25 mM Tris-HCl, 192 mM glycine, 0.1% SDS, 20% methanol.
5. TBS-T buffer: 150 mM NaCl, 50 mM Tris-HCl pH 7.4, Tween 20 0.1%.
6. Rabbit α-NOTCH1 (C-20)-R (Santa Cruz): 1:1000 dilution in TBS-T, 5% nonfat milk.
7. Rabbit α-NICD1 cleaved Val1744 (Cell Signaling): 1:500 dilution in TBS-T, 5% BSA.
8. Rabbit α-HES1 antibody (Abcam): 1:1000 dilution in TBS-T, 5% BSA.
9. Goat α-rabbit (HRP)-conjugated (Pierce): 1:1000 dilution in TBS-T, 5% nonfat milk.
10. Dehybridization buffer: 100 mM mercaptoethanol, SDS 2%, 62.5 mM Tris-HCl pH 6.7.

2.5 Confocal Microscopy

1. Fixing solution: 3% formaldehyde in DPBS, or FIX & PERM Cell Permeabilization Kit (Invitrogen).
2. Permeabilization solution: 0.2% TX-100 in DPBS, or with Invitrogen FIX & PERM Cell Permeabilization Kit.
3. Blocking solution: 10% fetal bovine serum in DPBS.
4. Rabbit α-NOTCH1 (C-20)-R (Santa Cruz): 1:1000 dilution in DPBS-5% fetal bovine serum.
5. Secondary fluorescent antibody (Alexa Fluor Red goat anti-rabbit, Invitrogen): 1:2000 dilution in DPBS-5% fetal bovine serum.

3 Methods

3.1 Analysis of the Transcription of NOTCH Receptor Target Genes, Hes1[1] y Hey1[1], by RT-qPCR

The steps to be followed are as follows.

1. Seed control and stable or transient transfectant cells (2×10^6) that overexpress or downregulate NOTCH genes expression in 75 cc flasks.
2. Grow cells in the appropriate medium in the presence of antibacterial antibiotics and, if applicable, the selection antibiotic of the expression vector used. Incubate cells under the appropriate temperature and CO_2 conditions and in the presence of the selective antibiotic when necessary. These cultures can be

treated with alpha- or gamma-secretase inhibitors at different concentrations from the beginning of the culture.

3. Grow all the different types of cells to compare at as close to confluence as possible. Besides, it is important to collect all cells to be compared at the same cell confluency since the expression of these genes' changes with such confluency.
4. To obtain total RNA, wash cultures once with DPBS and treat them with 2 mL of Trypsin-EDTA or collect them by a cell scraper. Centrifuge cells at 16,000 × *g* for 5 min in a table centrifuge. Subsequently, they can be washed with DPBS, and the samples can be frozen at −80 °C until further processing.
5. Silica adsorption chromatography columns can be used to obtain total RNA using the RNeasy Minikit (Qiagen) following the manufacturer's instructions, including the treatment of samples with DNase (Promega).
6. The concentration of total RNA in each sample can be determined by ultraviolet-visible spectrophotometry, measuring the absorbance ratio 260 nm/280 nm. For this purpose, a NanoDrop One equipment can be used following the manufacturer's instructions.
7. The RevertAid H Minus First Strand cDNA Synthesis Kit (Thermo Fisher Scientific) is used for cDNA synthesis from total RNA. The reaction mixture is prepared according to the manufacturer's instructions. One microgram of total RNA is mixed with 1 μL of oligo-dT and incubated at 70 °C for 5 min to denature the RNA and allow the oligo-dT to bind to the polyA tail of the mRNAs. Subsequently, to carry out the synthesis process, reaction buffer, a mixture of dNTPs, RiboLock to inhibit RNases, and reverse transcriptase enzyme is added. Incubate this mixture at 42 °C for 60 min to allow cDNA elongation and finally at 70 °C for 10 min. The cDNA synthesis is carried out on a T100 thermal cycler (Bio-Rad).
8. The Power Up SYBR Green Master Mix fluorochrome (Thermo Fisher Scientific) is used for RT-qPCR analysis. To perform RT-qPCR experiments the StepOne Plus kit (Applied Biosystems) can be used and samples are analyzed in triplicate. The specific oligonucleotides of analyzed genes are diluted to a final concentration of 7.5 M and the cDNA obtained at a ratio of 1:10 or 1:100 with nuclease-free water. Next, prepare a mixture per reaction well containing 5 μL of SYBR Green reagent and 0.4 μL of each of the oligonucleotides. Finally, in a 96-well qPCR plate, 5.8 μL of this mixture along with 4.2 μL of the diluted cDNA is placed per reaction well. The RT-qPCR reaction conditions are as follows: an initial stage at 95 °C for 20 s allowing denaturation and loss of DNA secondary

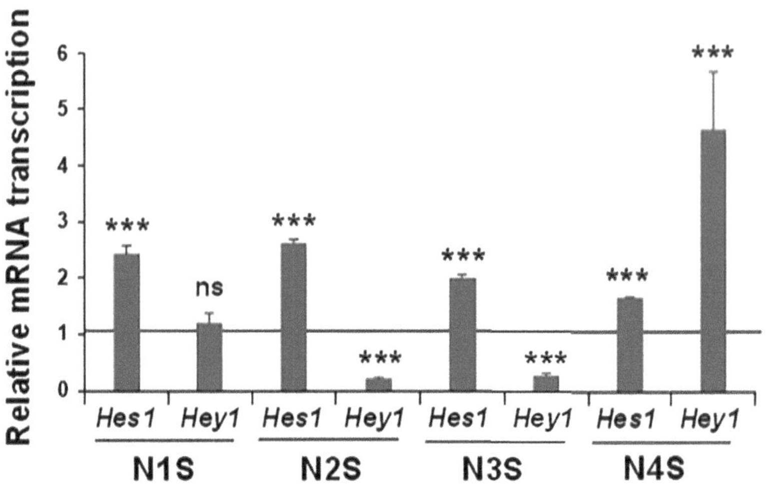

Fig. 1 NOTCH signaling in mesenchymal C3H10T1/2 cells stably overexpressing each one of the four NOTCH receptors [39]. qRT-PCR analysis of the relative *Hes1* and *Hey1* mRNA transcription levels in stably *Notch* gene-transfected pools. Stable *Notch1* transfectant (N1S); stable *Notch2* transfectant (N2S); stable *Notch3* transfectant (N3S), and stable *Notch4* transfectant (N4S). Data in qRT-PCR assays were normalized to *Rplp0* mRNA transcription levels. The fold activation or inhibition in all assays was measured relative to levels in non-differentiated empty vector–transfected cells, set arbitrarily at 1 (horizontal black line). Data are shown as the mean ± SD of at least three biological assays performed in triplicate. The statistical significance of Student's *t*-test results is indicated (***$p \leq 0.001$). Nonstatistical significance is indicated by ns

structures, followed by 40 cycles consisting of a first stage at 95 °C for 3 s, together with a second stage at 60 °C for 30 s allowing oligonucleotide binding and extension by the DNA polymerase. The results of the amplifications can be obtained using StepOne 2.3 software, with the parameters recommended by the manufacturer, always determining the melting curve for each gene. In all cases the expression of the constitutive gene *Rplp0* is analyzed to compare the relative expression levels of the different samples. Figure 1 represents one of these RT-qPCR assays.

3.2 Analysis of the Transcription of NOTCH Receptor Target Genes, Hes1 y Hey1, by Western Blot (See Note 1)

The steps to be followed are as follows.

1. To obtain protein extracts, cell cultures are washed once with DPBS, collected by treatment with trypsin/EDTA or with a cell scraper and DPBS. Cells are then centrifuged at 16,000 × *g* for 5 min. Subsequently, they can be washed again with DPBS, and samples can be frozen at −20 °C until further processing.
2. Cell pellets are vortexed after resuspending them in 100 μL of lysis buffer.

3. The samples are incubated on ice for 30 min and centrifuged at 16,000 × g for 10 min at 4 °C and then the supernatant is transferred to a new clean Eppendorf tube.
4. The protein concentration of cell extracts is determined by the bicinchoninic acid (BCA) method. In this method, the absorbance is proportional to the amount of protein present in the solution and can be determined spectrophotometrically at 562 nm by comparison with a known protein standard such as bovine serum albumin (1 mg/mL). The protein samples to be titrated are loaded in 96-well plate in triplicate. All protein samples are loaded in 10 μL of a final reaction volume, and 200 μL/well of a reagent composed of BCA and copper sulfate, mixed in the ratio recommended by the distributor (Sigma), is added. The plate is incubated at 37 °C for 30 min in an IN30 oven and the absorbance is determined on an ASYS UVM340 plate reader (Biochrom), following the manufacturer's instructions.
5. An appropriate amount of protein sample (50–100 g) from cell extract supernatants is diluted in 4× protein loading buffer and boiled for 5 min in a Tembloc thermoblock (Selecta) at 95 °C. To estimate molecular weight, the Prestained Protein Molecular Weight Marker or Plus Prestained Protein Molecular Weight Marker (Thermo Fisher Scientific) is used by diluting 5 μL in protein loading buffer.
6. Protein separation is performed by electrophoresis under denaturing conditions in discontinuous polyacrylamide gels (acrylamide–bisacrylamide mix [37.5:1] [Serva]) formed by a 4% packing gel and a separation gel at concentrations of 10–15%, depending on the size of the protein to be determined. Electrophoresis is carried out in electrophoresis buffer pH 8.3 by using the Mini-PROTEAN 3 Cell system (Bio-Rad). Thirty milliampere per gel are applied and electrophoresis is stopped when deemed appropriate depending on the size of the proteins to be analyzed.
7. Proteins are transferred to nitrocellulose membranes (Amersham Protran, GE Healthcare Life Sciences, Fisher Scientific) using the Mini Trans-Blot Electrophoretic Transfer Cell system (Bio-Rad), according to the instructions provided by the manufacturer. Transfer is carried out at 450 mA for 3 h in transfer buffer.
8. The membranes are blocked in a solution composed of skimmed milk powder or 5% bovine serum albumin (Sigma) in TBS-T buffer. Blocking is carried out for 1 h with gentle agitation. The membranes are then incubated under the same conditions with the primary antibodies overnight at 4 °C. After the incubation time with the primary antibody, the membranes

are washed three times with TBS-T for 15 min and incubated with the corresponding horseradish peroxidase (HRP)-conjugated secondary antibody for 1 h at room temperature.

9. Subsequently, the membranes are washed three times with TBS-T for 15 min and incubated with the chemiluminescence reagents of the Western Blotting Luminol Reagent (Santa Cruz) or Pierce ECL 2 Western blotting substrate kit (Thermo Fisher Scientific).
10. The emitted chemiluminescence can be analyzed by exposure of orthochromatic photographic films (CP-BU M, AGFA), which are not sensitive to high wavelengths such as red colors, so they can be manipulated under red illumination without veiling. Curix 60 developer (AGFA) can be used, and exposure times varied from 1 s to 15 min.
11. Upon completion, when it is necessary to incubate the membranes with other antibodies, they are treated with dehybridization buffer at 50–60 °C for 30 min in a Hybridization Oven/Shaker (Amersham Biosciences) and then washed extensively with TBS-T.
12. For densitometric quantification of the signals detected, photographic films can be scanned with the HP Officejet Pro-8600 scanner and converted to grayscale. The signals of the different proteins can be quantified with Quantity One Quantitation Software 4.6.5. (Basic) (Bio-Rad). The intensity of the signals per square millimeter can be determined using the Volume Rect Tool. To normalize data, densitometric analysis of tubulin protein signals in each sample can be used. Statistical analysis of densitometric signals in Western blot assays can be performed by calculating the average of different experiments and their standard deviation by using *t*-Student test. Figure 2 represents one of these Western blot assays to analyze de expression of HES1 protein.

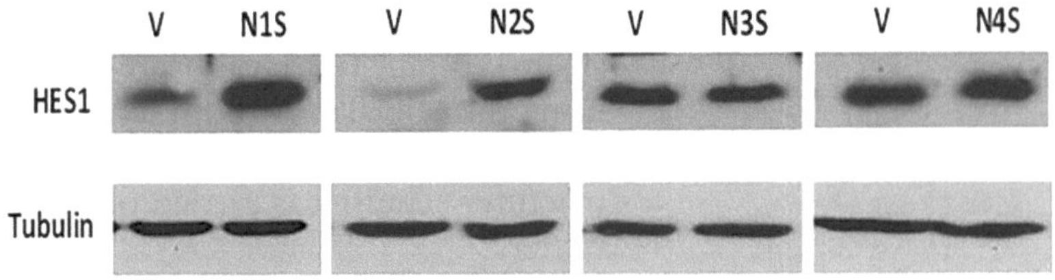

Fig. 2 Representative Western blot assay of HES1 protein expression in confluent 3T3-L1 preadipocytes stably overexpressing each one of the four NOTCH receptors. Stable *Notch1* transfectant (N1S); stable *Notch2* transfectant (N2S); stable *Notch3* transfectant (N3S), and stable *Notch4* transfectant (N4S). The expression of alpha-tubulin was used as a loading and quality control in Western blot

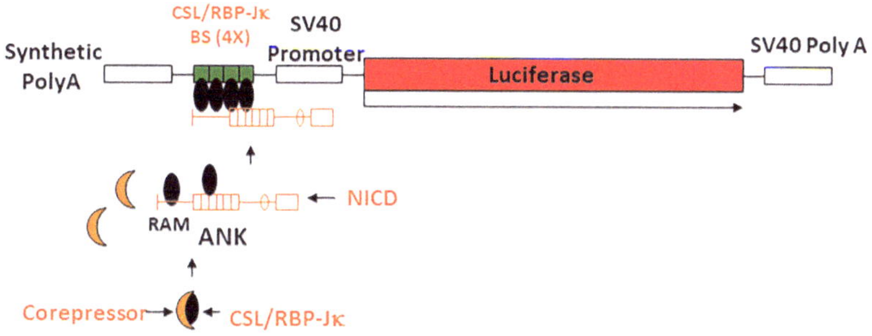

Fig. 3 pGLucWT, a luciferase construct derived from pGL3 vector (Promega) that contains four CSL/RBP-Jκ binding sites (BS) cloned upstream SV40 promoter. CSL/RBP-Jκ factor binds active NICD and modulates SV40 luciferase promoter

3.3 Analysis of the NOTCH Transcriptional Activity by Luciferase Assays to Determine the Global or Specific Activation of Notch Receptors Under a Given Treatment or in Response to the Modification of Gene Expression

To analyze NOTCH transcriptional activity by luciferase assays, the plasmid named pGLuc-WT (*see* **Note 2**) can be used (Fig. 3) [1].

The steps to be followed are as follows.

1. 20,000 cells/well are seeded in 12-well plates from the control cell line of interest C3H10T1/2 or from each of the possible stable transfectants generated.
2. A mixture of 50 μL of Opti-MEM, 2.5 μL of Superfect transfection reagent (Qiagen) for each μg of DNA, 0.5 μg of pGLuc-WT plasmid, 30 ng of pRLTK[1] plasmid and, if applicable, the expression plasmids to be analyzed at different concentrations, is added to 50% confluence cells after 24 h of culture.
3. To confirm the correct functioning of the luciferase activity assay system in the cell line, both positive and negative controls can be included in each assay. As a positive control, the effect of the expression of the active intracellular domain of NOTCH1 on the luciferase activity of cells can be studied. For this purpose, cells at 50% confluence are transiently transfected with a mixture of 50 μL of Opti-MEM, 0.5 μg of pGLuc-WT plasmid, 30 ng of pRLTK plasmid, 1 μg of empty vector plasmid or vector expressing active NICD1, together with 2.5 μL of Superfect transfection reagent. For the negative control, cells at 50% confluency are transiently transfected with a mixture of 50 μL of Opti-MEM, 0.5 μg of pGLuc-WT plasmid, 30 ng of pRLTK plasmid and 2.5 μL of Superfect transfection reagent for each μg of DNA used for transfection, and incubated throughout the assay with DMSO (dimethyl sulfoxide) or the NOTCH signaling inhibitor, DAPT, at a final concentration of 10 μM.
4. In all cases, cell samples are collected 48 h after transfection by aspirating the culture medium from each well, washing once

with DPBS and resuspending the cells in Passive Lysis Buffer 5× luciferase lysis buffer (Dual-Luciferase Reporter Assay System [Promega]).

5. Plates can be frozen at −80 °C until further processing.
6. The homogenized extract from each well is transferred to an Eppendorf tube and centrifuged at 16,000 × *g* for 5 min.
7. Finally, 20 μL of each sample is loaded in triplicate into a 96-well plate suitable for reading on a plate luminometer. The relative luciferase and renilla activity in each sample can be determined with an Orion II microplate luminometer (Berthold) according to the manufacturer's instructions.
8. The luciferase activity data in each sample is normalized to the renilla activity data. Statistical analysis in luciferase assays can be performed by calculating the average of different experiments and their standard deviation using *t*-Student test. Figure 4 is a representative luciferase assay showing transcriptional activity of each of the four overexpressed NOTCH receptors.

3.4 Determination the Specific Activation of Each NOTCH Receptor by Western Blot with Antibodies That Recognize the Active Forms of Each NOTCH Receptor

The steps to be followed in this section are the same as those used in Subheading 3.2, but in this section is important to control the conditions of collection of the different type of cultured confluent cells to compare. In this section, the effect of DAPT inhibitor (*see* **Note 3**), and EDTA-containing solutions (*see* **Note 4**) in cell cultures is analyzed.

The effect of different concentrations of DAPT on NOTCH activation, as well as the time that DAPT is active in cell cultures without the need to change the medium, is analyzed as follows:

1. Incubate cells from the beginning of the assay with DMSO, used as the control sample, or DAPT, at a final concentration of 10 μM for 24, 48, 72, or 96 h, and with different concentrations of DAPT (5, 2.5, 1 or 0.1 μM) for 24 h.
2. To obtain protein extracts to determine the inhibition of active NICD by DAPT in a Western blot assay, we can proceed as described previously in Subheading 3.2. Figure 5 shows the effect of DAPT in NOTCH1 inhibition.

   When analyzing NOTCH receptors activation, the way in which the samples are collected is very important, since small changes in this process lead to changes in NOTCH activation, as showed in Fig. 6. The following steps in this protocol allow us to analyze the effect of time and temperature in the activation of NOTCH receptors with EDTA-containing solutions.
3. Wash cell cultures once with DPBS and collect them with a cell scraper after treatment with trypsin/EDTA or EDTA/DPBS solutions at different times and temperatures.

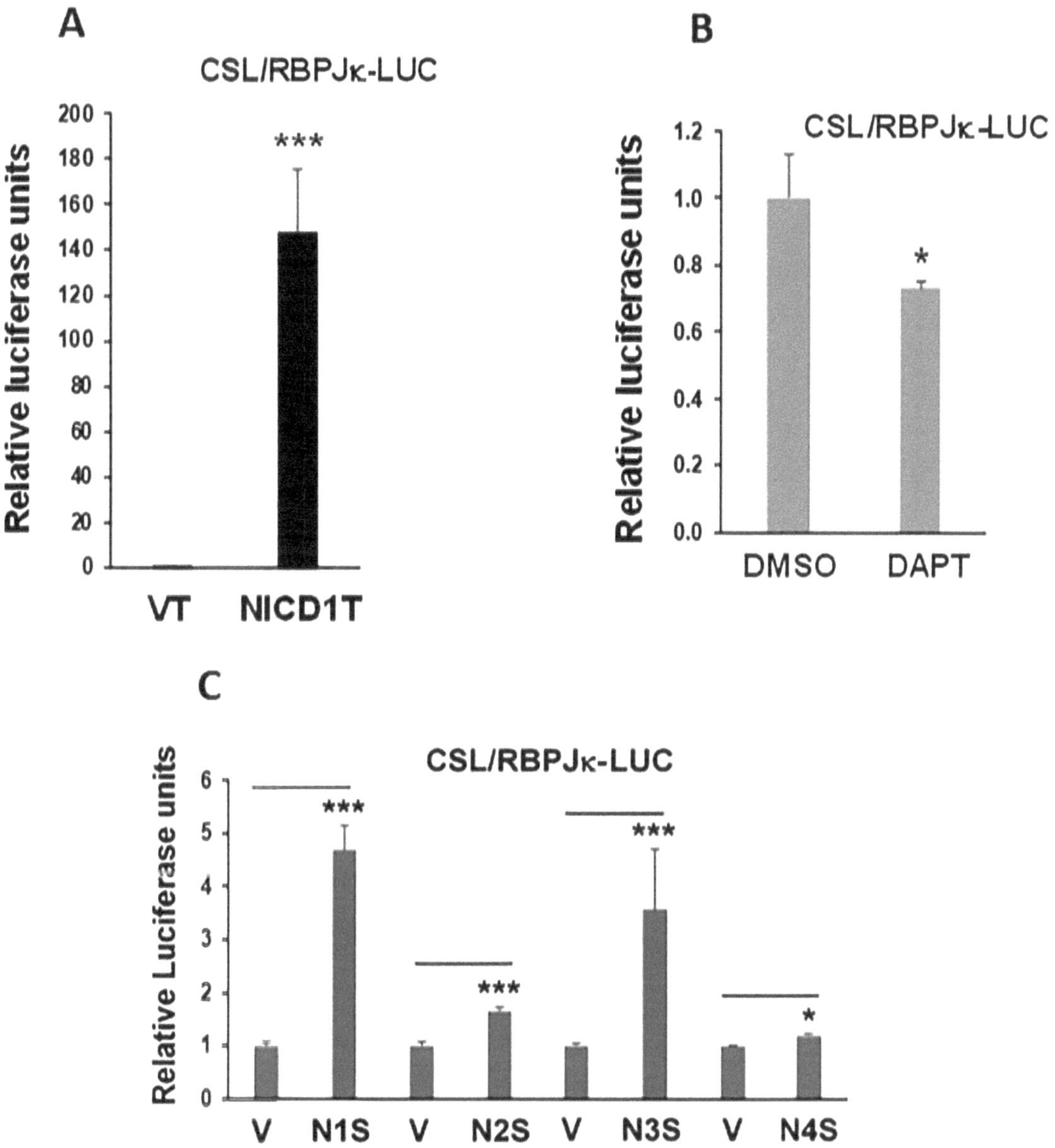

Fig. 4 Transcriptional activation of NOTCH signaling (CSL/RBPJκ-LUC) in mesenchymal C3H10T1/2 cells [39]. Transcriptional activity of cells that were transiently overexpressing active NICD1 (**a**) or treated or not with DAPT (**b**) as measured by luciferase assays. (**c**) Transcriptional activation of NOTCH signaling in mesenchymal C3H10T1/2 cells stably overexpressing each one of the complete NOTCH receptors, as measured by luciferase assays. Stable *Notch1* transfectant (N1S); stable *Notch2* transfectant (N2S); stable *Notch3* transfectant (N3S), and stable *Notch4* transfectant (N4S). The relative luciferase activities were always normalized with renilla values. The fold activation or inhibition in all assays was measured relative to the levels of control cells, set arbitrarily at 1 (V [stably-empty-vector-transfected cells]); VT [transiently-empty-vector-transfected cells]; DMSO [DMSO-treated cells]). Data are shown as the mean ± SD of at least three biological assays performed in triplicate. The statistical significance of Student's *t*-test results is indicated (*$p \leq 0.05$, ***$p \leq 0.001$)

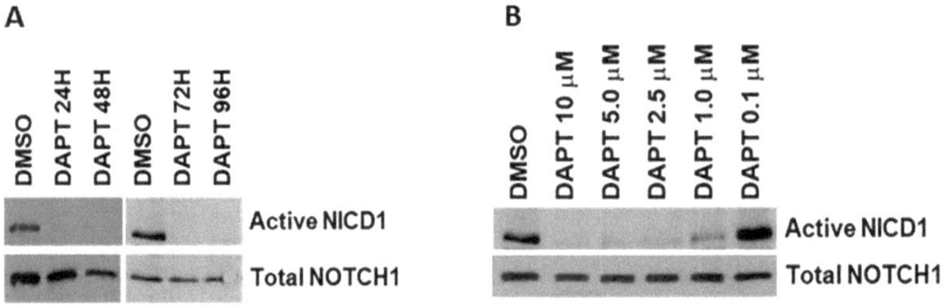

Fig. 5 Representative Western blot assays that show the levels of the active form of NOTCH1 receptor (NICD1) in SK-MEL-2 melanoma cells. (**a**) Cells treated with 10 μM DAPT for 24, 48, 72, or 96 h. (**b**) Cells treated with 5.0, 2.5, 1.0, or 0.1 μM DAPT for 24 h. To detect the active NICD1 form, cleaved NOTCH1 (Val1744) antibody was used. The expression of total NOTCH1 was detected with rabbit α-NOTCH1 (C-20)-R antibody and was used as a loading and quality control in the Western blot assays. DMSO-treated cells were used as the control

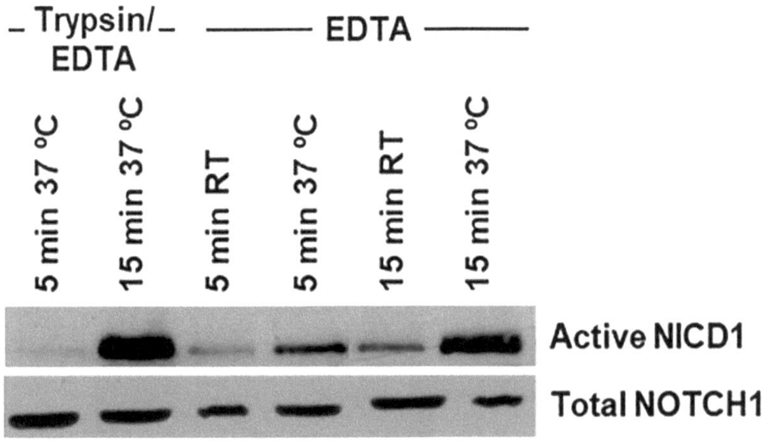

Fig. 6 Representative Western blot assay that shows the levels of the active form of NOTCH1 (NICD1) in SK-MEL-2 melanoma cells. Cells were treated with trypsin/EDTA for 5 and 15 min at 37 °C, and with EDTA for 5 or 15 min at room temperature (RT) or 37 °C. To detect the active NICD1 form, cleaved NOTCH1 (Val1744) antibody was used. The expression of total NOTCH1 was detected with rabbit α-NOTCH1 (C-20)-R antibody and was used as a loading and quality control in the Western blot assays

4. Cells are then centrifuged at 16,000 × *g* for 5 min. Subsequently, they can be washed again with DPBS, and samples can be frozen at −20 °C until further processing.
5. To obtain protein extracts to analyze the levels of active NICD in Western blot assays, we can proceed as described previously in Subheading 3.2. Figure 6 shows the effect of these EDTA-containing solutions in NOTCH1 processing.

3.5 Detection of the Intracellular Domain of Each NOTCH Receptor into the Cell Nucleus by Confocal Microscopy Using the Appropriate Antibodies That Recognize the Intracellular Domain of the Receptor

These are the steps to detect the activation of NOTCH receptors by confocal microscopy:

1. Seed cells directly in six-well plates on glass coverslips.
2. Fix with 3% formaldehyde in PBS, 10 min, wash 3× with DPBS, or with Invitrogen FIX & PERM Cell Permeabilization Kit.
3. Permeabilize 5 min with 0.2% TX-100 in PBS, wash with DPBS, or with Invitrogen FIX & PERM Cell Permeabilization Kit.
4. Block nonspecific junctions with 10% fetal bovine serum in DPBS for 0.5–1 h at room temperature inside a humid chamber.
5. Incubate with the corresponding primary NOTCH antibody (dilute in 5% fetal bovine serum in DPBS) for 1 h at room temperature or overnight at 4 °C. Wash with 3× PBS.
6. Incubate with secondary fluorescent antibody diluted to the appropriate concentration in DPBS-5% serum. One hour at room temperature. If precipitates are present, centrifuge before use.
7. Wash with DPBS (3×, 5 min) and mount in slides with fluorescence mounting medium such as Fluoromount G.
8. In the ultimate wash, DNA is labeled using Hoechst 33342 or DAPI at 0.5 μg/mL.
9. Visualize under confocal microscopy and quantify the fluorescence signal of the intracellular domain of NOTCH receptors in cell nuclei by using ImageJ program. Figure 7 shows representative pictures of this kind of assay by using NOTCH1 antibody as the primary antibody.

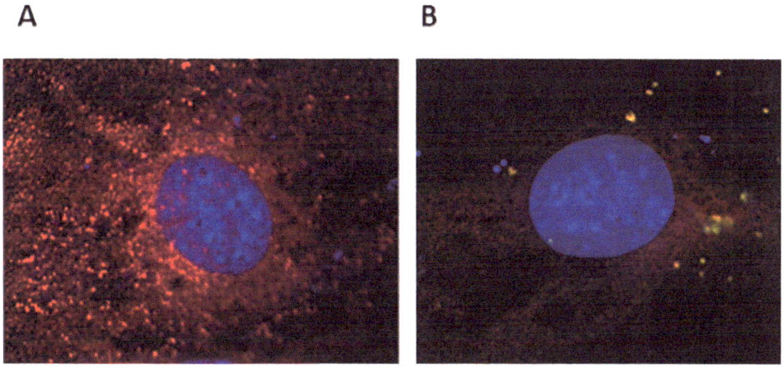

Fig. 7 Confocal microscopy assay of 3T3-L1 preadipocyte cells. Cells were firsts labelled with the rabbit α-NOTCH1 (C-20)-R antibody and then with a secondary fluorescent antibody (Alexa Fluor Red goat anti rabbit) that recognizes de primary antibody. The primary antibody recognizes the intracellular region of NOTCH1 receptor, which can be activated and migrates to the nucleus. Nuclei is stained with DAPI. The difference between (**a**, **b**) panels is the overexpression of NOTCH1 receptor in (**a**)

4 Notes

1. *Hes* genes have a cyclic expression and their proteins have a very short half-life. These proteins repress the transcription of their genes and ensure that their activity is time limited. HEY factors block the transcriptional activity of NOTCH by preventing its binding to DNA, thus exerting negative feedback and modulating signaling [23, 24].
2. Signaling pathways other than the NOTCH receptor signaling pathway can also modulate the expression of *Hes* and *Hey* genes. For this reason, the most suitable vector for analyzing transcriptional activation of NOTCH receptors is a plasmid like pGLuc-WT.
3. DAPT (DMSO solution) inhibits the gamma-secretase complex, which processes NOTCH receptors to generate the active NICD form.
4. EDTA can sequester the calcium that holds the NOTCH heterodimer in the cell membrane. It is important to control the structure of the receptor so that the extracellular and intracellular regions do not separate and affect their correct processing to generate the active intracellular domains that are detected by specific antibodies of NOTCH receptors. When collecting the samples, the time and temperature of incubation with trypsin/EDTA 0.025% solution or EDTA at a final concentration of 5 mM must be controlled to avoid the incorrect interpretation of the results.

Acknowledgments

We want to thank to our research colleagues Dra. María-Milagros Rodríguez-Cano, Dra. María-Julia González-Gómez, and Dr. Jorge Laborda for their help and advice, and to Mss. María de los Ángeles Ballesteros for her inestimable technical support.

References

1. Baladron V, Ruiz-Hidalgo M, Nueda M, Diaz-Guerra M, Garcia-Ramirez J, Bonvini E, Gubina E, Laborda J (2005) dlk acts as a negative regulator of Notch1 activation through interactions with specific EGF-like repeats. Exp Cell Res 303(2):343–359. https://doi.org/10.1016/j.yexcr.2004.10.001
2. Borggrefe T, Oswald F (2009) The Notch signaling pathway: transcriptional regulation at Notch target genes. Cell Mol Life Sci 66(10):1631–1646. https://doi.org/10.1007/s00018-009-8668-7
3. Artavanis-Tsakonas S, Matsuno K, Fortini ME (1995) Notch signaling. Science 268(5208):225–232
4. Artavanis-Tsakonas S, Rand MD, Lake RJ (1999) Notch signaling: cell fate control and signal integration in development. Science 284(5415):770–776

5. Bi P, Yue F, Karki A, Castro B, Wirbisky SE, Wang C, Durkes A, Elzey BD, Andrisani OM, Bidwell CA, Freeman JL, Konieczny SF, Kuang S (2016) Notch activation drives adipocyte dedifferentiation and tumorigenic transformation in mice. J Exp Med 213(10):2019–2037. https://doi.org/10.1084/jem.20160157
6. Yang X, Klein R, Tian X, Cheng HT, Kopan R, Shen J (2004) Notch activation induces apoptosis in neural progenitor cells through a p53-dependent pathway. Dev Biol 269(1): 81–94. https://doi.org/10.1016/j.ydbio.2004.01.014. S0012160604000636 [pii]
7. Zhang XP, Zheng G, Zou L, Liu HL, Hou LH, Zhou P, Yin DD, Zheng QJ, Liang L, Zhang SZ, Feng L, Yao LB, Yang AG, Han H, Chen JY (2008) Notch activation promotes cell proliferation and the formation of neural stem cell-like colonies in human glioma cells. Mol Cell Biochem 307(1-2):101–108. https://doi.org/10.1007/s11010-007-9589-0
8. Koch U, Radtke F (2007) Notch and cancer: a double-edged sword. Cell Mol Life Sci 64(21): 2746–2762. https://doi.org/10.1007/s00018-007-7164-1
9. Maillard I, Adler SH, Pear WS (2003) Notch and the immune system. Immunity 19(6): 781–791. S107476130300325X [pii]
10. Robey E (1997) Notch in vertebrates. Curr Opin Genet Dev 7(4):551–557. S0959-437X(97)80085-8 [pii]
11. Weinmaster G (2000) Notch signal transduction: a real rip and more. Curr Opin Genet Dev 10(4):363–369. S0959-437X(00)00097-6 [pii]
12. Kadesch T (2004) Notch signaling: the demise of elegant simplicity. Curr Opin Genet Dev 14(5):506–512. https://doi.org/10.1016/j.gde.2004.07.007. S0959-437X(04)00118-2 [pii]
13. Laborda J (2014) Un siglo de NOTCH. http://cienciaes.com/quilociencia/2014/03/09/un-siglo-de-notch/. Accessed 15 marzo 2016
14. Lindsell CE, Boulter J, diSibio G, Gossler A, Weinmaster G (1996) Expression patterns of Jagged, Delta1, Notch1, Notch2, and Notch3 genes identify ligand-receptor pairs that may function in neural development. Mol Cell Neurosci 8(1):14–27. https://doi.org/10.1006/mcne.1996.0040. S1044-7431(96)90040-8 [pii]
15. D'Souza B, Meloty-Kapella L, Weinmaster G (2010) Canonical and non-canonical Notch ligands. Curr Top Dev Biol 92:73–129. https://doi.org/10.1016/S0070-2153(10)92003-6. S0070-2153(10)92003-6 [pii]
16. D'Souza B, Miyamoto A, Weinmaster G (2008) The many facets of Notch ligands. Oncogene 27(38):5148–5167. https://doi.org/10.1038/onc.2008.229. onc2008229 [pii]
17. Kumar R, Juillerat-Jeanneret L, Golshayan D (2016) Notch antagonists: potential modulators of cancer and inflammatory diseases. J Med Chem 59(17):7719–7737. https://doi.org/10.1021/acs.jmedchem.5b01516
18. Kopan R, Ilagan MX (2009) The canonical Notch signaling pathway: unfolding the activation mechanism. Cell 137(2):216–233. https://doi.org/10.1016/j.cell.2009.03.045. S0092-8674(09)00382-1 [pii]
19. Laborda J, Sausville E, Hoffman T, Notario V (1993) dlk, a putative mammalian homeotic gene differentially expressed in small cell lung carcinoma and neuroendocrine tumor cell line. J Biol Chem 268:3817–3820
20. Nueda M, Baladrón V, García-Ramírez J, Sánchez-Solana B, Ruvira M, Rivero S, Ballesteros M, Monsalve E, Díaz-Guerra M, Ruiz-Hidalgo M, Laborda J (2007) The novel gene EGFL9/Dlk2, highly homologous to Dlk1, functions as a modulator of adipogenesis. J Mol Biol 367:1270–1280
21. Sanchez-Solana B, Luisa Nueda M, Desamparados Ruvira M, Jose Ruiz-Hidalgo M, Maria Monsalve E, Rivero S, Javier Garcia-Ramirez J, Diaz-Guerra MJM, Baladron V, Laborda J (2011) The EGF-like proteins DLK1 and DLK2 function as inhibitory non-canonical ligands of NOTCH1 receptor that modulate each other's activities. BBA-Mol Cell Res 1813(6):1153–1164. https://doi.org/10.1016/j.bbamcr.2011.03.004
22. Traustadóttir G, Jensen C, Thomassen M, Beck H, Mortensen S, Laborda J, Baladrón V, Sheikh S, Andersen D (2016) Evidence of non-canonical NOTCH signaling: delta-like 1 homolog (DLK1) directly interacts with the NOTCH1 receptor in mammals. Cell Signal 28:246–254
23. Fior R, Henrique D (2009) "Notch-off": a perspective on the termination of Notch signalling. Int J Dev Biol 53(8–10):1379–1384. https://doi.org/10.1387/ijdb.072309rf. 072309rf [pii]
24. Andersson ER, Sandberg R, Lendahl U (2011) Notch signaling: simplicity in design, versatility in function. Development 138(17): 3593–3612. https://doi.org/10.1242/dev.063610. 138/17/3593 [pii]
25. Robey EA, Bluestone JA (2004) Notch signaling in lymphocyte development and function. Curr Opin Immunol 16(3):360–366. https://

doi.org/10.1016/j.coi.2004.03.009. S0952791504000469 [pii]
26. Kopan R (2002) Notch: a membrane-bound transcription factor. J Cell Sci 115(Pt 6): 1095–1097
27. Falix F, Aronson D, Lamers W, Gaemers I (2012) Possible roles of DLK1 in the Notch pathway during development and disease. Biochim Biophys Acta 1822:988–995
28. Guo H, Lu Y, Wang J, Liu X, Keller E, Liu Q, Zhou Q, Zhang J (2014) Targeting the Notch signaling pathway in cancer therapeutics. Thorac Cancer 5:473–486
29. Fortini ME (2001) Notch and presenilin: a proteolytic mechanism emerges. Curr Opin Cell Biol 13(5):627–634. S0955-0674(00) 00261-1 [pii]
30. Bray S, Furriols M (2001) Notch pathway: making sense of suppressor of hairless. Curr Biol 11(6):R217–R221. S0960-9822(01) 00109-9 [pii]
31. Nichols JT, Miyamoto A, Weinmaster G (2007) Notch signaling--constantly on the move. Traffic 8(8):959–969. https://doi.org/10.1111/j.1600-0854.2007.00592.x. TRA592 [pii]
32. Kanwar R, Fortini ME (2004) Notch signaling: a different sort makes the cut. Curr Biol 14(24):R1043–R1045. https://doi.org/10.1016/j.cub.2004.11.041. S0960982204009303 [pii]
33. Fortini ME (2009) Notch signaling: the core pathway and its posttranslational regulation. Dev Cell 16(5):633–647. https://doi.org/10.1016/j.devcel.2009.03.010. S1534-5807(09) 00131-2 [pii]
34. Bray S, Bernard F (2010) Notch targets and their regulation. Curr Top Dev Biol 92: 253–275. https://doi.org/10.1016/S0070-2153(10)92008-5. S0070-2153(10)92008-5 [pii]
35. Saint Just Ribeiro M, Wallberg AE (2009) Transcriptional mechanisms by the coregulator MAML1. Curr Protein Pept Sci 10(6): 570–576. CPPS-24 [pii]
36. Nam Y, Sliz P, Song L, Aster JC, Blacklow SC (2006) Structural basis for cooperativity in recruitment of MAML coactivators to Notch transcription complexes. Cell 124(5): 973–983. https://doi.org/10.1016/j.cell.2005.12.037. S0092-8674(06)00122-X [pii]
37. Iso T, Kedes L, Hamamori Y (2003) HES and HERP families: multiple effectors of the Notch signaling pathway. J Cell Physiol 194(3): 237–255. https://doi.org/10.1002/jcp.10208
38. Fischer A, Gessler M (2007) Delta-Notch--and then? Protein interactions and proposed modes of repression by Hes and Hey bHLH factors. Nucleic Acids Res 35(14):4583–4596. https://doi.org/10.1093/nar/gkm477. gkm477 [pii]
39. Rodriguez-Cano MM, Gonzalez-Gomez MJ, Sanchez-Solana B, Monsalve EM, Diaz-Guerra MM, Laborda J, Nueda ML, Baladron V (2020) NOTCH receptors and DLK proteins enhance brown adipogenesis in mesenchymal C3H10T1/2 cells. Cells 9(9):2032. https://doi.org/10.3390/cells9092032

Chapter 8

Somatic Clonal Analyses Using FLP/FRT and MARCM System to Understand Notch Signaling Mechanism and Its Regulation

Vartika Sharma, Nalani Sachan, Mousumi Mutsuddi, and Ashim Mukherjee

Abstract

Notch signaling regulates an array of developmental decisions and has been implicated in a multitude of diseases, including cancer over the past a few decades. The simplicity and versatility of the Notch pathway in *Drosophila* make it an ardent system to study Notch biology, its regulation, and functions. In this chapter, we highlight the use of two powerful techniques, namely, FLP/FRT and MARCM in the study of Notch signaling. These mosaic analysis techniques are powerful tools to analyze gene functions in different biological processes. The section briefly explains the principle and the protocols with suitable examples.

Key words FLP/FRT, Mosaic analysis, MARCM, Notch signaling, *Drosophila*

1 Introduction

Notch signaling is an evolutionarily conserved signaling pathway that regulates a wide array of developmental processes, including cell-fate specification, cell proliferation, and cell death, in order to maintain tissue homeostasis throughout development [1, 2]. The pleiotropic role of Notch signaling is intricately linked with the complexity of its regulation in different cellular contexts. Activation of the Notch trans-membrane receptor by either of its ligands, Delta or Serrate (Jagged in mammals), triggers a series of proteolytic cleavages resulting in the release of Notch intracellular domain (Notch-ICD), that translocates to the nucleus and directly interacts with the DNA binding protein CSL [CBF-1/Su(H)/LAG-1] and the coactivator Mastermind to promote transcription [3, 4]. Aberrant Notch signaling is often associated with many inherited disorders [Alagille syndrome, CADASIL], diseases, and a multitude of

Dongyu Jia (ed.), *Notch Signaling Research: Methods and Protocols*,
Methods in Molecular Biology, vol. 2472, https://doi.org/10.1007/978-1-0716-2201-8_8,

cancer [5]; thus it is a prerequisite to understand the mechanism of signaling regulation as well as its signaling outcomes.

Our understanding of genetic mechanisms that underlie any biological processes relies heavily on loss-of-function studies. The loss-of-function study allows us to analyze the phenotypes caused by these perturbations, and thus the wild-type functions of the particular gene can be assessed. Over the years, many Notch and its pathway mutants have been identified and characterized that potentially abolishes the functions of the Notch pathway [6]. Since most of the genes are required at various developmental stages, their complete knockdown results in embryonic lethality, making the study quite impossible. Thus, a mosaic study where only a proportion of tissue is homozygous mutant in an otherwise wild-type background was the need of the hour. The emergence of the powerful tools, namely, FLP/FRT system and its derivative MARCM system aided in overcoming this barrier.

In this chapter, we will discuss the basics and methodology to generate loss-of-function and gain-of-function clones by using the FLP/FRT technique and we will also lay our emphasis on the MARCM technique, where we can generate positively marked homozygous mutant clones in otherwise unlabeled heterozygous wild-type tissue.

1.1 FLP/FRT System

This powerful technique for the generation of genetic mosaics in *Drosophila* is based on the FLP (flippase)/FRT (FLP recognition target) system, initially adapted from yeast (*Saccharomyces cerevisiae*). The system basically works as a site-specific recombination technology that targets the recombination enzyme FLP to specific DNA sequences designated as FRT sites [7, 8]. The controlled induction of this mitotic recombination is usually performed by expressing FLP under an inducible or tissue-specific promoter. *hsp70* (Heat shock promoter) is the most common promoter being used to control the pulses of FLP expression in response to temperature modulations. It is to be noted that FLP is not expressed at 18 °C, and exhibits a relatively low expression at 25 °C but is highly expressed at temperatures above 30 °C [8]. Modulating the duration of heat shock, the temperature, or the stage of development can control both the recombination frequency and the clone induction in the *Drosophila* tissues. Exposure to heat shock thus potentially catalyzes the recombination between two FRT sites during G2 phase of the cell cycle. The mitotic recombination between the two FRT sites on homologous chromosome arms generates clones of cells homozygous for mutations that lie distal to the FRT sites in otherwise wild-type heterozygous tissue. Upon chromatid segregation during mitosis, the cells that underwent recombination show homozygous mutant patches/clones in an otherwise wild-type background (Fig. 1a). Thus the technique bypasses the potentially detrimental effects of

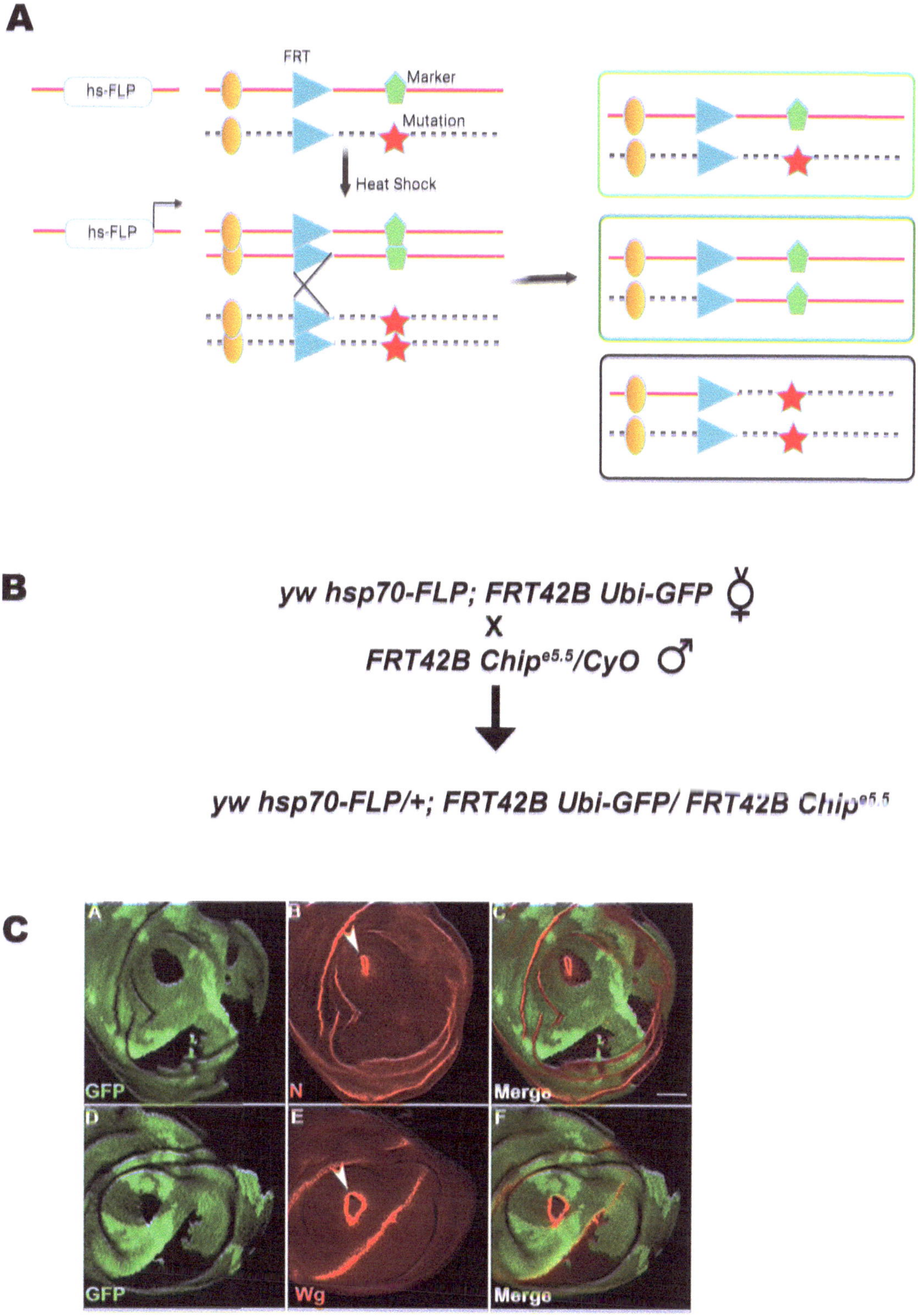

Fig. 1 Generation of mutant somatic clones using FLP/FRT system. (**a**) A schematic diagram showing FLP/FRT-mediated mitotic recombination. The yeast recombinase Flippase is under the control of the *hsp70* promoter, *hs-FLP*. Heat-shock induces FLP expression which in turn triggers the recombination between the two FRT

mutations in other tissues or in early developmental stages that are more susceptible to lethality. Hereby with the help of two examples, we demonstrate the use of FLP/FRT to generate loss-of-function and gain-of-function clones in the *Drosophila* wing imaginal discs.

1.2 Mosaic Analysis with a Repressible Cell Marker (MARCM)

To study the effect of homozygous mutations in a more robust and sensitive manner, MARCM analysis is deployed. As we have discussed in the previous section that FLP/FRT system generates mutant clones which remain unmarked in a positively marked wild-type background [7, 8]. However, for particular cell types, such as neurons, marking mutant clonal cells is more useful in order to understand their complex processes from the numerous densely packed surrounding cells. The MARCM technique overcomes this barrier by allowing one to label homozygous mutant cells in otherwise unlabeled heterozygous surrounding cells via mitotic recombination. MARCM was first used to visualize neuronal projections using fluorescent markers, but it is now widely used in all tissue types to generate positively labeled mutant clones. MARCM uses a cocktail of the yeast GAL80 repressor protein which inhibits the activity of GAL4 transcription factor, *GAL4-UAS* binary expression system [9], and FLP/FRT system to generate genetically labeled clones. In the MARCM system, GAL80 represses the activity of GAL4 proteins under normal conditions so that eventually expression of any UAS construct present in the genome is blocked. Since both GAL4 and GAL80 are kept under tubulin 1 promoter, they have ubiquitous expression. The *tub Gal80* transgene is kept distal to an FRT in one chromosome and the other homologous chromosome has the same FRT with mutation of the gene of interest distal to the FRT. FLP induced recombination results in three types of daughter cells, one is like the parent cell carrying a copy of *tub-GAL80* and a copy of mutation, the other is the wild-type cells that carries two copies of *tub-GAL80,* and the third one, which is homozygous mutant cells without *tub-GAL80* (*see* Fig. 3a). Loss of *GAL80* relieves repression of *GAL4* and allows for expression of GAL4-driven reporter gene specifically in mutant clonal cells [10, 11]. Figure 3 systematically explains the MARCM genetics. A "MARCM ready" flies constituting of FLP recombinase, an FRT site, *tub-GAL80* (tubulin 1 promoter), *tub-GAL4,* and a *UAS-marker.* When this MARCM ready flies are crossed to a line carrying

Fig. 1 (continued) sites on homologous chromosomes. Segregation of recombinant chromosomes at mitosis results in homozygous mutant and homozygous wild-type cells along with heterozygous ones. Note that the homozygous mutant clones are marked by the absence of marker (GFP or RFP are the commonly used markers). (**b**) An example of a genetic cross to obtain *Chip*$^{e5.5}$ mutant clone using FLP/FRT system in *Drosophila* larval tissue. (**c**) *Chip*$^{e5.5}$ loss-of-function effect on the expression of Notch, and its target Wg. Note that the *Chip*$^{e5.5}$ mutant clones are marked by the absence of GFP. *Chip*$^{e5.5}$ clones in the dorsal compartment of the wing imaginal disc show an ectopic expression of Notch and its target Wg (marked by arrowhead). Scale Bar in **c** (**a–f**: 50 μm). (Figure **c** reproduced from ref. 12 with permission from Elsevier)

corresponding FRT and mutations, marked homozygous clones can be visualized (see the procedure for details). In order to overexpress another gene in the null mutant clonal cells, *UAS-transgene* can be combined along with mutant lines which allows one to visualize the loss-of-function effect of one gene in the background of the gain-of-function effect of another gene.

We have used these elegant techniques to investigate the functional implications of a good number of novel interacting partners involved in the regulation of Notch signaling [12–19].

2 Materials

2.1 General Equipment

1. Basic equipment to perform *Drosophila* handling (stereomicroscope, light source, CO_2/Ether, anesthetizer, fly food, vials, etc.).
2. 18 °C and 25 °C incubators to maintain *Drosophila* crosses.
3. 37 °C water bath for heat shock (if using heat-shock promoter for FLP expression).
4. Imaging microscope and software (confocal/fluorescence microscope).

2.2 Loss-of-Function Clones in Wing Imaginal Discs

2.2.1 Drosophila *Stocks*

Fly stocks are readily available from the Bloomington Stock Center (http://www.bdsc.indiana.edu/) or Vienna *Drosophila* Resource Center (http://stockcenter.vdrc.at). The list below is specific for an example of a loss-of-function clone generation in wing imaginal discs using FRT42B insertion carrying chromosomes shown here.

1. *yw hsp70-FLP; FRT42B Ubi-GFP.*
2. *FRT42B Chip* $^{e5.5}$/CyO.

2.3 Gain-of-Function Clones in Wing Imaginal Discs

2.3.1 Drosophila *Stocks*

The following flies were used for gain-of-function clone generation using FLP/FRT technique:

1. *yw hsp70-FLP; Act-FRT y + FRT-GAL4, UAS-GFP/CyO.*
2. *UAS-HA-TRAF6; UAS-FLAG-Dx.*

2.4 MARCM Clones in Larval Tissue

2.4.1 Drosophila *Stocks*

MARCM fly stocks are readily available from the Bloomington Stock Center (http://www.flybase.org) or Vienna *Drosophila* Resource Center (http://stockcenter.vdrc.at). Some of the important MARCM stocks available at Bloomington Stock Center are listed in Table 1.

The list below is specific for an example of loss-of-function clone generation using MARCM technique in the *Drosophila* larval brain shown here.

1. *yw hsp70-FLP Tub-GAL4 UAS-GFP;+; FRT82B Tub-GAL80.*
2. *UAS-Notch-ICD; FRT82B imp-*$\alpha3$ ^{D93}*/TM6B.*

Table 1
List of MARCM fly stocks available at Bloomington Stock Center

S. No.	Bloomington stock number	MARCM component
1.	5132	FRT and tubP-GAL80 recombinant for X: FRT19A (1C2)
2.	5191	FRT and tubP-GAL80 recombinant for 3L: FRT80D (75E1)
3.	5192	FRT and tubP-GAL80 recombinant for 2L: FRT40A (31E3)
4.	5140	FRT and tubP-GAL80 recombinant for 2R: FRT42B (47A7)
5.	5190	FRT and tubP-GAL80 recombinant for 3L: FRT80B (75E1)
6.	5135	FRT and tubP-GAL80 recombinant for 3R: FRT82B (insertion unknown)
7.	5138	*tubP-GAL4*
8.	5580	FLP driven by a tissue-specific promoter: *ey-FLP*
9.	8862	FLP driven by a heat-shock promoter: *hs-FLP*
10.	5136	UAS-mCD8::GFP on X chromosome
11.	5137	UAS-mCD8::GFP on second chromosome
12.	5130	UAS-mCD8::GFP on third chromosome
13.	7118	UAS-myr::mRFP on second chromosome
14.	7119	UAS-myr::mRFP on third chromosome
15.	28832	*hs-FLP* and UAS-mCD8::GFP on X chromosome

3 Methods

3.1 Loss-of-Function Clones in Wing Imaginal Discs

1. To generate homozygous mosaic clones of *Chip* in wing imaginal discs, virgin flies carrying *FLP, FRT* sites, and suitable marker (*yw hsp70-FLP; FRT42B Ubi-GFP*) were crossed with males carrying the mutant allele (*FRT42B Chip*$^{e5.5}$/*CyO*). For stock details (*see* **Note 1**).
2. Heat shock was given at 37 °C for 45 min at 24 h after egg laying (AEL).
3. The vials were then transferred to a 25 °C incubator for proper development of the larvae.
4. Third instar larvae were then dissected out and stained for Notch and Wingless (Wg).
5. The *Chip* mutant clones were identified by the absence of GFP expression corresponding to which the status of Notch and Notch downstream target Wg were checked. An ectopic

expression of Notch and its target Wg was observed in *Chip*$^{e5.5}$ clones in the dorsal compartment of the wing imaginal disc (Fig. 1c [12]).

3.2 Gain-of-Function Clone in Wing Imaginal Discs

1. To generate gain-of-function clones in wing discs, virgin flies carrying the specific genes under the control of *UAS* promoter (*UAS-HA-TRAF6; UAS-FLAG-Dx*) were crossed with male flies with a FLP-out construct and a marker as mentioned above (*yw hsp70-FLP; Act-FRT y + FRT-GAL4, UAS-GFP/ CyO*) (*see* **Note 2**).
2. 24 h after egg-laying, the first instar larvae (F1 generation) were subjected to heat shock in a 37 °C water bath for 1 h.
3. The vials were then transferred to a 25 °C incubator for proper growth and development of the larvae.
4. Third instar larvae were then dissected out and stained for Notch target, Wg. The *TRAF6* and *Deltex* gain-of-function clones were identified by the presence of GFP expression corresponding to which the status of Notch target, Wg was checked.
5. The GFP positive cells show an elevated expression of Notch target, Wg, upon *TRAF6* and *Deltex* overexpression (Fig. 2c).

3.3 MARCM Clones in Larval Tissue

1. To generate *imp-α3*^{D93} mutant clones in Notch overexpression background, virgin flies of *UAS-Notch-ICD; FRT82B imp-α3*^{D93}*/TM6B* was crossed with male *yw hsp70-FLP Tub--GAL4 UAS-GFP; FRT82B Tub-GAL80* flies (*see* **Note 3**).
2. A day after egg-laying, the first instar larvae (F1 generation) were subjected to heat shock in a 37 °C water bath for 45 min.
3. The vials were then transferred to a 25 °C incubator for proper growth and development of the larvae.
4. Following, the third instar larvae were then dissected out and stained for Notch. The *imp-α3*^{D93} mutant clones in Notch overexpression background were identified by the presence of GFP expression corresponding to which the status of Notch was checked (*see* **Notes 4–9** for additional details on MARCM).

4 Notes

1. *yw hsp70-FLP; FRT42B Ubi-GFP* stock was obtained from Bloomington Stock Center. This line harbors FLP enzyme sequence under the control of heat shock promoter on the first chromosome that catalyzes the double-strand DNA breaks and recombinations at FRT sites. The second chromosome harbors an FRT site on which the enzyme flippase will act to

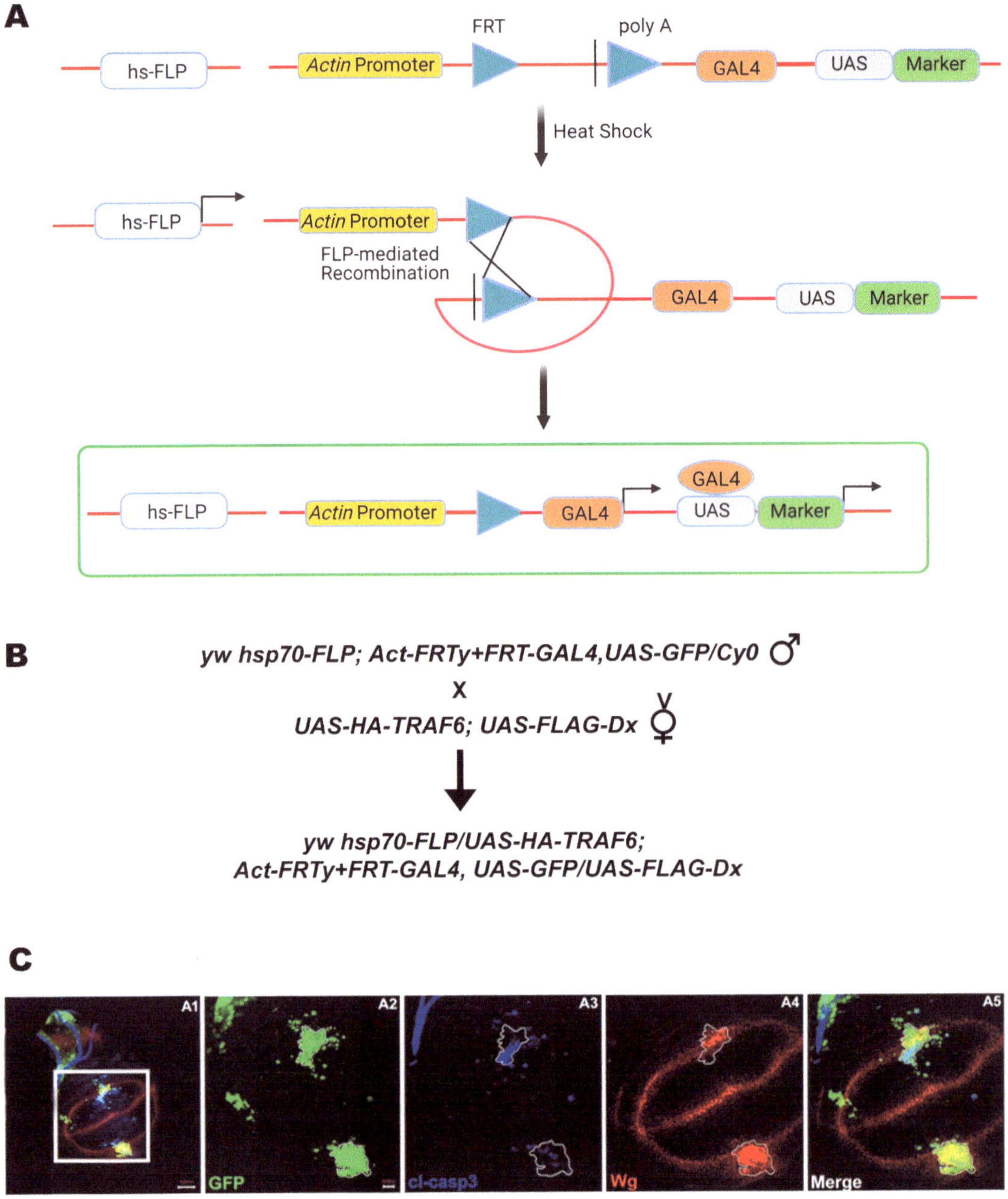

Fig. 2 Generation of *TRAF6* and *Deltex* gain-of-function clones using FLP/FRT technique. (**a**) A schematic diagram showing FLP/FRT mediated mitotic recombination to generate gain-of-function clones. The expression of *GAL4* by the *Actin* promoter is disrupted by the presence of polyA site (transcription terminator). The site-specific recombination between the tandem FRT sites, induced by flippase results in the removal of the transcription terminator, thereby *GAL4* is induced under *Actin* promoter and consequently induces the expression of the transgenes under the control of *GAL4* inducible promoter, *UAS*-marker. (**b**) A genetic cross to obtain *TRAF6* and *Deltex* gain-of-function clones. (**c**) The GFP positive cells that mark the clonal area, where an elevated expression of Notch target, Wg was observed upon *TRAF6* and *Deltex* overexpression, parallel to which an ectopic caspase expression was observed in the same GFP positive cells. Scale Bar in **c** (A1: 50 μm; A2–A5: 20 μm)

generate recombinations. Note that the FRT sites are generally located close to the centromere on a chromosome arm. Ubi-GFP serves as a marker, and hereby, the mutant cells are marked by the absence of the green fluorescence protein (GFP).

FRT42B Chip$^{e5.5}$*/CyO* is a null allele of *Chip* (*Chip*$^{e5.5}$) harbors an FRT42B insertion on the same chromosome arm.

2. Gain-of-function clones utilize the *GAL4/UAS* system along with FLP/FRT, where *GAL4* expression causes transcriptional activation of a marker gene under the control of the *UAS* promoter [9]. *yw hsp70-FLP; Act-FRT y + FRT-GAL4, UAS-GFP/CyO* line harbors a FLP-out construct consisting of a constitutive promoter, followed by an FRT site, a marker gene with a poly-A site (transcriptional terminator), and a second FRT site. Upon heat shock, the FRT cassette is released thereby expressing the *UAS-transgene* in an *Act-GAL4* dependent manner. The *UAS-GFP* serves as a marker that is constitutively expressed in those cells that are positive for the mutation (*see* Fig. 2a).

   UAS-HA-TRAF6; UAS-FLAG-Dx transgenic line contains HA-epitope tagged *TRAF6* and FLAG-epitope tagged *Deltex* coding sequence downstream to the *UAS* promoter.

3. MARCM ready stock (*yw hsp70-FLP Tub-GAL4 UAS-GFP; FRT82B Tub-GAL80*) features a constitutively actively *Tub-GAL4.* Note that the GAL4 can be anywhere in the genome except on the same chromosome arm harboring *Tub-GAL80. Tub-GAL80* on the other hand represses the functions of *Tub-GAL4* under normal conditions, thereby repressing the expression of the marker (*UAS-GFP*).

   UAS-Notch-ICD; FRT82B imp-$\alpha 3^{D93}$*/TM6B* transgenic line contains *Notch-ICD* downstream to *UAS* on the second chromosome. On the third chromosome, *imp* mutant harbors an FRT site upstream to the mutation. The line enables us to visualize *imp* loss-of-function effect in *Notch-ICD* overexpression background.

   The cytoplasmic localization of Notch in *imp-*$\alpha 3^{D93}$ mutant cells in Notch overexpression background suggests that *imp* null clones restrict the nuclear translocation of Notch and it was determined through this experiment that Importin-α3 protein is essential for nuclear translocation of Notch-ICD (Fig. 3c [13]).

4. Flies carrying the *P[ry+, hs-neo, FRT]* element were selected by their resistance to G418 (Geneticin). To prepare fly medium containing G418, a few holes were made in fly food already poured in vials and 0.2 ml of 25 mg/ml freshly made G418 solution was added per 10 ml of food medium, allowed to

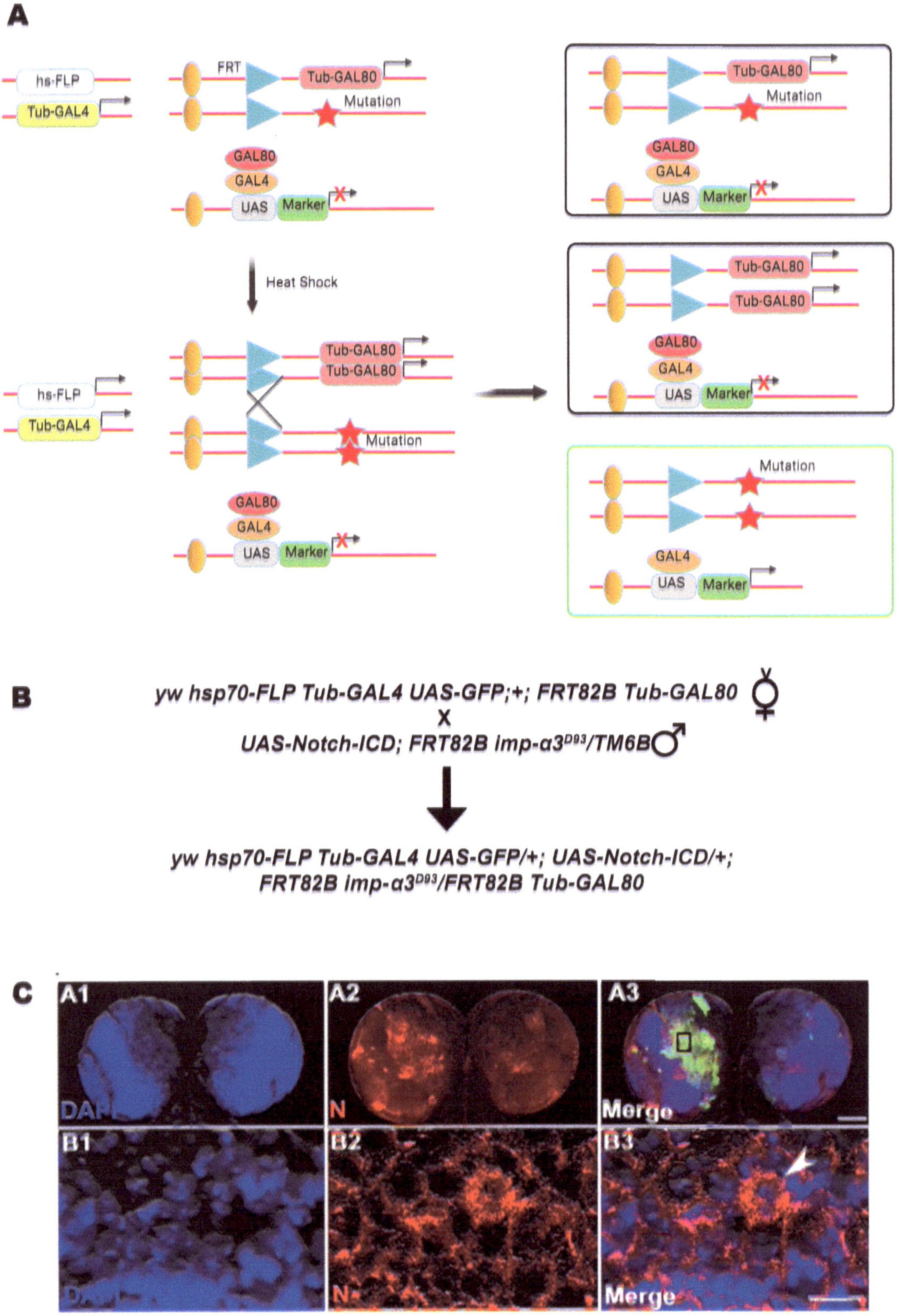

Fig. 3 Generation of mutant clones using MARCM technique. (**a**) A schematic representation of the MARCM technique. The GAL80 protein represses the GAL4 transcription factor, thus GAL4-dependent expression of

air-dry for several hours before using them [8]. These food vials containing G418 can be stored at 18 °C for 2 weeks.

5. To increase the efficiency of clone generation through MARCM, G2 phase can be extended by allowing flies to develop at lower temperatures after induction of FLP expression.
6. On the contrary, if the number of clones are more, a reduction in the length of heat shock or a reduction in the number of FLP copies will serve the purpose to reduce the number of clones.
7. It is worth considering that the MARCM ready flies might be weak due to the presence of multiple genetic insertions; therefore it is advisable to take extra care of these stocks and to ensure periodically that they retain their transgenes.
8. For heat shock, vials should be submerged in a circulating water bath at 37 °C for a uniform heat induction throughout the food vials.
9. It is always advisable to prepare a fresh fixative (3% Paraformaldehyde in PBS) each time for immunostaining of larval tissues.

Acknowledgments

We extend our sincere thanks to the Bloomington Stock Centre and Developmental studies Hybridoma Bank (DSHB) for fly stocks and some of the antibodies used in our studies. We also acknowledge the confocal facility of DBT-BHU-ISLS, Banaras Hindu University for microscopy assistance. This work was supported by grants from Department of Science and Technology (DST), Government of India and Institute of Eminence Scheme, Banaras Hindu University, India. Fellowship support to Vartika Sharma was provided by the Council of Scientific and Industrial Research (CSIR), Government of India. Artwork is created with BioRender (BioRender.com).

Fig. 3 (continued) UAS-marker (GFP or RFP) is blocked. On the contrary, cells lacking GAL80 relieve repression of GAL4 and allows expression of GAL4-driven *UAS*-marker. FLP-induced site-specific mitotic recombination between FRT sites gives rise to daughter cells exhibiting ubiquitous expression of GAL80 under *tubulin* promoter and repress GAL4 and consequently repress the expression of *UAS*-marker. On the contrary, in homozygous mutant daughter cells, the GAL4 transcription factor is active due to the absence of GAL80, thereby allowing the expression of the specific marker under the UAS promoter. (**b**) A scheme of genetic cross to obtain *imp-α3*^{D93}mutant clone in *Drosophila* larval tissue in which *UAS-Notch-ICD* is overexpressed. (**c**) *imp-α3*^{D93} loss-of-function effect on Notch localization in *UAS-Notch-ICD* overexpression background. It clearly shows that absence of Importin-α3 inhibits the translocation of Notch-ICD into the nucleus. Note, the arrowhead points the cytoplasmic localization of Notch in *imp-α3*^{D93} mutant cells. Scale bar in **c** (A1–A3: 100 μm; B1–B3: 10 μm). (Figure **c** reproduced from ref. 13 with permission from PLoS One)

References

1. Andersson ER, Sandberg R, Lendahl U (2011) Notch signaling: simplicity in design, versatility in function. Development 138:3593–3612
2. Artavanis-Tsakonas S, Rand MD, Lake RJ (1999) Notch signaling: cell fate control and signal integration in development. Science 284:770–776
3. Fortini ME (2009) Notch signaling: the core pathway and its posttranslational regulation. Dev Cell 16:633–647
4. Kopan R, Ilagan MXG (2009) The canonical Notch signaling pathway: unfolding the activation mechanism. Cell 137:216–233
5. Siebel C, Lendahl U (2017) Notch signaling in development, tissue homeostasis, and disease. Physiol Rev 97:1235–1294
6. Zacharioudaki E, Bray SJ (2014) Tools and methods for studying Notch signaling in *Drosophila melanogaster*. Methods 68:173–182
7. Golic KG, Lindquist S (1989) The FLP recombinase of yeast catalyzes site-specific recombination in the *Drosophila* genome. Cell 59:499–509
8. Xu T, Rubin GM (1993) Analysis of genetic mosaics in developing and adult *Drosophila* tissues. Development 117:1223–1237
9. Brand AH, Perrimon N (1993) Targeted gene expression as a means of altering cell fates and generating dominant phenotypes. Development 118:401–415
10. Lee T, Luo L (1999) Mosaic analysis with a repressible cell marker for studies of gene function in neuronal morphogenesis. Neuron 22:451–461
11. Lee T, Luo L (2001) Mosaic analysis with a repressible cell marker (MARCM) for *Drosophila* neural development. Trends Neurosci 24:251–254
12. Sachan N, Mishra AK, Mutsuddi M et al (2015) Chip physically interacts with Notch and their stoichiometry is critical for Notch function in wing development and cell proliferation in *Drosophila*. Biochim Biophys Acta Gen Subj 1850:802–812
13. Sachan N, Mishra AK, Mutsuddi M et al (2013) The *Drosophila* importin-α3 is required for nuclear import of Notch in vivo and it displays synergistic effects with Notch receptor on cell proliferation. PLoS One 8:e68247
14. Mukherjee A, Veraksa A, Bauer A et al (2005) Regulation of Notch signalling by non-visual β-arrestin. Nat Cell Biol 7:1191–1201
15. Mishra AK, Sachan N, Mutsuddi M et al (2014) TRAF6 is a novel regulator of Notch signaling in *Drosophila melanogaster*. Cell Signal 26:3016–3026
16. Paul MS, Dutta D, Singh A et al (2018) Regulation of Notch signaling in the developing *Drosophila* eye by a T-box containing transcription factor, Dorsocross. Genesis 56(10): e23251
17. Singh A, Dutta D, Paul MS et al (2018) Pleiotropic functions of the chromodomain-containing protein hat-trick during oogenesis in *Drosophila melanogaster*. G3 (Bethesda) 8:1067–1077
18. Singh A, Paul MS, Dutta D et al (2019) Regulation of Notch signaling by the chromatin-modeling protein Hat-trick. Development 146:dev170837
19. Sharma V, Mutsuddi M, Mukherjee A (2021) Deltex cooperates with TRAF6 to promote apoptosis and cell migration through Eiger-independent JNK activation in *Drosophila*. Cell Biol Int 45:686–700

Chapter 9

Analyzing the Interaction of RBPJ with Mitotic Chromatin and Its Impact on Transcription Reactivation upon Mitotic Exit

Kostiantyn Dreval, Robert J. Lake, and Hua-Ying Fan

Abstract

The sequence-specific transcription factor RBPJ, also known as CSL (CBF1, Su(H), Lag1), is an evolutionarily conserved protein that mediates Notch signaling to guide cell fates. When cells enter mitosis, DNA is condensed and most transcription factors dissociate from chromatin; however, a few, select transcription factors, termed bookmarking factors, remain associated. These mitotic chromatin-bound factors are believed to play important roles in maintaining cell fates through cell division. RBPJ is one such factor that remains mitotic chromatin associated and therefore could function as a bookmarking factor. Here, we describe how to obtain highly purified mitotic cells from the mouse embryonal carcinoma cell line F9, perform chromatin immunoprecipitation with mitotic cells, and measure the first run of RNA synthesis upon mitotic exit. These methods serve as basis to understand the roles of mitotic bookmarking by RBPJ in propagating Notch signals through cell division.

Key words RBPJ, Notch signaling, Mitotic bookmarking, Chromatin immunoprecipitation, Nascent RNA transcription, Purification of mitotic cells, Mouse embryonal carcinoma cells

1 Introduction

The faithful propagation of transcription programs through cell division is essential for the maintenance of cell identity and lineage fidelity. During mitosis, the chromosomes condense, transcription ceases, and the majority of the transcriptional machinery dissociates from mitotic chromatin [1]. Nonetheless, most transcription programs are faithfully propagated through cell division. A critical question concerns how the memory of a specific transcription program is maintained through mitosis. Epigenetic marks, such as specific histone modifications, are believed to "bookmark" certain genes to maintain transcriptional memory [2]. More recently, sequence-specific transcription factors that are selectively retained on mitotic chromatin have emerged as potential players in mitotic

Dongyu Jia (ed.), *Notch Signaling Research: Methods and Protocols*,
Methods in Molecular Biology, vol. 2472, https://doi.org/10.1007/978-1-0716-2201-8_9,

chromatin bookmarking [3–7]; however, significant questions remain. These include the mechanisms of selective mitotic chromatin retention and the extent to which their mitotic chromatin association contributes to cell identity maintenance. Answers to these questions will offer insights into fundamental mechanisms of developmental regulation and may open up novel avenues to develop mitosis-based protocols for cellular reprograming as well as cancer therapy.

We have found that the sequence-specific transcription factor, RBPJ, is retained on the mitotic chromatin of mouse embryonal carcinoma F9 cells [8, 9], which have many characteristic features of embryonic stem cells (ES). Importantly, RBPJ is the major downstream effector of the Notch signaling pathway, which controls a wide variety of cell-fate choices at virtually all stages of development [10–15]. RBPJ plays an important role in ES cells by maintaining pluripotency and self-renewal [16]. Moreover, a large body of evidence has revealed that aberrant Notch activation plays a key role in carcinogenesis and tumor progression [17–30]. Consequently, the Notch singling pathway has emerged as a therapeutic target of great interest for cancer treatment [31–35]. Modulating the interaction between RBPJ and mitotic chromatin may provide a means to intervene in Notch-related cancers.

Here, we describe how to enrich mitotic cells (Subheading 3.1), determine mitotic index (Subheading 3.2), examine chromatin occupancy sites of RBPJ in mitotically enriched cells by chromatin immunoprecipitation followed by quantitative PCR (ChIP-qPCR) (Subheading 3.3). In addition, we will describe an approach to manipulate mitotic chromatin RBPJ occupancy by decreasing levels of its binding partner, histone deacetylase I (HDAC1), using short hairpin RNA (shRNA) (Subheading 3.4). Lastly, we will show how to measure the first-run of RNA synthesis upon mitotic exit to correlate changes in mitotic chromatin RBPJ binding with changes in nascent transcript production (Subheading 3.5) [36].

2 Materials

2.1 Mitotic Cell Enrichment and Mitotic Index Determination

1. Murine embryonal carcinoma F9 cells are maintained in Dulbecco's Modified Eagle Medium (DMEM) supplemented with 10% fetal bovine serum (FBS) and 1% glutamine. F9 cells divide every 8–12 h and therefore require subculture every 36–48 h at dilutions of 1:10–1:20.
2. 0.25% Gelatin in water.
3. 1 mg/ml Nocodazole in DMSO (1000×).
4. Phosphate buffered saline (PBS).

5. 0.25% Trypsin–EDTA.
6. 10% Formaldehyde (methanol free).
7. 1 M Glycine.
8. Wash Buffer I: 0.25% Triton X-100, 10 mM EDTA, 10 mM HEPES (pH 6.5).
9. Microscope slides and 20 × 20 mm cover glasses (#1.5).
10. Cytospin funnels and filter paper.
11. Humidified chamber.
12. Immunostaining Solution: 1× PBS, 0.5% Triton X-100, 10% goat serum.
13. Immunostaining Wash Buffer: 1× PBS, 0.5% Triton X-100.
14. Anti-phospho-histone H3 (Ser10) antibody (*see* **Note 1**).
15. Anti-Phospho-RNA pol II CTD (Ser2) antibody.
16. 1 mg/ml 2-(4-Amidinophenyl)-6-indolecarbamidine (DAPI) in water (1000×).
17. ProLong Gold antifade mountant.

2.2 Chromatin Immunoprecipitation

1. PBS.
2. Wash Buffer I: 0.25% Triton X-100, 10 mM EDTA, 10 mM HEPES (pH 6.5).
3. Wash Buffer II: 200 mM NaCl, 1 mM EDTA, 10 mM HEPES (pH 6.5).
4. ChIP Lysis Buffer: 0.1% deoxycholate, 1 mM EDTA, 50 mM HEPES (pH 7.5), 140 mM NaCl, 1% Triton X-100.
5. High Salt Wash Buffer: 0.1% deoxycholate, 1% Triton X-100, 1 mM EDTA, 50 mM HEPES (pH 7.5), 500 mM NaCl.
6. LiCl Wash Buffer: 250 mM LiCl, 1% deoxycholate, 1% Triton X-100, 1 mM EDTA, 10 mM Tris (pH 8.0).
7. TE/TX Wash Buffer: 10 mM Tris (pH 8.0), 1 mM EDTA, 0.1% Triton X-100.
8. Elution Buffer I: 1% SDS, 10 mM EDTA, 50 mM Tris (pH 8.0).
9. Elution Buffer II: 0.67% SDS, 1 mM EDTA, 10 mM Tris (pH 8.0).
10. DNA purification spin columns.
11. Anti-RBPJ antibody.
12. Protein A agarose beads (*see* **Note 2**).
13. Beads Blocking Solution: 1 × PBS, 5% BSA, 0.1% Tween 20.

2.3 shRNA-Mediated Protein Reduction

1. 293T cells are maintained in DMEM supplemented with 10% FBS and 1% glutamine.
2. PBS.

3. 0.25% Trypsin–EDTA.
4. 1 mg/ml Polyethylenimine (PEI) in water. Heat to 65 °C to dissolve completely. Aliquot and store at −20 °C (*see* **Note 3**).
5. 0.45 μm PVDF filters.
6. Bleach.
7. Third-generation lentiviral packaging system: pRSV-Rev, pMDLg/pRRE, and pCMV-VSV-g.
8. HDAC1 MISSION shRNA plasmid DNA: TRCN0000229438 (Sigma).
9. Control MISSION shRNA plasmid DNA: SHC002 (Sigma).

2.4 Measurement of Nascent Transcript Production During Mitotic Exit

1. 1 mg/ml Nocodazole in DMSO (1000×).
2. TRIzol Reagent, chloroform, isopropanol, ethanol.
3. Standard cDNA synthesis kit.

3 Methods

3.1 Cell Synchronization and Mitotic Cell Enrichment

F9 cells are maintained on gelatin-coated dishes. Upon completion of this protocol, ~3 × 10^5 mitotic F9 cells (more than 95% pure) are expected from a 100 mm dish.

1. For F9 cell maintenance, coat 100 mm dishes with gelatin by adding 5 ml of a 0.25% gelatin solution and allow to sit for at least 10 min at room temperature. Remove gelatin and rinse with PBS.
2. Have a dish of F9 cells that are ~90–95% confluent ready to use.
3. Aspirate media and rinse cells with PBS.
4. Add 2 ml of trypsin–EDTA and leave for 1 min at room temperature. Remove trypsin–EDTA and incubate cells for ~5 min at room temperature until all cells detach.
5. Add 10 ml DMEM and pipet cell suspension up and down ten times with the pipette tip touching the plate bottom.
6. Check with a microscope for complete cell dissociation. If cells are not completely dissociated from each other, continue to pipet up and down until they are. This step is important to remove cell clumps that easily detach from the dish and will result in asynchronous cell contamination of the mitotic cell preparation.
7. Combine 6 ml of fresh DMEM with 3 ml of the F9 cell suspension and add to a gelatin coated dish (**step 1**) which will result in a final cell density of ~1.5 × 10^4 cells/cm^2.

8. Twenty-four hours later, remove media and wash cells once with PBS.
9. Add 5 ml of complete media containing 1 μg/ml nocodazole and incubate for 4 h.
10. Mitotic cells are loosely attached to the plate relative to asynchronous cells, so they can be collected by gently dislodging with the 5 ml of medium. Set the speed of the pipettor to its lowest. Pipet the media very gently on top of the cells in a crisscross pattern. Rotate the dish and repeat two more times. It is important to note that nonmitotic F9 cells also adhere loosely, so pipetting needs to be very gentle and conducted a minimal number of times to achieve high mitotic cell enrichment.
11. Pellet mitotic cells by centrifugation at 150 × *g* for 3 min. Discard supernatant and resuspend the cell pellet in 8 ml of complete media containing 1 μg/ml nocodazole.
12. Add 800 μl of 10% formaldehyde to achieve a final concentration of 1%.
13. Place cells on a rocking platform at lowest speed and incubate for 10 min at room temperature.
14. Quench formaldehyde by adding 2.5 ml of 1 M glycine. Return to the rocking platform and incubate for another 10 min.
15. Centrifuge cells at 150 × *g* for 3 min. Remove supernatant.
16. Resuspend cells in Wash Buffer I to a final concentration of ~1 × 10^6 cells/ml.
17. Transfer 200 μl to a microcentrifuge tube, which will be used to determine mitotic index as described in Subheading 3.2. Store the remaining cells at 4 °C while determining mitotic index (*see* **Note 4**).

3.2 Cyto-centrifugation and Immunostaining to Determine Mitotic Index

1. Assemble a cytospin funnel by placing the disposable filter paper insert into the rectangular plastic base of the cytospin funnel.
2. Insert a labeled microscope slide into the holder, and place the assembled cytospin funnel into the cytospin.
3. Pipet 20 μl of Wash Buffer I into the smaller funnel.
4. To the larger funnel, pipet 50–100 μl of the cell suspension.
5. Centrifuge at 125 × *g* for 3 min using medium acceleration.
6. Place the microscope slide with cells facing up in a humidified chamber.
7. Pipet 20 μl of primary antibody, diluted in immunostaining solution, directly onto the area where the cells have attached. Carefully cover the cell area with a cover glass.

8. Close the chamber and incubate at room temperature for 1 h.
9. Wash slides three times with Immunostaining Wash Buffer, using either a slide staining chamber or a 50 ml tube, and remove excess solution by touching the edges of the slide to a paper towel.
10. Pipet 20 μl of a fluorochrome-labeled secondary antibody onto the area containing cells, diluted in immunostaining solution that contains DAPI (1 μg/ml), cover with a cover glass.
11. Close the chamber and incubate at room temperature for 1 h.
12. Wash slides three times in Immunostaining Wash Buffer and remove excess solution.
13. Apply 10 μl of Prolong-Gold antifade mountant, and cover with a cover glass. Avoid air bubbles by dropping the cover glass onto the slide from a 45° angle. Make sure that the mounting solution has solidified before imaging.
14. Image cells and calculate the mitotic cell index by dividing the number of mitotic cells to the total number of cells (DAPI stained) in a field of view (Fig. 1). Mitotic chromatin is stained by a phospho-histone H3 (Ser10) antibody but not stained by an anti-phospho-RNA pol II CTD (Ser2) antibody. Examine at least 5 independent fields per slide (>100 cells).

3.3 RBPJ Chromatin Immunoprecipitation

1. Pellet the stored mitotic cells (Subheading 3.1) by centrifugation at 150 × *g* for 3 min.
2. Discard supernatant and resuspend the cell pellet in 1 ml of Wash Buffer II.
3. Place cells on a rotating platform and rotate for 5 min.

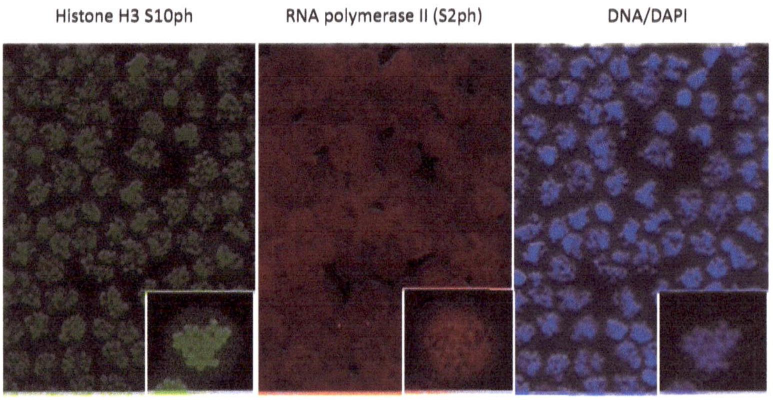

Fig. 1 Mitotic cell index determination. Cells were immunostained with antibodies against phospho-histone H3 (Ser10) (green) and phospho-RNA pol II CTD (Ser2) (red). DNA was counterstained with DAPI (blue). The mitotic cell population in this preparation was >99%

4. Pellet cells in a refrigerated microcentrifuge at 21,000 × *g* for 5 min.
5. Discard supernatant and resuspend cell pellet in 1 ml of ChIP Lysis Buffer.
6. Sonicate cells on ice using a 0.125″ diameter tapered probe (Branson) for 12 cycles with settings of 30 s on, 90 s off and amplitude of 40%.
7. Centrifuge sonicated lysate in a refrigerated microcentrifuge at 21,000 × *g* for 15 min.
8. Transfer solubilized chromatin to a new microcentrifuge tube.
9. Aliquot 400 μl of lysate into a microcentrifuge tube and add anti-RBPJ antibody.
10. Aliquot another 400 μl of lysate into a microcentrifuge tube, which will be used as beads-only control, which does not receive antibody.
11. Place tubes and remaining lysate (which will be used for input control) at 4 °C overnight.
12. Coat protein A agarose beads with BSA and DNA by combining 500 μl of net protein A agarose beads with 1 ml of Beads Blocking Solution and 150 μl of 10 mg/ml salmon sperm DNA (previously sonicated to a size range of 200 to 1000 bps). Incubate the mix on a rotator overnight at 4 °C.
13. Centrifuge lysate at 21,000 × *g* for 30 min at 4 °C to remove any precipitate that may have formed during the overnight incubation.
14. Transfer supernatant to a new tube and add 20 μl of blocked protein A beads (prepared in **step 12**) (*see* **Note 5**).
15. Rotate at 4 °C for 3 h.
16. Pellet beads by centrifugation at 750 × *g* for 30 s at room temperature.
17. Wash beads twice with 500 μl of ChIP Lysis Buffer (*see* **Note 6**).
18. Wash beads once with 500 μl of High-Salt Wash Buffer.
19. Wash beads once with 500 μl of LiCl Wash Buffer.
20. Wash beads three times with TE/TX.
21. Add 100 μl of Elution Buffer I to beads and incubate at 65 °C for 15 min.
22. Pellet beads and collect supernatant.
23. Add 150 μl of Elution Buffer II to the beads, centrifuge and combine with first elution.
24. Reverse-crosslink eluted chromatin by incubating at 65 °C for 16 h.

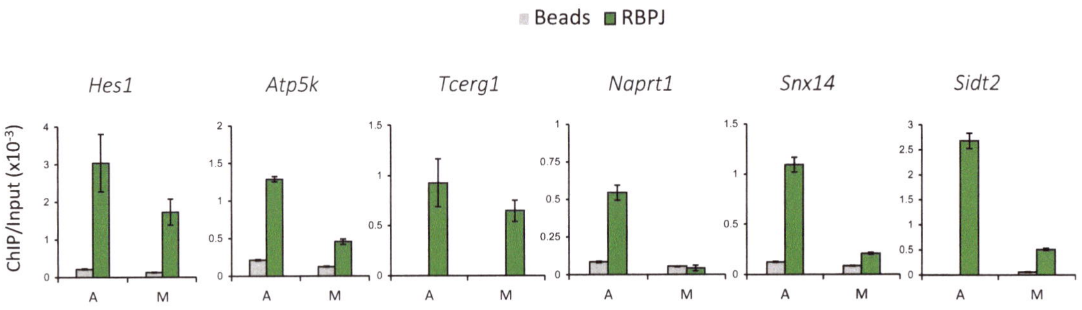

Fig. 2 RBPJ chromatin occupancy in asynchronous and mitotic F9 cells. RBPJ ChIP was carried out in asynchronous and mitotic-enriched F9 cells. Quantitative PCR was used to compare RBPJ chromatin occupancy between asynchronous (**a**) and mitotic (M) cells. Bars represent ChIP signals normalized to signals from input DNA

Table 1
Transfection mix for packaging lentivirus in 293T cells[a]

Plasmid DNA	Length	% DNA in the transfection mix	DNA mass
pRSV-Rev	4174 bp	16%	3.2 μg
pMDLg/pRRE	8895 bp	33%	6.6 μg
pCMV-VSV-g	6363 bp	24%	4.8 μg
pLKO.1 (either control or HDAC1-targeting shRNA)	7086 bp	27%	5.4 μg
Total	26,518 bp	100%	20 μg

[a] DNA amount required for transfecting one 100 mm plate of 293T cells

25. Purify reverse-crosslinked DNA with spin columns according to manufacturer's protocol.
26. The purified ChIPed DNA is ready for qPCR analysis (Fig. 2) (*see* **Note 7**).

3.4 shRNA-Mediated Knockdown of HDAC1

1. Prepare transfection mix: Add 1 ml of serum-free DMEM to each of two microcentrifuge tubes (one tube for control shRNA and a second for shRNA targeting HDAC1). Add equimolar ratios of plasmids according to the values provided in Table 1 (*see* **Note 8**).
2. Vortex for 10 s.
3. Add 47.5 μl of a 1 mg/ml PEI solution to each tube.
4. Immediately vortex for 10 s.
5. Incubate transfection mix for at least 10 min at room temperature.

6. Plate 293T cells to a confluency of ~80% while transfection mix is forming.
7. Add transfection mix to 293T cells.
8. The following day, remove media and add 6 ml of fresh DMEM and place cells back in the incubator (*see* **Note 9**).
9. On the same day **step 8** is performed, plate F9 cells to a confluency of ~10% (~5 × 10^3 cells/cm^2). One 100 mm dish of cells prepared in this manner will yield three 100 mm dishes of F9 cells with knocked-down HDAC1 upon completion of this protocol.
10. Twenty-four hours after **steps 8** and **9**, collect lentivirus-containing medium from transfected 293T cells, and filter through 0.45 μm PVDF filter unit (~6 ml) to remove contaminating 293T cells (*see* **Note 9**). Add 6 ml of fresh DMEM to the filtered medium.
11. Remove medium from the plate containing F9 cells (**step 11**) and add the ~12 ml of lentivirus-containing medium (*see* **Note 9**). Place cells in an incubator.
12. After 24 h, remove lentivirus-containing medium, add 10 ml of fresh DMEM, and place cells back in an incubator (*see* **Note 9**).
13. Twenty-four hours later, split infected F9 cells 1:3 and incubate for additional 24 h (72 h post infection) (*see* **Note 9**).
14. Use 1 plate to determine the level of knockdown by western blot analysis, comparing control shRNA expressing cells to HDAC1 shRNA expressing cells (Fig. 3a), (*see* **Note 10**). An example of the effect of HDAC1 knockdown on the association of RBPJ with mitotic chromatin is shown in Fig. 3.

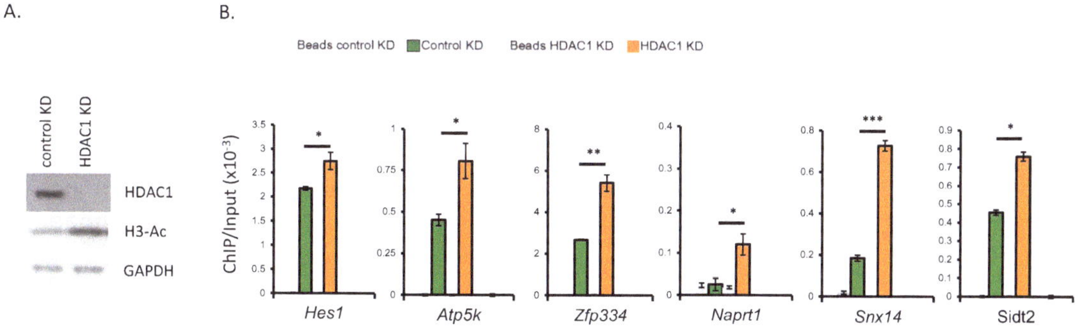

Fig. 3 HDAC1 negatively regulates RBPJ occupancy of mitotic chromatin in F9 cells. (**a**) Western blot revealing the extent of HDAC1 knockdown and its impact on histone H3 acetylation. (**b**) Comparison of RBPJ occupancy in mitotic-enriched F9 cells treated with HDAC1 or control shRNA

3.5 Measuring the First Run of RNA Synthesis After Release from Mitosis Arrest

1. Collect mitotic cells from the two remaining plates, using the method described in Subheading 3.1.
2. Resuspend cell pellet in 10 ml of DMEM without nocodazole and transfer to a 50 ml tube.
3. Immediately transfer 500 μl of the cell suspension to a microcentrifuge tube (time point 0) and place the 50 ml tube with the cap loose in an incubator.
4. Immediately pellet cells in the microcentrifuge tube with a high-speed pulse, remove medium and snap freeze pellet in liquid nitrogen (or a dry ice/ethanol bath).
5. Remove 500 μl of the cell suspension every 30 min over a course of 4 h (mix cell suspension before removing an aliquot). Prepare frozen cell pellets at each time point as described in **step 3**.
6. Continue with RNA purification using Trizol once all cell pellets have been collected.
7. Prepare cDNA using standard protocols. For each time point, prepare two reactions for analysis: one reaction for cDNA synthesis (RT) and one reaction to be incubated without reverse transcriptase, which controls for DNA-contamination (no RT).
8. Perform qPCR analysis of the cDNA to assess relative nascent transcript abundance using primers that span exon-intron boundaries. Signals derived from nascent transcripts are normalized to signals derived from beta-actin mRNA measured by using primers that lie within exons and span introns (Fig. 4). Store unused cDNA at −20 °C. During calculations, relative nascent transcript levels are calculated by subtracting the signal obtained in the no-RT reaction from the signal obtained in the RT reaction for each time point.

4 Notes

1. Anti-phospho-Ser/Thr-Pro MPM-2 antibody can also be used as a mitosis marker.
2. Agarose beads can be substituted with Protein A magnetic beads and used according to manufacturer's recommendations.
3. Commercially available transfection reagents can also be used, although these are more expensive.
4. Unless otherwise specified, proceed between each step of this protocol as soon as possible. Avoid storing cells for prolonged periods.

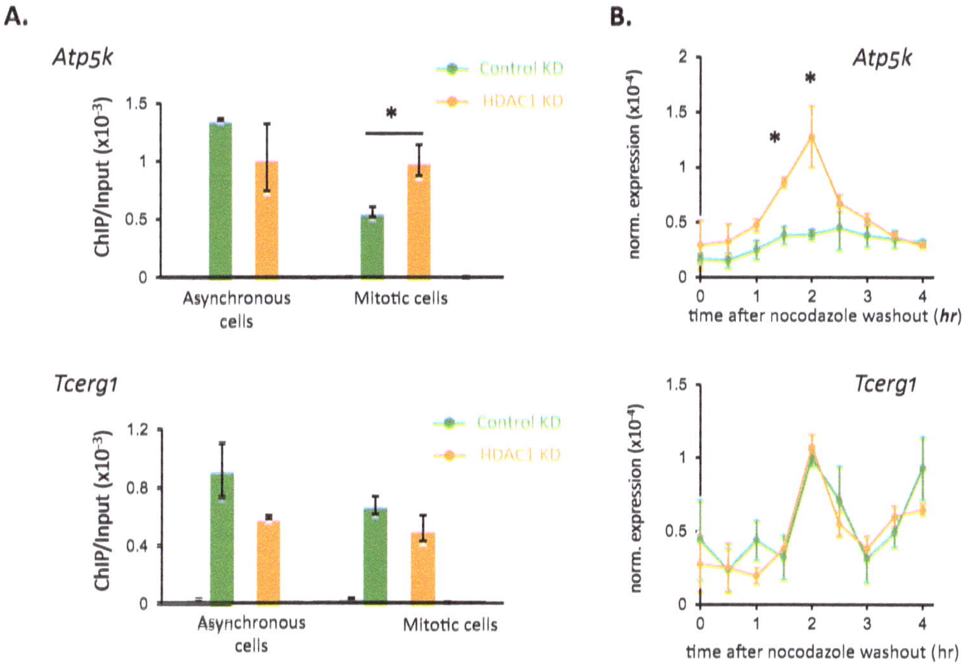

Fig. 4 Mitotic chromatin RBPJ occupancy enhances nascent transcript production upon mitotic exit. (**a**) Quantitative PCR analysis showing the impact of HDAC1 on RBPJ-chromatin association in asynchronous and mitotic cells at two loci. (**b**) Normalized nascent transcript levels at different times after release from mitotic arrest. Data are presented as means ± SEM. *: $p \leq 0.05$

5. If the immunoprecipitated chromatin will be used in next-generation sequencing, protein A agarose beads should not be coated with salmon sperm DNA.
6. Unless otherwise specified, keep all reagents and samples at 4 °C or on ice.
7. A single dish of asynchronous F9 cells can be collected similarly as a comparative control.
8. If using a lentiviral expression vector other than pLKO.1, calculate new ratios, substituting the size of your lentiviral construct for that of pLKO.1.
9. Immediately after use, decontaminate all pipettes, pipette tips, filter units, dishes and any other supplies that come in contact with lentivirus with 10% bleach before discarding.
10. Time of viral treatment is empirically determined by monitoring the KD efficiency, which may be different in other target cells.

Acknowledgments

This work was supported by American Heart Association 17GRNT33400020 to H.Y.F., and Cancer Center Support Grant P30CA118100. Support for microscopy imaging was provided by the University of New Mexico Cancer Center Fluorescence Microscopy Shared Resource, funded by NCI 2P30 CA118100 and NIGMS 5P50 GM085273.

References

1. Martinez-Balbas MA, Dey A, Rabindran SK, Ozato K, Wu C (1995) Displacement of sequence-specific transcription factors from mitotic chromatin. Cell 83(1):29–38. https://doi.org/10.1016/0092-8674(95)90231-7
2. Kouskouti A, Talianidis I (2005) Histone modifications defining active genes persist after transcriptional and mitotic inactivation. EMBO J 24(2):347–357. https://doi.org/10.1038/sj.emboj.7600516
3. Zaidi SK, Young DW, Montecino MA, Lian JB, van Wijnen AJ, Stein JL, Stein GS (2010) Mitotic bookmarking of genes: a novel dimension to epigenetic control. Nat Rev Genet 11(8):583–589. https://doi.org/10.1038/nrg2827
4. Kadauke S, Blobel GA (2013) Mitotic bookmarking by transcription factors. Epigenetics Chromatin 6(1):6. https://doi.org/10.1186/1756-8935-6-6
5. Michelotti EF, Sanford S, Levens D (1997) Marking of active genes on mitotic chromosomes. Nature 388(6645):895–899. https://doi.org/10.1038/42282
6. Caravaca JM, Donahue G, Becker JS, He X, Vinson C, Zaret KS (2013) Bookmarking by specific and nonspecific binding of FoxA1 pioneer factor to mitotic chromosomes. Genes Dev 27(3):251–260. https://doi.org/10.1101/gad.206458.112
7. Dey A, Nishiyama A, Karpova T, McNally J, Ozato K (2009) Brd4 marks select genes on mitotic chromatin and directs postmitotic transcription. Mol Biol Cell 20(23):4899–4909. https://doi.org/10.1091/mbc.E09-05-0380
8. Lake RJ, Tsai PF, Choi I, Won KJ, Fan HY (2014) RBPJ, the major transcriptional effector of Notch signaling, remains associated with chromatin throughout mitosis, suggesting a role in mitotic bookmarking. PLoS Genet 10(3):e1004204. https://doi.org/10.1371/journal.pgen.1004204
9. Fortini ME, Artavanis-Tsakonas S (1994) The suppressor of hairless protein participates in notch receptor signaling. Cell 79(2):273–282. https://doi.org/10.1016/0092-8674(94)90196-1
10. Artavanis-Tsakonas S, Rand MD, Lake RJ (1999) Notch signaling: cell fate control and signal integration in development. Science 284(5415):770–776. https://doi.org/10.1126/science.284.5415.770
11. Bray S, Bernard F (2010) Notch targets and their regulation. Curr Top Dev Biol 92:253–275. https://doi.org/10.1016/S0070-2153(10)92008-5
12. Das D, Lanner F, Main H, Andersson ER, Bergmann O, Sahlgren C, Heldring N, Hermanson O, Hansson EM, Lendahl U (2010) Notch induces cyclin-D1-dependent proliferation during a specific temporal window of neural differentiation in ES cells. Dev Biol 348(2):153–166. https://doi.org/10.1016/j.ydbio.2010.09.018
13. Kurpinski K, Lam H, Chu J, Wang A, Kim A, Tsay E, Agrawal S, Schaffer DV, Li S (2010) Transforming growth factor-beta and notch signaling mediate stem cell differentiation into smooth muscle cells. Stem Cells 28(4):734–742. https://doi.org/10.1002/stem.319
14. Lowell S, Benchoua A, Heavey B, Smith AG (2006) Notch promotes neural lineage entry by pluripotent embryonic stem cells. PLoS Biol 4(5):e121. https://doi.org/10.1371/journal.pbio.0040121
15. Jang J, Ku SY, Kim JE, Choi K, Kim YY, Kim HS, Oh SK, Lee EJ, Cho HJ, Song YH, Lee SH, Lee SH, Suh CS, Kim SH, Moon SY, Choi YM (2008) Notch inhibition promotes human embryonic stem cell-derived cardiac mesoderm differentiation. Stem Cells 26(11):2782–2790. https://doi.org/10.1634/stemcells.2007-1053
16. Mason MJ, Fan G, Plath K, Zhou Q, Horvath S (2009) Signed weighted gene co-expression network analysis of transcriptional regulation in murine embryonic stem cells. BMC Genomics

10:327. https://doi.org/10.1186/1471-2164-10-327
17. Bhagat TD, Zou Y, Huang S, Park J, Palmer MB, Hu C, Li W, Shenoy N, Giricz O, Choudhary G, Yu Y, Ko YA, Izquierdo MC, Park AS, Vallumsetla N, Laurence R, Lopez R, Suzuki M, Pullman J, Kaner J, Gartrell B, Hakimi AA, Greally JM, Patel B, Benhadji K, Pradhan K, Verma A, Susztak K (2017) Notch pathway is activated via genetic and epigenetic alterations and is a therapeutic target in clear cell renal cancer. J Biol Chem 292(3):837–846. https://doi.org/10.1074/jbc.M116.745208
18. Aoki S, Mizuma M, Takahashi Y, Haji Y, Okada R, Abe T, Karasawa H, Tamai K, Okada T, Morikawa T, Hayashi H, Nakagawa K, Motoi F, Naitoh T, Katayose Y, Unno M (2016) Aberrant activation of Notch signaling in extrahepatic cholangiocarcinoma: clinicopathological features and therapeutic potential for cancer stem cell-like properties. BMC Cancer 16(1):854. https://doi.org/10.1186/s12885-016-2919-4
19. Su Q, Xin L (2016) Notch signaling in prostate cancer: refining a therapeutic opportunity. Histol Histopathol 31(2):149–157. https://doi.org/10.14670/HH-11-685
20. Wu G, Wilson G, George J, Qiao L (2015) Modulation of Notch signaling as a therapeutic approach for liver cancer. Curr Gene Ther 15(2):171–181
21. Yuan X, Wu H, Han N, Xu H, Chu Q, Yu S, Chen Y, Wu K (2014) Notch signaling and EMT in non-small cell lung cancer: biological significance and therapeutic application. J Hematol Oncol 7:87. https://doi.org/10.1186/s13045-014-0087-z
22. Flores AN, McDermott N, Meunier A, Marignol L (2014) NUMB inhibition of NOTCH signalling as a therapeutic target in prostate cancer. Nat Rev Urol 11(9):499–507. https://doi.org/10.1038/nrurol.2014.195
23. Ou JP, Lin HY, Su KY, Yu SL, Tseng IH, Chen CJ, Hsu HC, Chan DC, Sophia Chen YL (2012) Potential therapeutic role of Z-Isochaihulactone in lung cancer through induction of apoptosis via notch signaling. Evid Based Complement Alternat Med 2012:809204. https://doi.org/10.1155/2012/809204
24. Han J, Hendzel MJ, Allalunis-Turner J (2011) Notch signaling as a therapeutic target for breast cancer treatment? Breast Cancer Res 13(3):210. https://doi.org/10.1186/bcr2875
25. Al-Hussaini H, Subramanyam D, Reedijk M, Sridhar SS (2011) Notch signaling pathway as a therapeutic target in breast cancer. Mol Cancer Ther 10(1):9–15. https://doi.org/10.1158/1535-7163.MCT-10-0677
26. Villaronga MA, Bevan CL, Belandia B (2008) Notch signaling: a potential therapeutic target in prostate cancer. Curr Cancer Drug Targets 8(7):566–580
27. Rizzo P, Miao H, D'Souza G, Osipo C, Song LL, Yun J, Zhao H, Mascarenhas J, Wyatt D, Antico G, Hao L, Yao K, Rajan P, Hicks C, Siziopikou K, Selvaggi S, Bashir A, Bhandari D, Marchese A, Lendahl U, Qin JZ, Tonetti DA, Albain K, Nickoloff BJ, Miele L (2008) Cross-talk between notch and the estrogen receptor in breast cancer suggests novel therapeutic approaches. Cancer Res 68(13):5226–5235. https://doi.org/10.1158/0008-5472.CAN-07-5744
28. Shi W, Harris AL (2006) Notch signaling in breast cancer and tumor angiogenesis: cross-talk and therapeutic potentials. J Mammary Gland Biol Neoplasia 11(1):41–52. https://doi.org/10.1007/s10911-006-9011-7
29. Jackstadt R, van Hooff SR, Leach JD, Cortes-Lavaud X, Lohuis JO, Ridgway RA, Wouters VM, Roper J, Kendall TJ, Roxburgh CS, Horgan PG, Nixon C, Nourse C, Gunzer M, Clark W, Hedley A, Yilmaz OH, Rashid M, Bailey P, Biankin AV, Campbell AD, Adams DJ, Barry ST, Steele CW, Medema JP, Sansom OJ (2019) Epithelial NOTCH signaling rewires the tumor microenvironment of colorectal cancer to drive poor-prognosis subtypes and metastasis. Cancer Cell 36(3):319–336.e317. https://doi.org/10.1016/j.ccell.2019.08.003
30. Stylianou S, Clarke RB, Brennan K (2006) Aberrant activation of notch signaling in human breast cancer. Cancer Res 66(3):1517–1525. https://doi.org/10.1158/0008-5472.CAN-05-3054
31. Xiao YF, Yong X, Tang B, Qin Y, Zhang JW, Zhang D, Xie R, Yang SM (2016) Notch and Wnt signaling pathway in cancer: crucial role and potential therapeutic targets (review). Int J Oncol 48(2):437–449. https://doi.org/10.3892/ijo.2015.3280
32. Yuan X, Wu H, Xu H, Xiong H, Chu Q, Yu S, Wu GS, Wu K (2015) Notch signaling: an emerging therapeutic target for cancer treatment. Cancer Lett 369(1):20–27. https://doi.org/10.1016/j.canlet.2015.07.048
33. Previs RA, Coleman RL, Harris AL, Sood AK (2015) Molecular pathways: translational and therapeutic implications of the Notch signaling pathway in cancer. Clin Cancer Res 21(5):955–961. https://doi.org/10.1158/1078-0432.CCR-14-0809

34. Shao H, Huang Q, Liu ZJ (2012) Targeting Notch signaling for cancer therapeutic intervention. Adv Pharmacol 65:191–234. https://doi.org/10.1016/B978-0-12-397927-8.00007-5
35. Miele L, Miao H, Nickoloff BJ (2006) NOTCH signaling as a novel cancer therapeutic target. Curr Cancer Drug Targets 6(4):313–323
36. Dreval K, Lake RJ, Fan HY (2019) HDAC1 negatively regulates selective mitotic chromatin binding of the Notch effector RBPJ in a KDM5A-dependent manner. Nucleic Acids Res 47(9):4521–4538. https://doi.org/10.1093/nar/gkz178

Chapter 10

Assessing the Roles of Potential Notch Signaling Components in Instructive and Permissive Pathways with Two *Drosophila* Pericardial Reporters

Manoj Panta, Andrew J. Kump, Kristopher R. Schwab, and Shaad M. Ahmad

Abstract

The highly conserved Notch signaling pathway brings about the transcriptional activation of target genes via either instructive or permissive mechanisms that depend on the identity of the specific target gene. As additional components of the Notch signaling pathway are identified, assessing whether each of these components are utilized exclusively by one of these mechanisms (and if so, which), or by both, becomes increasingly important. Using RNA interference-mediated knockdowns of the Notch component to be tested, reporters for two Notch-activated pericardial genes in *Drosophila melanogaster*, immunohistochemistry, and fluorescence microscopy, we describe a method to determine the type of signaling mechanism—instructive, permissive, or both—to which a particular Notch pathway component contributes.

Key words Notch signaling pathway, Notch permissive signaling mechanism, Notch instructive signaling mechanism, Gene regulation, Transcriptional regulation, Cardiac cell subtype-specific gene expression, *Drosophila* embryonic heart, Heart development, Enhancer–reporter constructs

1 Introduction

Cell–cell signaling via the Notch receptor is a highly conserved signaling pathway that regulates numerous metazoan developmental processes such as cell proliferation, differentiation, and apoptosis [1–6]. Notch signaling, initiated by ligand binding to Notch receptors, results in a proteolytic cleavage that releases the Notch intracellular domain (NICD) from the cell membrane. NICD then enters the nucleus and binds with the transcription factor CSL to activate the transcription of relevant target genes [2–6]. In the absence of Notch signaling, CSL associates with corepressors to form a repressor complex that binds dynamically and rapidly on and off the enhancers of target genes to prevent their transcription [4, 6–12]. During Notch signaling, the interaction of NICD with

Dongyu Jia (ed.), *Notch Signaling Research: Methods and Protocols*,
Methods in Molecular Biology, vol. 2472, https://doi.org/10.1007/978-1-0716-2201-8_10,

CSL forms an activator complex that competes dynamically with the CSL repressor complex for CSL binding motifs on the enhancers. NICD enhances both the recruitment of the NICD-CSL activator complex and its dwell time at the enhancer, thereby favoring its presence over the CSL repressor complex [12]. This allows for the transcription of target genes to occur by one of two mechanisms: (1) a Notch-permissive process in which alleviating repression through the competitive displacement of the CSL repressor complex is sufficient to initiate transcription due to the presence of other local activators that are then free to exert their positive effects, or (2) a Notch-instructive process, which necessitates the activating effect of the NICD-CSL activator complex that allows transcription initiation to occur [4, 6, 10–12]. Thus, as additional components of the Notch signaling pathway are identified, the assessment of whether these components are utilized exclusively by the Notch-permissive or the Notch-instructive mechanism, or by both, becomes important in describing Notch activity.

In the fruit fly *Drosophila melanogaster*, Notch signaling brings about cell type–specific gene expression in different cardiac cell types [13–20]. The *Drosophila* heart consists of two groups of cells: two paired inner rows of cardial cells (CCs) surrounded by pericardial cells (PCs) (Fig. 1). The CC nuclei express *Myocyte enhancer factor 2* (*Mef2*) while the PC nuclei express both *Zn finger homeodomain 1* (*zfh1*) and *Holes in muscle* (*Him*) [13, 20–24]. A subset of PCs, the Eve-PCs, also express *even-skipped* (*eve*) in addition to *zfh1* [25, 26]. The Delta ligand expressed by the CCs binds to the Notch receptor in the adjacent PCs (i.e., all PCs other than the Eve-PCs, henceforth referred to as core-PCs) to activate transcription of *Him* and *zfh1* in the latter, *Him* in a permissive manner by displacing the CSL repressor complex with the NICD-CSL complex on the *Him* enhancer (Fig. 2a), and *zfh1* in an instructive manner that requires the inductive activating effect of the NICD-CSL activator complex on the *zfh1* enhancer (Fig. 2d) [13, 14].

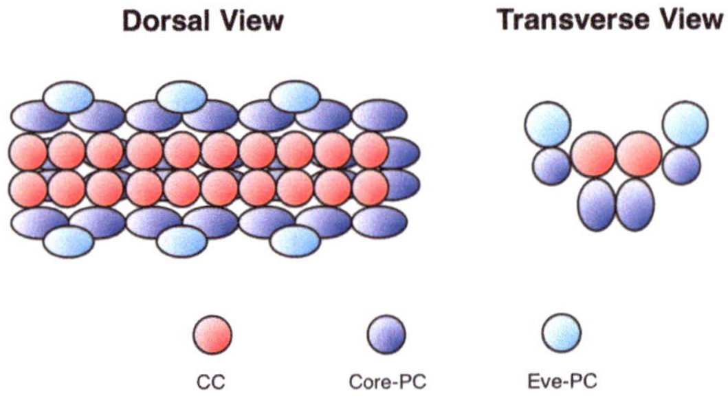

Fig. 1 Relative positions of the distinct cardiac cell nuclei in the *Drosophila* heart. The cardial cells (CCs) express *Mef2*, the core pericardial cells (Core-PCs) express both *zfh1* and *Him*, and the Eve pericardial cells (Eve-PCs) express *zfh1*, *Him*, and *eve*

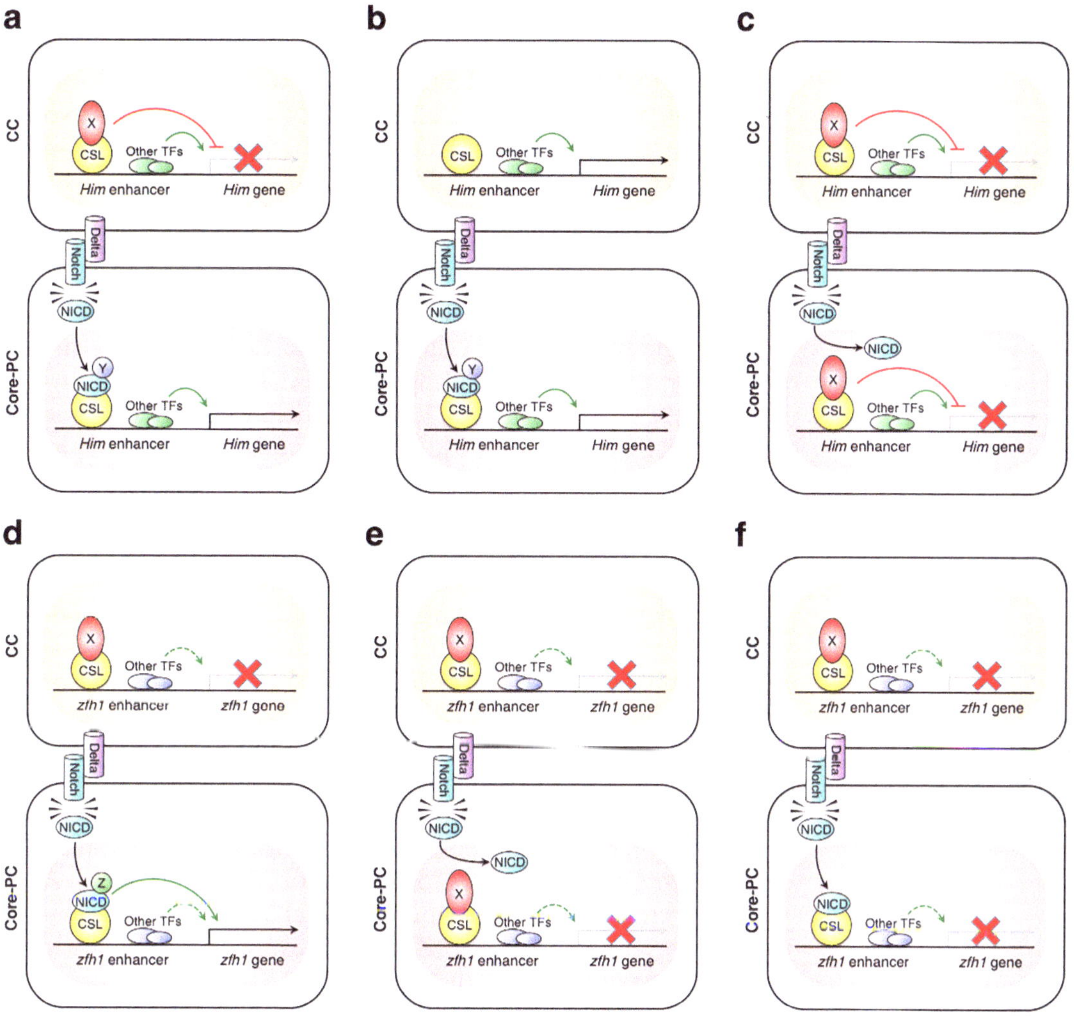

Fig. 2 Schematic of the Notch signaling pathway driving *Him* and *zfh1* expression in the *Drosophila* heart and the effect of knocking down components of the pathway. (**a**) *Him* is repressed in CCs by binding of the CSL repressor complex to its enhancer. In PCs, the competitive displacement of the CSL repressor complex by the NICD-CSL complex alleviates this repression and allows *Him* to be transcribed. (**b**) If the pathway component (X in the figure) being assayed is required for repression by the CSL repressor complex, for example, a corepressor, then its knockdown will result in *Him* being also ectopically expressed in CCs. (**c**) If, instead, the pathway component (Y in the figure) is required for the NICD-CSL complex to competitively displace the CSL repressor complex, for example, by stabilizing the binding of NICD to CSL, then its knockdown will result in *Him* expression being eliminated in core-PCs. (**d**) *zfh1* expression in core-PCs is achieved only through the instructive effect of the NICD-CSL activator complex binding to the *zfh1* enhancer. (**e**) Thus, knockdown of a component (Z in the figure) required to bring about this transcription by stabilizing the NICD-CSL activator complex would result in transcription being abrogated in core-PCs. (**f**) Alternatively, Z could be a component recruited or used by the NICD-CSL activator complex, such as a coactivator or member of a histone acetyltransferase complex, to promote transcription. In this case too, knockdown of Z would result in the elimination of *zfh1* expression in core-PCs

We have taken advantage of the distinct Notch signaling mechanisms for *Him* and *zfh1* to devise a simple RNA interference (RNAi) knockdown-based assay to examine whether newly discovered Notch components are utilized exclusively by the Notch-permissive or the Notch-instructive mechanism, or by both. Briefly, we have created transgenic fly lines with enhancer–reporter constructs for *Him* (*Him*^{WT}*-lacZ*) and *zfh1* (*zfh1*^{WT}*-lacZ*) that express the β-galactosidase reporter exclusively in PCs in embryos, that is, in the same expression pattern as the endogenous genes [13, 14]. The Notch signaling pathway component to be tested will be knocked down specifically in the cardiac mesoderm (the precursor of the heart) using the GAL4-UAS system by the *TinD-GAL4* driver which we have also included in these previously described fly lines [27, 28]. If the component functions in the permissive Notch signaling pathway solely by repressing the transcription of target genes in the absence of NICD, then its knockdown will result in the *Him*^{WT}*-lacZ* reporter being ectopically expressed in the CCs (Fig. 2b). If, instead, the component acts in the permissive Notch signaling pathway in concert with the NICD-CSL complex to bring about transcription of the target genes solely by displacing the CSL repressor complex, then its knockdown will result in the *Him*^{WT}*-lacZ* reporter expression disappearing in the core-PCs (Fig. 2c). Finally, if the component is required to activate transcription of target genes in the instructive Notch signaling pathway through the inductive effect of the NICD-CSL activator complex, then its knockdown will lead to the disappearance of *zfh1*^{WT}*-lacZ* reporter expression in the core-PCs (Fig. 2e, f). Additionally, knockdown of components that result in both *Him*^{WT}*-lacZ* reporter and *zfh1*^{WT}*-lacZ* reporter expression disappearing in core-PCs would indicate that those components are used in both the Notch permissive and Notch instructive pathways. Thus, examining the effect of the knockdown of individual Notch signaling pathway components on the cardiac expression of these reporters can serve to identify the specific category of Notch signaling mechanisms that they are involved in (summarized in Table 1).

The protocol for this assay involves acquiring the relevant fly lines (transgenic lines with the enhancer–reporter constructs and the *TinD-GAL4* driver from us as well as controls and transgenic lines with UAS-RNAi constructs for the pathway components to be tested), crossing these lines to obtain embryos of the desired genotype, dechorionating, fixing, and immunostaining these embryos with relevant primary and secondary antibodies, and then examining the embryos using fluorescent microscopy.

Table 1
Identification of the specific Notch signaling mechanism, permissive or instructive, a Notch signaling pathway component is involved in, based on the effect of its knockdown on the expression of *Him* and *zfh1* pericardial reporters

Effect of component knockdown	Interpretation
Ectopic expression of β-galactosidase driven by the *Him* enhancer in CCs	Component functions in the Notch-permissive signaling pathway to repress transcription of target genes in the absence of NICD
Elimination of *Him* enhancer-driven β-galactosidase expression in the core-PCs	Component functions in the Notch-permissive signaling pathway to enable the NICD-CSL complex to competitively displace the CSL repressor complex at the target gene enhancer
Elimination of *zfh1* enhancer-driven β-galactosidase expression in the core-PCs	Component functions in the Notch-instructive signaling pathway to activate transcription of target genes through the inductive effect of the NICD-CSL activator complex

2 Materials

Use distilled and deionized water.

2.1 Drosophila *Lines*

1. *Him*^{WT}*-lacZ; TinD-GAL4 UAS-Dcr-2*: This fly line can be obtained from the authors (*see* **Note 1**). We also intend to submit this line to the Bloomington *Drosophila* Stock Center (https://bdsc.indiana.edu) for ease of access and dissemination.
2. *zfh1*^{WT}*-lacZ; TinD-GAL4 UAS-Dcr-2*: This fly line can also be obtained from us, and we intend to submit this line too to the Bloomington *Drosophila* Stock Center (*see* **Note 1**).
3. Canton-S or Oregon-R flies: These wild-type flies can be obtained from any *Drosophila* stock center or almost any research laboratory using *Drosophila*.
4. Transgenic fly lines containing UAS-RNAi constructs for Notch signaling pathway components to be tested. Appropriate RNAi lines can be obtained from the Bloomington *Drosophila* Stock Center, the Vienna *Drosophila* Resource Center (https://stockcenter.vdrc.at), the National Institute of Genetics—Japan (https://www.nig.ac.jp/nig), and individual investigators (*see* **Note 2**).

2.2 Fly Crosses, Egg Laying, and Embryo Collection

1. Basic apparatus for rearing, anesthetizing, and sexing flies: Plastic or polypropylene bottles with fly food for culturing *Drosophila*, an incubator capable of holding these bottles and adjustable to temperatures between 25 °C and 29 °C, apparatus for anesthetizing flies, and dissecting microscopes for sexing flies.

2. Small embryo collection cages: These can be purchased from commercial vendors or made in the laboratory (*see* **Note 3**).
3. Embryo collection plates: Heat 1 L of water until lukewarm in a pan or beaker on a stirring hotplate and gradually add in and stir 32 g of bacteriological agar such that it does not form clumps. Next, add in 189 g of molasses and heat the stirring mixture until it comes to a boil. Turn off the heat once it is boiling. Once the stirring mixture cools down to 60 °C, add 15 mL of tegosept solution (18 g of methyl 4-hydroxybenzoate in 750 mL of reagent alcohol), stir, and pour into 60 mm × 15 mm petri dishes until the bottom of the dishes are completely covered. Let the agar cool down until solidified, cover the petri dishes, and store at 4 °C.
4. Yeast paste: Mix dry yeast to a small quantity of water (approximately 1 g of yeast to 1.3 mL of water) until it has the consistency of cream cheese. Store at 4 °C.
5. Egg baskets: 100 μm cell strainers designed to fit 50 mL conical tubes.
6. Small fine-haired paint brushes.
7. Squirt bottle containing water.

2.3 Dechorionation

1. Petri dishes (60 mm × 15 mm).
2. 50% Bleach (Clorox) in water.
3. Squirt bottle containing water.
4. Squirt bottle containing 0.1% Triton X-100 in water.

2.4 Embryo Fixation

1. Scintillation vials.
2. Pasteur pipettes.
3. Small fine-haired paint brushes.
4. 5× PEM solution: 0.5 M PIPES, 10 mM $MgSO_4$, 5 mM EGTA. Adjust pH to 6.9 with KOH.
5. 16% EM grade formaldehyde.
6. Heptane.
7. Methanol.
8. Platform shaker.
9. 1.5 mL Microcentrifuge tubes.

2.5 Antibody Staining

1. 10× phosphate-buffered saline (PBS) stock solution: 1.37 M NaCl, 27 mM KCl, 100 mM Na_2HPO_4, 18 mM KH_2PO_4, pH 7.4 (*see* **Note 4**).
2. PBST solution: 1× PBS, 0.2% Tween 20. Make by diluting appropriate volumes of 10× PBS stock solution and 10% Tween 20 in water.

3. Normal goat serum (NGS).
4. Rabbit anti-Mef2 antibody (*see* **Note 5**).
5. Guinea pig anti-Zfh1 antibody (*see* **Note 6**).
6. Mouse monoclonal anti-β-galactosidase antibody (*see* **Note 7**).
7. Alexa Fluor 568 goat anti-rabbit IgG (H + L) cross-adsorbed secondary antibody (*see* **Note 8**).
8. Alexa Fluor 647 goat anti-guinea pig IgG (H + L) highly cross-adsorbed secondary antibody (*see* **Note 8**).
9. Alexa Fluor 488 goat anti-mouse IgG (H + L) cross-adsorbed secondary antibody (*see* **Note 8**).
10. Vectashield antifade mounting medium: available from Vector Laboratories.

2.6 Microscopy

1. Fluorescence microscope with filter sets capable of imaging Alexa Fluor 488, 568, and 647 antibodies: confocal microscopes or microscopes with apotomes capable of taking Z-stacks are preferred.
2. Secure Seal Imaging Spacers (0.12 mm depth, 20 mm diameter).
3. Slides and coverslips.

3 Methods

Carry out all procedures at room temperature unless otherwise specified.

3.1 Fly Crosses and Egg Laying

1. Before setting up an embryo collection cage, put a small dollop of yeast paste on the molasses-agar surface at the center of each embryo collection plate (Fig. 3a).

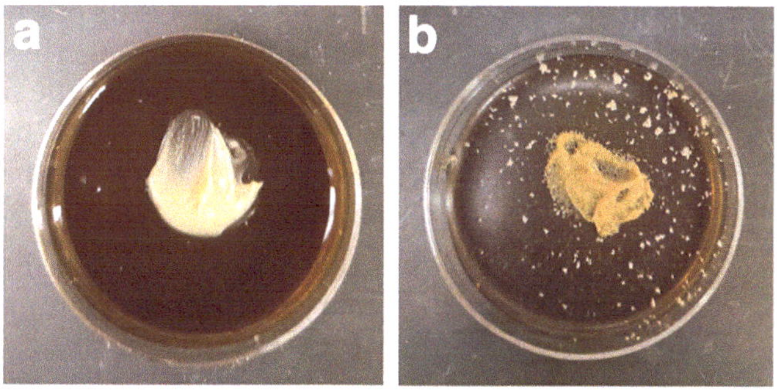

Fig. 3 Embryo collection plates before and after egg laying. (**a**) Plate immediately prior to being used in a cage. Note the small dollop of yeast paste at the center of the plate. (**b**) Plate after a 3 h egg laying period in the cage

2. An individual cage is set up by placing anesthetized flies of the appropriate genotypes (about 200 virgin females and 100 recently eclosed males) in the cage, and sealing the cage opening with the yeast paste containing embryo collection plate. Invert the cage such that the embryo collection plate is on the bottom only after all the anesthetized flies have regained consciousness and have begun moving. For each fly line containing a UAS-RNAi construct for a candidate Notch signaling pathway component to be assessed, set up the four following simultaneous fly crosses in separate embryo cages in a 29 °C incubator (*see* **Notes 9–11**):
   (a) *Him*^{WT}*-lacZ; TinD-GAL4 UAS-Dcr-2* virgin females × wild-type males. This cross will produce embryos that allow the normal *Him* enhancer-driven transcription to be assessed as a baseline/control for the assay.
   (b) *Him*^{WT}*-lacZ; TinD-GAL4 UAS-Dcr-2* virgin females × males bearing the UAS-RNAi construct for the Notch pathway component to be tested. This cross will produce embryos that allow the effect of the RNAi-mediated knockdown on Notch-permissive *Him* enhancer-driven transcription to be assessed.
   (c) *zfh1*^{WT}*-lacZ; TinD-GAL4 UAS-Dcr-2* virgin females × wild-type males. This cross will produce embryos that allow the normal *zfh1* enhancer-driven transcription to be assessed as a baseline/control for the assay.
   (d) *zfh1*^{WT}*-lacZ; TinD-GAL4 UAS-Dcr-2* virgin females × males bearing the UAS-RNAi construct for the Notch pathway component to be tested. This cross will produce embryos that allow the effect of the RNAi-mediated knockdown on Notch-instructive *zfh1* enhancer-driven transcription to be assessed.
3. Change the embryo collection plate on each cage with a new plate (that has a fresh dab of yeast paste) every 8 h on the first day (Day 1). With a little practice, this is achieved by inverting the cage, tapping it on the bench to make the flies fall to the bottom, quickly replacing the old embryo collection plate with a new one before any flies can escape, and inverting it again such that the plate is on the bottom. Discard the old embryo collection plates.
4. From Day 2 onward, use the following procedure: first change the embryo collection plate, and discard the previous plate. After 3 h, change the plate again, but this time cover the removed plate and keep it at 29 °C in the incubator. This plate should now have many visible *Drosophila* eggs (white specks) laid on it during that 3 h period (Fig. 3b). After 12.5 h at 29 °C (which allows all the embryos to age to stage 16, a point at which the effect of the Notch pathway

component on the reporter expression in the heart can be conveniently assessed), begin the dechorionation and fixing process described in the following sections (*see* **Note 12**).

3.2 Fix Solution

1. While the embryos are maturing during the 12.5 h period in the incubator, prepare the fix solution by adding the following in the order they are listed to each scintillation vial.
   - (a) 1.375 mL water.
   - (b) 0.5 mL 5× PEM.
   - (c) 0.625 mL 16% EM grade formaldehyde.
   - (d) 2.5 mL heptane.

   Four (4) scintillation vials with fix solution are required for each fly line containing a UAS-RNAi construct for the candidate Notch signaling pathway component being assessed.
2. After adding the heptane, close the lids tightly and shake the vials vigorously by hand to saturate the heptane with formaldehyde.

3.3 Embryo Collection and Dechorionation

1. After 12.5 hours in the incubator, recover the embryo collection plates. The small white specks scattered on the surface of the plate are embryo-containing eggs. Add enough water to a plate to cover the eggs, and then use a paint brush to dissolve the remaining yeast paste (if any) and gently detach the eggs from the molasses-agar surface.
2. Once the yeast paste is dissolved, and most of the eggs have been detached from the molasses-agar surface, pour the mix of water, dissolved yeast paste, and eggs into a labeled egg basket. Repeat for the remaining genotypes (each in its separate embryo collection plate) into separate labeled egg baskets. The egg baskets serve as filtering devices to collect the embryo-containing eggs and remove dissolved yeast and water. Wash the eggs in the baskets with squirts of water until the water runs clear.
3. The following bleach treatment with vigorous washing removes the outer chorion layer of the embryo. Place the egg baskets in a petri dish partially filled with 50% bleach for 5 min such that the level of the bleach solution is just below the rim of the baskets. During these 5 min, use a pasteur pipette to repeatedly suck up bleach solution from the petri dish and expel it inside the baskets to continually rinse the embryos.
4. Immediately wash the embryos in the egg baskets, first by squirting with 0.1% Triton X-100, and then extensively with water to remove any traces of bleach or detergent.

3.4 Embryo Fixation

1. Blot the exterior of the egg baskets with a disposable laboratory tissue or a paper towel to remove excess moisture. Then use fine paint brushes to transfer the embryos from the egg baskets to the fix solutions in separate labeled scintillation vials. The embryos will easily come off the paint brushes and settle at the interface between the two phases. A single layer of embryos is desired at this interface. Perform this step quickly to prevent desiccation of the dechorionated embryos.
2. Close the lids tightly, secure the vials on their side on the platform shaker, and run the shaker at 125 rpm for 20 min.
3. After 20 min, gently release and remove the vials from the shaker. Moving the vials too vigorously at this point increases the risk of embryos getting stuck on the lids or the sides of the vials.
4. Draw off embryos from the interface using a pasteur pipette. Discard the lower aqueous phase in the pasteur pipette, and transfer the embryos to a labeled 1.5 mL microcentrifuge tube. Repeat for each genotype (each in its separate scintillation vial).
5. Wash embryos three times with heptane. Each wash is achieved by adding 1 mL of heptane gently to the embryos in the tube, then using a pasteur pipette to remove as much of the heptane as possible without drawing out the embryos.
6. Add 0.6 mL of heptane followed by 0.6 mL of methanol to the embryos in each tube. Hold tubes in a fist and shake vigorously for 60 s to facilitate devitellinization (removal of the opaque vitelline membrane) of the embryos.
7. Let the tubes stand vertically in a rack for 30 s. Devitellinized embryos will sink to the bottom of the tube.
8. Use a pasteur pipette to remove the top layer (heptane) and the interface (which will contain empty vitelline membranes and embryos which were not devitellinized).
9. Wash embryos five times with methanol. Each wash is achieved by adding 1 mL of methanol gently to the embryos in the tube, then using a pasteur pipette to remove as much of the methanol as possible without drawing out the embryos.
10. Add 1 mL of methanol to the embryos and store at −20 °C for antibody staining (*see* **Note 13**).

3.5 Antibody Staining

1. Take out the microcentrifuge tubes of embryos that were fixed and stored at −20 °C. Place the tubes vertically in a rack such that the embryos settle to the bottom.
2. For each tube of fixed embryos, prepare 1 mL of 50% methanol in PBST.

3. Carefully aspirate the supernatant (methanol) from the tubes without disturbing or removing the fixed embryos, add 1 mL of 50% methanol in PBST to each tube, and rock the tubes at room temperature for 5 min to begin the process of rehydrating the embryos (*see* **Note 14**).
4. Place the microcentrifuge tubes vertically in a rack. After the embryos have settled to the bottom of the tube, aspirate the supernatant (50% methanol) without disturbing the embryos.
5. Add 1 mL of PBST to the microcentrifuge tubes and rock them at room temperature for 5 min. Then place them vertically in a rack to allow the embryos to settle to the bottom, and remove the supernatant (PBST) without disturbing the embryos.
6. Repeat **step 5** three (3) more times. These washes should remove all traces of methanol from the embryos.
7. While repeating **step 5**, prepare 2% NGS in PBST.
8. After the final aspiration of PBST from the tubes in **step 6**, add 1 mL of 2% NGS in PBST to each tube of embryos, and rock at room temperature for 1 h. The NGS in this and subsequent steps acts as a blocking agent to prevent nonspecific binding of antibodies to the embryo.
9. During **step 8**, prepare the primary antibody cocktail. Add all three primary antibodies (rabbit anti-Mef2 antibody, guinea pig anti-Zfh1 antibody, and mouse monoclonal anti-β-galactosidase antibody) to their prescribed dilutions in 2% NGS in PBST, and rock for at least 10 min to create the primary antibody cocktail (*see* **Note 15**).
10. After 1 h of incubation in the blocking solution (**step 8**), place the microcentrifuge tubes vertically in a rack and allow embryos to settle to the bottom of the tube. Aspirate the supernatant (2% NGS in PBST) without disturbing the embryos, add 0.5 mL of the primary antibody cocktail to each tube of embryos, and then rock the tubes at 4 °C overnight.
11. Prepare fresh 2% NGS in PBST the next day.
12. Place the tubes of embryos that had been rocking overnight at 4 °C in the primary antibody cocktail vertically in a rack such that the embryos settle to the bottom. Remove the supernatant (the primary antibody cocktail) (*see* **Note 16**).
13. Add 1 mL of 2% NGS in PBST to the microcentrifuge tubes and rock them at room temperature for 5 min. Then place the tubes vertically in a rack to allow the embryos to settle to the bottom, and aspirate the supernatant (2% NGS in PBST) without disturbing the embryos.

14. Repeat **step 13** seven (7) more times. These washes will remove all traces of the primary antibody cocktail from the tubes while continuing to prevent any nonspecific binding of the antibodies to the embryo.
15. During the wash process, that is, while repeating **step 13**, prepare the secondary antibody cocktail. Add all three secondary antibodies (Alexa Fluor 568 goat anti-rabbit IgG (H + L) cross-adsorbed secondary antibody, Alexa Fluor 647 goat anti-guinea pig IgG (H + L) highly cross-adsorbed secondary antibody, and Alexa Fluor 488 goat anti-mouse IgG (H + L) cross-adsorbed secondary antibody) each to a 1:300 dilution in 2% NGS in PBST in a tube. Cover the tube completely with aluminum foil to prevent photobleaching of the antibodies and rock for at least 10 min.
16. After the final aspiration of 2% NGS in PBST from the tubes in **step 14**, add 0.5 mL of the secondary antibody cocktail to each tube of embryos. Immediately cover each tube completely with aluminum foil to prevent photobleaching.
17. Rock these covered tubes containing the embryos in the secondary antibody cocktail either at room temperature for 1 h or at 4 °C overnight.
18. Place these covered tubes that had been rocking in the secondary antibody cocktail vertically in a rack such that the embryos settle to the bottom. Aspirate the supernatant (the secondary antibody cocktail) without disturbing the embryos.
19. Add 1 mL of PBST to the microcentrifuge tubes and rock them at room temperature for 5 min. Then place them vertically in a rack to allow the embryos to settle to the bottom, and remove the supernatant (PBST).
20. Repeat **step 19** seven (7) more times. These washes will remove all traces of the secondary antibody cocktail from the tubes. Ensure that the tubes remain covered with aluminum foil at all times to prevent photobleaching.
21. After the final removal of PBST from the tubes in **step 20**, add 2–3 drops of Vectashield antifade mounting medium to each tube of embryos. Ensure that the tubes are still completely covered with aluminum foil and store them at 4 °C until microscopy.

3.6 Mounting and Microscopy

1. Peel and attach a Secure Seal Imaging Spacer on a clean microscope slide (Fig. 4a) (*see* **Note 17**).
2. Use a micropipette to transfer 10–15 μL of Vectashield with embryo suspension from the relevant microcentrifuge tube to the slide surface at the center of the hole in the spacer (Fig. 4b, c).

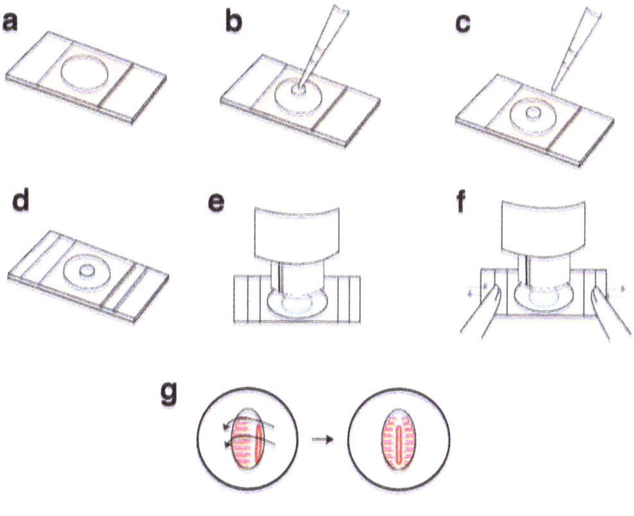

Fig. 4 Sequential steps in mounting and microscopy. (**a**) Attach a Secure Seal Imaging Spacer to the slide. (**b, c**) Transfer 10–15 μL of Vectashield with embryos suspended in it to the slide surface at the center of the hole in the spacer. (**d**) Gently place a coverslip on the spacer such that it also touches the drop of Vectashield. (**e**) Examine the embryos under low magnification. (**f**) Rotate embryos to bring them to the correct orientation by gently sliding the coverslip. (**g**) Schematic showing the rotation of embryos to the correct orientation by gentle sliding of the coverslip

3. Gently place a coverslip on the spacer such that it also touches the drop of Vectashield containing the embryos (Fig. 4d). Try not to let air bubbles form in the Vectashield during this process.
4. Examine the fluorescent antibody-labeled embryos on the slide under low magnification (e.g., with a 10× objective) in a fluorescence microscope with filters set to detect Alexa Fluor 568 staining (Fig. 4e). The anti-Mef2 antibody-labeled nuclei of the somatic muscles and/or the CCs of the tubular heart should be clearly visible (Fig. 5). *Drosophila* Stage 16 embryos have a convex ventral surface, and a relatively flatter dorsal surface where the heart is located. Embryos may thus lie rotated along their long axis in a manner that makes viewing and imaging of the heart difficult (Fig. 5a). Gently slide the coverslip the tiniest degree along the spacer surface (Fig. 4f) while viewing an embryo through the eyepiece to rotate it such that the dorsal surface and the characteristic tubular heart now directly faces the objective (Figs. 4g and 5b, c).
5. Switch to higher magnification (20× or 40× objectives) and examine/take images of the heart with filters set to detect first Alexa Fluor 568 staining (for anti-Mef2-labeled CCs), then Alexa Fluor 647 staining (for anti-Zfh1-labeled PCs), and finally Alexa Fluor 488 (for β-galactosidase reporter

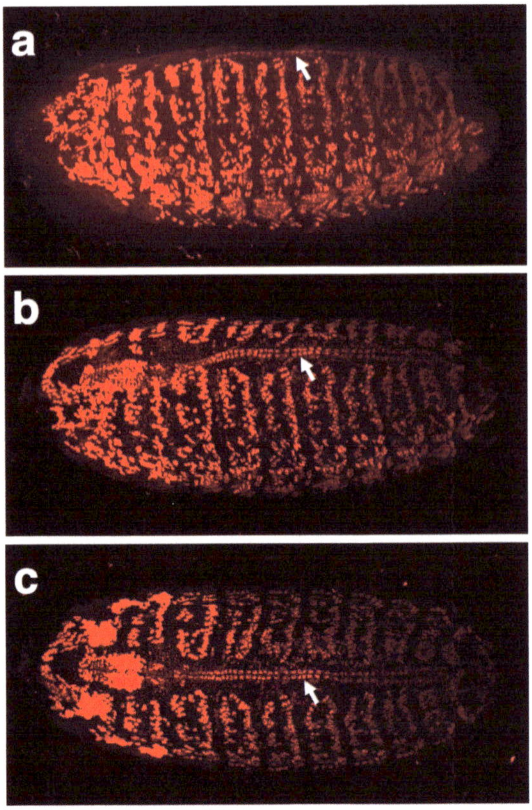

Fig. 5 Rotation of embryos to the correct orientation for imaging. Mef2-expressing nuclei (red) of the somatic muscles and heart in Stage 16 embryos observed under low magnification using a fluorescent microscope with filters set to detect Alexa Fluor 568 staining. (**a**) Embryo in an orientation that precludes clear imaging of the tubular heart (arrow). (**b**) Gentle sliding of the coverslip showing the gradual rotation of the same embryo. (**c**) The same embryo rotated to the proper orientation for detailed analysis of the heart at higher magnification and with all three fluorophores

expression). If the fluorescence microscope is a confocal microscope or equipped with an apotome, capture Z-stacks of sufficient depth that accommodate all the Mef2-labeled CCs and Zfh1-labeled PCs of the heart. Otherwise, take images along multiple different focal planes in all three filter set channels to ensure that every CC and PC is captured in focus. Examine 20–30 embryonic hearts for each genotype, paying particular attention to whether β-galactosidase reporter expression driven by the *Him* enhancer is ectopically present in CCs or missing in core-PCs in Notch pathway component knockdown embryos compared to controls. Similarly, note whether β-galactosidase reporter expression driven by the *zfh1* enhancer is missing in core-PCs in Notch pathway component knockdown embryos compared to controls.

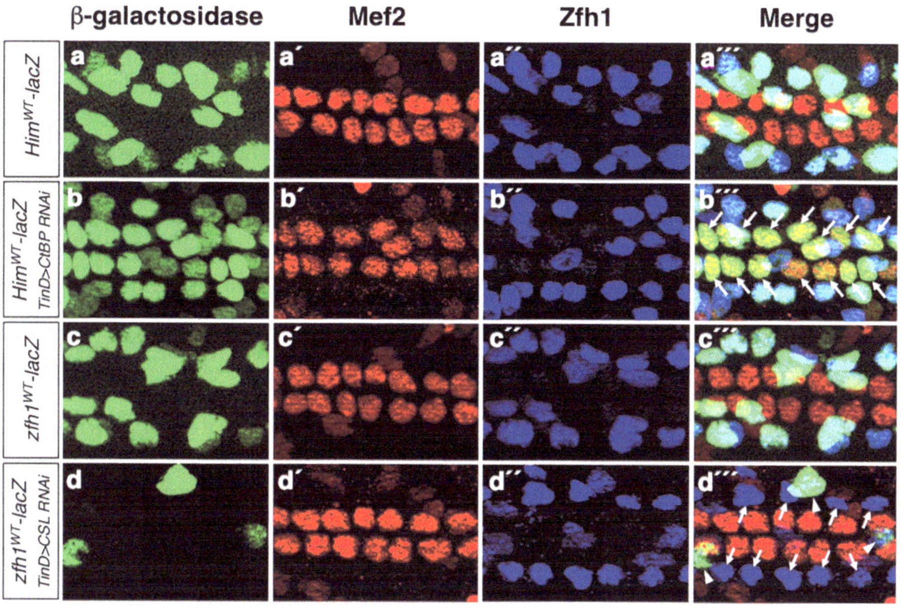

Fig. 6 Examples of RNAi-mediated knockdown of Notch signaling pathway components on reporter expression driven by the Notch-permissive *Him* and the Notch-instructive *zfh1* enhancers. (**a–d**‴) *lacZ* reporter activity (β-galactosidase, green) driven by wild-type *Him* or *zfh1* enhancers in Stage 16 embryos. All CCs express Mef2 (red) while PCs are marked by Zfh1 (blue). (**a-a**‴) In an otherwise wild-type embryo, β-galactosidase, driven by the *Him* enhancer is present only in PCs. (**b-b**‴) Knockdown of a component required for repressing the transcription of target genes in the absence of NICD, in this case CtBP, a corepressor that forms the repressor complex with CSL, results in derepression in the CCs (arrows) of β-galactosidase driven by the *Him* enhancer. (**c-c**‴) β-galactosidase, driven by the *zfh1* enhancer is also present only in PCs in otherwise wild-type embryos. (**d-d**‴) Knockdown of a component required for instructive transcriptional activation by the NICD-CSL activator complex, in this case CSL itself, eliminates *zfh1* enhancer-driven β-galactosidase expression in the core-PCs (arrows). β-galactosidase expression will still be detected in the Eve-PCs (arrowheads). Note that a very similar elimination of *Him* enhancer-driven β-galactosidase expression in the core-PCs would also be seen if the knocked down component functioned in the Notch-permissive signaling pathway to enable the NICD-CSL complex to competitively displace the CSL repressor complex at the target gene enhancer

3.7 Analysis and Interpretation

1. Examine the hearts of embryos obtained from the cross: *Him*^{WT}*-lacZ; TinD-GAL4 UAS-Dcr-2* virgin females × wild-type males. These are the controls for the *Him* enhancer–reporter construct. Ensure that β-galactosidase is expressed in each and every Zfh1-labeled PC but not in a single Mef2-labeled CC as in Fig. 6a-a‴.
2. Examine the hearts of embryos obtained from the cross: *Him*^{WT}*-lacZ; TinD-GAL4 UAS-Dcr-2* virgin females × males bearing the UAS-RNAi construct for the Notch pathway component to be tested. First assess whether β-galactosidase is expressed in one or more Mef2-labeled CCs in addition to all Zfh1-labeled PCs. If ectopic expression of β-galactosidase is seen in some CCs in addition to all PCs as in Fig. 6b-b‴, then

the component knocked down by RNAi functions in the Notch-permissive signaling pathway to repress transcription of target genes in the absence of NICD (Table 1) (*see* **Note 18**). If no ectopic expression of β-galactosidase is detected in CCs, assess whether β-galactosidase is expressed in each and every Zfh1-labeled PC, as in the control in **step 1**. If β-galactosidase reporter expression is missing in some or all PCs, similar to what is seen in Fig. 6d-d‴, then the component knocked down by RNAi functions in the Notch-permissive signaling pathway to enable the NICD-CSL complex to competitively displace the CSL repressor complex at the target gene enhancer (Table 1).

3. Examine the hearts of embryos obtained from the cross: *zfh1*^{WT}*-lacZ; TinD-GAL4 UAS-Dcr-2* virgin females × wild-type males. These are the controls for the *zfh1* enhancer–reporter construct. Ensure that β-galactosidase is expressed in each and every Zfh1-labeled PC but not in a single Mef2-labeled CC as in Fig. 6c-c‴.
4. Examine the hearts of embryos obtained from the cross: *zfh1*^{WT}*-lacZ; TinD-GAL4 UAS-Dcr-2* virgin females × males bearing the UAS-RNAi construct for the Notch pathway component to be tested. Assess whether β-galactosidase is expressed in each and every Zfh1-labeled PC, as in the control in **step 3**, or whether β-galactosidase expression is missing in some PCs. If β-galactosidase reporter expression is missing in some or all PCs, as in Fig. 6d-d‴, then the component knocked down by RNAi functions in the Notch-instructive signaling pathway to activate transcription of target genes through the inductive effect of the NICD-CSL activator complex (Table 1) (*see* **Note 19**).
5. Based on the expectations described above and in Table 1 and the effect of the Notch pathway component knockdown on reporter expression driven by the *Him* and *zfh1* enhancers, determine whether the component is utilized by the Notch-permissive and/or the Notch-instructive mechanism.

4 Notes

1. In addition to the enhancer reporter constructs (*Him*^{WT}*-lacZ* or *zfh1*^{WT}*-lacZ*) and the cardiac mesoderm-specific driver (*TinD-GAL4*), these lines also contain the transgenic construct *UAS-Dcr-2*. Inclusion of this *UAS-Dcr-2* construct ensures that the long dsRNAs used in the earlier generation of UAS-RNAi constructs also provide efficient knockdowns. The cardiac mesoderm-specific driver, *TinD-GAL4*, ensures that

the Notch signaling pathway component being assessed is not knocked down elsewhere in the embryo where it may be essential.

2. Use the RSVP tool (https://fgr.hms.harvard.edu/rsvp) tool to identify and obtain existing UAS-RNAi stocks for the Notch signaling pathway component to be knocked down from the Bloomington Drosophila Stock Center, the Vienna Drosophila Resource Center, or the National Institute of Genetics—Japan [29]. We recommend acquiring all available UAS-RNAi stocks for a particular component and performing separate series of crosses with them, since they will differ in knockdown efficiency. We also recommend performing and comparing reverse transcription quantitative real-time PCR (RT-qPCR) on total RNA collected from Stage 13–16 control and knockdown embryos (9–15.5 h after egg laying) from these series of crosses to determine RNAi-mediated knockdown efficiency. If the Notch signaling pathway component was identified in a system other than flies, the DIOPT tool (https://www.flyrnai.org/diopt) can be used to identify the *Drosophila* ortholog of the component in order to secure the relevant UAS-RNAi stocks [30]. Finally, FlyBase (http://flybase.org) can be used to identify UAS-RNAi fly lines constructed by and available from individual investigators that may not be present in the previously mentioned stock centers.
3. Small embryo collection cages can also be made from 100 mL polypropylene beakers whose open ends fit snugly over 60 mm × 15 mm petri dishes. Heat the tip of a syringe needle in the flame of a Bunsen burner and use it to make a number of air holes in the side of the beaker. Make certain that the holes are too small to prevent the flies from escaping. Prepare cages by dropping anesthetized flies into such a beaker, closing the opening with an embryo collection plate, using rubber bands or adhesive tape to keep the embryo collection plate in position, and inverting the cage such that the plate is at the bottom when the flies begin to stir (Fig. 7). We find making cages in this manner considerably more economical than purchasing commercial cages.
4. While 10× PBS pH 7.4 is relatively easy to prepare in the laboratory, it is also available from many vendors.
5. We use and recommend the rabbit anti-Mef2 antibodies that we requested and obtained as gifts from either Dr. Roger Jacobs in the Department of Biology at McMaster University or Dr. Bruce Paterson at the National Cancer Institute, NIH [31, 32]. Use at a 1:1000 dilution.
6. We use and recommend the guinea pig anti-Zfh1 antibody that we requested and obtained as a gift from Dr. James Skeath in

Fig. 7 Small embryo collection cage prepared as described in **Note 3**

the Department of Genetics at Washington University School of Medicine in St. Louis [33]. Use at a 1:1000 dilution.

7. We use and recommend the mouse monoclonal anti-β-galactosidase antibody available from Promega (catalog no. Z3783). Use at a 1:500 dilution.
8. The Alexa Fluor conjugated secondary antibodies from ThermoFisher Scientific that we suggest are effective at distinguishing Mef2, Zfh1, and β-galactosidase staining with the filter sets we use. Investigators are welcome to use different sets of fluorescent secondary antibodies that work best with their specific microscopy setups. We do recommend using a fluorescent secondary antibody for the anti-Mef2 primary antibody that is easily visible through the eyepiece (i.e., not Alexa Fluor 647 or any fluorophore in the "far-red" range) since we use the Mef2 staining for proper rotation and orientation of the embryos.
9. We recommend performing the crosses at 29 °C since the GAL4-UAS system is most efficient at that temperature. However, for certain genotypes, this temperature can lead to very few or no embryos surviving to Stage 16. In such cases, lower the temperature to a level that allows sufficient embryos to survive.
10. While the reciprocal crosses can be performed, we have empirically found that we obtain more efficient knockdowns by crossing males with the UAS-RNAi construct to virgin females bearing the *TinD-GAL4* driver, the *UAS-Dcr-2* construct, and the enhancer–reporter construct.
11. The crosses described in this method assume that the male flies bearing the UAS-RNAi constructs are homozygous viable. However, in some instances, the insertion of the UAS-RNAi

construct into the chromosome may be lethal when homozygous, resulting in the construct being carried over a balancer. In such cases, first put the UAS-RNAi construct over an embryonic *lacZ*-expressing balancer such as *CyO, ftz-lacZ* or *TM3, ftz-lacZ* and then perform the crosses with males bearing the construct over these marked balancers. Identify the knockdown embryos as those lacking the marked balancer since they fail to express β-galactosidase in the same pattern as the balancer.

12. Note that multiple embryo plate collections can be obtained from the same cage in a day. After the first 3 h collection, simply perform another 3 h collection on the second plate, allow it to age for 12.5 h before dechorionation and fixation, and then repeat again if necessary. Also note that the 12.5 h of aging is specific for embryos being raised at 29 °C. If, due to reduced embryo survival, the crosses and aging of the embryos are being performed at 25 °C, age the embryos for 13 h to reach Stage 16.
13. While the embryos can be used immediately for antibody staining, we have found that storing them overnight at −20 °C improves immunostaining.
14. For this and many of the subsequent wash steps, we find it easiest to simply aspirate the supernatant with a micropipette tip at the end of a pasteur pipette attached to a vacuum line (with appropriate safety trap containers and filters). However, a micropipette or even a pasteur pipette with a rubber bulb can also be used to remove the supernatant.
15. The primary antibody cocktail consists of 1:1000 dilutions of both the rabbit anti-Mef2 and guinea pig anti-Zfh1, and 1:500 dilutions of the Promega mouse anti-β-galactosidase we recommended. Please determine the appropriate dilutions empirically if the primary antibodies are obtained from other sources.
16. Since we have limited amounts of some of the primary antibodies, we remove the primary antibody cocktail with a micropipette, store it at 4 °C, and reuse it for subsequent immunostainings.
17. The Secure Seal Imaging Spacer both prevents the coverslip from flattening the embryos and allows adjustment of the orientation of the embryos for imaging by slight sliding of the coverslip on the spacer. Temporary spacers can also be prepared in the laboratory by placing three strips of Scotch Magic Tape on top of one another on a glass sheet. Cut off small pieces of this triple layered tape strip with a razor blade and place on both sides of the Vectashield drop on the slide. Use these tape pieces as bridge supports on which to place the coverslip.

18. Ectopic β-galactosidase need not be present in each and every CC of the heart since the *TinD-GAL4* driver does not function with identical efficiency in each and every heart cell. Reproducible β-galactosidase expression that is detected in even some CCs of the embryos with candidate component knockdowns but never in the CCs of the controls is sufficient to conclude that the knocked down component functions in the Notch-permissive signaling pathway to repress transcription of target genes in the absence of NICD.
19. Since disruption of the Notch-instructive signaling pathway does not affect *zfh1* enhancer-driven expression in the Eve-PCs, absence of β-galactosidase expression in the core-PCs is sufficient to conclude that the knocked down component functions in the Notch-instructive signaling pathway to activate transcription of target genes through the inductive effect of the NICD-CSL activator complex, even if β-galactosidase continues to be expressed in the Eve-PCs. This result is generally detected as the disappearance of β-galactosidase expression in the majority of the Zfh1-labeled PCs. However, if more specificity is required in the analysis, the Eve-PCs can be distinguished from the core-PCs as being the most dorsally positioned cells among the Zfh1-labeled PCs. Alternatively, staining with antibodies against the Eve protein can also serve to distinguish the Eve-PCs which express Eve from the core-PCs which do not.

Acknowledgments

This work was supported by the American Heart Association grant 16SDG31390005 to S.M.A. and a Rich and Robin Porter Cancer Research Center fellowship to A.J.K. We thank Sarah Hall for assistance in illustrating our figures.

References

1. Borggrefe T, Oswald F (2009) The Notch signaling pathway: transcriptional regulation at Notch target genes. Cell Mol Life Sci 66(10): 1631–1646. https://doi.org/10.1007/s00018-009-8668-7
2. Bray SJ (2006) Notch signalling: a simple pathway becomes complex. Nat Rev Mol Cell Biol 7(9):678–689. https://doi.org/10.1038/nrm2009
3. Bray SJ (2016) Notch signalling in context. Nat Rev Mol Cell Biol 17(11):722–735. https://doi.org/10.1038/nrm.2016.94
4. Falo-Sanjuan J, Bray SJ (2020) Decoding the notch signal. Develop Growth Differ 62(1): 4–14. https://doi.org/10.1111/dgd.12644
5. Kopan R, Ilagan MX (2009) The canonical Notch signaling pathway: unfolding the activation mechanism. Cell 137(2):216–233. https://doi.org/10.1016/j.cell.2009.03.045
6. Bray S, Bernard F (2010) Notch targets and their regulation. Curr Top Dev Biol 92: 253–275. https://doi.org/10.1016/S0070-2153(10)92008-5

7. Barolo S, Stone T, Bang AG, Posakony JW (2002) Default repression and Notch signaling: hairless acts as an adaptor to recruit the corepressors Groucho and dCtBP to suppressor of hairless. Genes Dev 16(15):1964–1976. https://doi.org/10.1101/gad.987402
8. Morel V, Lecourtois M, Massiani O, Maier D, Preiss A, Schweisguth F (2001) Transcriptional repression by suppressor of hairless involves the binding of a hairless-dCtBP complex in *Drosophila*. Curr Biol 11(10):789–792. https://doi.org/10.1016/s0960-9822(01)00224-x
9. Nagel AC, Krejci A, Tenin G, Bravo-Patino A, Bray S, Maier D, Preiss A (2005) Hairless-mediated repression of notch target genes requires the combined activity of Groucho and CtBP corepressors. Mol Cell Biol 25(23): 10433–10441. https://doi.org/10.1128/MCB.25.23.10433-10441.2005
10. Bray S, Furriols M (2001) Notch pathway: making sense of suppressor of hairless. Curr Biol 11(6):R217–R221
11. Barolo S, Posakony JW (2002) Three habits of highly effective signaling pathways: principles of transcriptional control by developmental cell signaling. Genes Dev 16(10):1167–1181. https://doi.org/10.1101/gad.976502
12. Gomez-Lamarca MJ, Falo-Sanjuan J, Stojnic R, Abdul Rehman S, Muresan L, Jones ML, Pillidge Z, Cerda-Moya G, Yuan Z, Baloul S, Valenti P, Bystricky K, Payre F, O'Holleran K, Kovall R, Bray SJ (2018) Activation of the Notch signaling pathway in vivo elicits changes in CSL nuclear dynamics. Dev Cell 44(5):611–623 e617. https://doi.org/10.1016/j.devcel.2018.01.020
13. Ahmad SM, Busser BW, Huang D, Cozart EJ, Michaud S, Zhu X, Jeffries N, Aboukhalil A, Bulyk ML, Ovcharenko I, Michelson AM (2014) Machine learning classification of cell-specific cardiac enhancers uncovers developmental subnetworks regulating progenitor cell division and cell fate specification. Development 141(4):878–888. https://doi.org/10.1242/dev.101709
14. Panta M, Kump AJ, Dalloul JM, Schwab KR, Ahmad SM (2020) Three distinct mechanisms, Notch instructive, permissive, and independent, regulate the expression of two different pericardial genes to specify cardiac cell subtypes. PLoS One 15(10):e0241191. https://doi.org/10.1371/journal.pone.0241191
15. Grigorian M, Mandal L, Hakimi M, Ortiz I, Hartenstein V (2011) The convergence of Notch and MAPK signaling specifies the blood progenitor fate in the *Drosophila* mesoderm. Dev Biol 353:105–118. https://doi.org/10.1016/j.ydbio.2011.02.024
16. Mandal L, Banerjee U, Hartenstein V (2004) Evidence for a fruit fly hemangioblast and similarities between lymph-gland hematopoiesis in fruit fly and mammal aorta-gonadal-mesonephros mesoderm. Nat Genet 36(15286786):1019–1023
17. Hartenstein AY, Rugendorff A, Tepass U, Hartenstein V (1992) The function of the neurogenic genes during epithelial development in the *Drosophila* embryo. Development 116(1295737):1203–1220
18. Han Z, Olson EN (2005) Hand is a direct target of tinman and GATA factors during *Drosophila* cardiogenesis and hematopoiesis. Development 132(15):3525–3536. https://doi.org/10.1242/dev.01899
19. Gajewski K, Choi CY, Kim Y, Schulz RA (2000) Genetically distinct cardial cells within the *Drosophila* heart. Genesis 28(1):36–43. https://doi.org/10.1002/1526-968X(200009)28:1<36::AID-GENE50>3.0.CO;2-4
20. Ward EJ, Skeath JB (2000) Characterization of a novel subset of cardiac cells and their progenitors in the *Drosophila* embryo. Development 127(11044409):4959–4969
21. Lai ZC, Fortini ME, Rubin GM (1991) The embryonic expression patterns of *zfh-1* and *zfh-2*, two *Drosophila* genes encoding novel zinc-finger homeodomain proteins. Mech Dev 34(2–3):123–134
22. Su MT, Fujioka M, Goto T, Bodmer R (1999) The *Drosophila* homeobox genes *zfh-1* and *even-skipped* are required for cardiac-specific differentiation of a *numb*-dependent lineage decision. Development 126(14):3241–3251
23. Elwell JA, Lovato TL, Adams MM, Baca EM, Lee T, Cripps RM (2015) The myogenic repressor gene *holes in muscles* is a direct transcriptional target of twist and tinman in the *Drosophila* embryonic mesoderm. Dev Biol 400(2):266–276. https://doi.org/10.1016/j.ydbio.2015.02.005
24. Bodmer R, Frasch M (2010) Development and aging of the *Drosophila* heart. In: Rosenthal N, Harvey RP (eds) Heart development and regeneration. Academic Press, London, UK, pp 47–86
25. Park M, Yaich LE, Bodmer R (1998) Mesodermal cell fate decisions in *Drosophila* are under the control of the lineage genes *numb, Notch,* and *sanpodo*. Mech Dev 75(1–2):117–126
26. Carmena A, Gisselbrecht S, Harrison J, Jimenez F, Michelson AM (1998)

Combinatorial signaling codes for the progressive determination of cell fates in the *Drosophila* embryonic mesoderm. Genes Dev 12(24): 3910–3922

27. Brand AH, Perrimon N (1993) Targeted gene expression as a means of altering cell fates and generating dominant phenotypes. Development 118(2):401–415
28. Yin Z, Xu XL, Frasch M (1997) Regulation of the twist target gene *tinman* by modular *cis*-regulatory elements during early mesoderm development. Development 124(24): 4971–4982
29. Perkins LA, Holderbaum L, Tao R, Hu Y, Sopko R, McCall K, Yang-Zhou D, Flockhart I, Binari R, Shim HS, Miller A, Housden A, Foos M, Randkelv S, Kelley C, Namgyal P, Villalta C, Liu LP, Jiang X, Huan-Huan Q, Wang X, Fujiyama A, Toyoda A, Ayers K, Blum A, Czech B, Neumuller R, Yan D, Cavallaro A, Hibbard K, Hall D, Cooley L, Hannon GJ, Lehmann R, Parks A, Mohr SE, Ueda R, Kondo S, Ni JQ, Perrimon N (2015) The transgenic RNAi project at Harvard Medical School: resources and validation. Genetics 201(3):843–852. https://doi.org/10.1534/genetics.115.180208
30. Hu Y, Flockhart I, Vinayagam A, Bergwitz C, Berger B, Perrimon N, Mohr SE (2011) An integrative approach to ortholog prediction for disease-focused and other functional studies. BMC Bioinformatics 12:357. https://doi.org/10.1186/1471-2105-12-357
31. Lilly B, Zhao B, Ranganayakulu G, Paterson BM, Schulz RA, Olson EN (1995) Requirement of MADS domain transcription factor D-MEF2 for muscle formation in *Drosophila*. Science 267(5198):688–693
32. Vanderploeg J, Vazquez Paz LL, MacMullin A, Jacobs JR (2012) Integrins are required for cardioblast polarisation in *Drosophila*. BMC Dev Biol 12:8. https://doi.org/10.1186/1471-213X-12-8
33. Tian X, Hansen D, Schedl T, Skeath JB (2004) Epsin potentiates *Notch* pathway activity in *Drosophila* and *C. elegans*. Development 131(23):5807–5815. https://doi.org/10.1242/dev.01459

Chapter 11

Image-Based Single-Molecule Analysis of Notch-Dependent Transcription in Its Natural Context

ChangHwan Lee, Tina Lynch, Sarah L. Crittenden, and Judith Kimble

Abstract

Notch signaling is crucial to animal development and homeostasis. Notch triggers the transcription of its target genes, which produce diverse outcomes depending on context. The high resolution and spatially precise assessment of Notch-dependent transcription is essential for understanding how Notch operates normally in its native context in vivo and how Notch defects lead to pathogenesis. Here we present biological and computational methods to assess Notch-dependent transcriptional activation in stem cells within their niche, focusing on germline stem cells in the nematode *Caenorhabditis elegans*. Specifically, we describe visualization of single RNAs in fixed gonads using single-molecule RNA fluorescence in situ hybridization (smFISH), live imaging of transcriptional bursting in the intact organism using the MS2 system, and custom-made MATLAB codes, implementing new image processing algorithms to capture the spatiotemporal patterns of Notch-dependent transcriptional activation. These methods allow a powerful analysis of in vivo transcriptional activation and its dynamics in a whole tissue. Our methods can be adapted to essentially any tissue or cell type for any transcript.

Key words smFISH, MS2, Confocal, Wide-field microscopy, Germline stem cells (GSC) single-molecule visualization, Transcriptional dynamics, Transcriptional bursting, RNA live imaging, Whole worm live imaging

1 Introduction

Notch signaling is a highly conserved cell–cell signaling pathway that is crucial to metazoan development and homeostasis [1–3]. Dysfunction or misregulation of Notch can lead to a variety of human pathologies, including cancer, congenital heart defects, and neurodegenerative diseases [4–10]. Notch signaling works by activating transcription of its target genes, whose products regulate key biological processes such as stem cell maintenance and tissue repair [11–14]. Physical interaction between a Notch ligand on a signaling cell and a Notch receptor on a receiving cell triggers cleavage of the receptor, allowing the Notch intracellular domain (NICD) to enter the nucleus and form a ternary complex with the

Dongyu Jia (ed.), *Notch Signaling Research: Methods and Protocols*,
Methods in Molecular Biology, vol. 2472, https://doi.org/10.1007/978-1-0716-2201-8_11,

DNA-binding protein, CSL/RBPJκ and the coactivator, MAML [2]. That complex then activates transcription of its target genes [15]. Therefore, Notch target transcripts are a hallmark of Notch activation. The transcriptional response to Notch signaling has been analyzed in diverse organisms, including *Drosophila* and zebrafish [16–20], and our recent work in *C. elegans* germ cells highlights the power of single-molecule analyses done at the resolution of nascent transcripts at individual Notch target loci in their native context [21–24]. We used single-molecule Fluorescence In Situ Hybridization (smFISH) to visualize both nascent transcripts in the nucleus and mature mRNAs in the cytoplasm at the single-RNA molecule resolution in an intact tissue [25–28]. Importantly, this method detects transcripts from the endogenous wild-type gene rather than relying on an artificial or indirect reporter, an approach that monitors the Notch transcriptional response in its authentic in vivo setting. In addition, we developed a computational method with custom-made image processing algorithms to automate the detection of RNA molecules in the whole tissue with high precision for their multidimensional spatial pattern analysis [21]. We developed these biological and computational methods to detect and analyze the Notch transcriptional response in the *C. elegans* germline, where Notch signaling is required to maintain germline stem cells (GSCs) [11, 29]. Recently, we successfully adapted our approach to other tissues in *C. elegans* (e.g., intestine and seam cells, unpublished) and to human and mouse cells in culture (e.g., induced pluripotent stem cells and cardiomyocytes [30]) with a few modifications and optimization steps in both biological and computational methods.

Our previous work with smFISH for the Notch target *sygl-1* revealed the probabilistic and stochastic nature of Notch-dependent transcription in GSCs [21–24]. The Notch-dependent transcriptional response is graded within the GSC pool, with its probability higher near the stem cell niche. However, not all GSCs, even GSCs adjacent to the niche, are transcriptionally active. Furthermore, the transcriptional activity (number of *sygl-1* nascent transcripts generated) at each locus is highly variable. There was no clear pattern or correlation with other critical factors such as cell cycle, cell size, or location to explain the variability in *sygl-1* transcription [21].

These results led us to the idea that the Notch-dependent transcription may occur in bursts. To test this idea, we developed a whole animal live imaging system in *C. elegans* using the MS2 system coupled to a worm immobilization method [24]. The MS2 system takes advantage of the strong and specific binding of the MS2 Coat Protein (MCP) to an array of tandem repeats of MS2 loops that are inserted into the target locus [31, 32]. We introduced 24 MS2 loops into a Notch target gene, *sygl-1,* as a transgene to visualize its transcripts in the nucleus. Combining the MS2 system, intact worm live imaging, and our computational method, we could trace *sygl-1* nascent transcripts for over 10 h. We thus were able to analyze the dynamics of Notch-dependent transcription at each locus in its natural setting. Notch indeed activates transcription in a bursty manner. Moreover, the strength of Notch modulates the transcriptional burst duration rather than the frequency, which is unexpected from previous in vitro studies about regulation of transcriptional dynamics [24, 33–35].

Here, we present our methods of single-RNA molecule visualization in fixed tissues (Subheading 3.1) and long-term whole animal live imaging using the MS2 system (Subheading 3.2). For the smFISH method, we include updates and greater depth than our previously published method [23]. We also present our computational methods to detect Notch-dependent transcription (Subheadings 3.1 and 3.2). The information of detected transcripts and nuclei such as their 3D coordinates and signal intensities can be used for further analyses as in our previous publications [21–23, 26, 30]. These methods can be used in any tissue or cell type for any transcripts after optimization.

2 Materials

2.1 smFISH in the *C. elegans* Germline

1. RNase-free microcentrifuge tubes (1.5 mL) (*see* **Note 1**).
2. Microscope slides.
3. Microscope cover glass with refractive index matching immersion oil for the microscope in use.
4. Aluminum foil.
5. Scalpel (Feather disposable scalpel #10).
6. RNase-free filtered tips (20, 200, and 1000 μL tips).
7. smFISH probe(s) conjugated with fluorophores (LGC BioSearch Technologies, https://www.biosearchtech.com/),
8. RNaseZap®.
9. ProLong Gold Antifade Reagent Mountant.
10. Nail polish (transparent).
11. Levamisole (or Tetramisole).
12. 37% formaldehyde
13. Ethanol (200 proof).
14. 4′,6-diamidino-2-phenylindole, dihydrochloride (DAPI)
15. TE buffer (pH 8.0, RNase-free): 10 mM Tris base, 1 mM EDTA.
16. PBST: RNase-free 1× phosphate-buffered saline (PBS; 137 mM NaCl, 2.7 mM KCl, 10 mM Na_2HPO_4, and 1.8 mM KH_2PO_4) + 0.1% Tween 20.

17. Permeabilization buffer: RNase-free 1× PBS + 0.1% Triton X-100.
18. Hybridization buffer (10 mL, RNase-free): 1 g dextran sulfate, 10 mg *E. coli* tRNA, 100 μL vanadyl ribonucleoside complex (200 mM), 40 μL nuclease-free BSA (50 mg/mL), 1 mL 20× SSC1 mL deionized formamide, 7.3 mL RNase-free H_2O (up to 10 mL volume). Bring all reagents to room temperature before making the hybridization buffer (*see* **Note 2**).
19. Wash buffer (RNase-free): 1 mL RNase-free 20× SSC, 1 mL deionized formamide, 8 mL RNase-free H_2O, 10 μL Tween 20. Bring all reagents to room temperature before making the wash buffer.

2.2 MS2 RNA Live Imaging

1. VALAP: Vaseline, lanolin, paraffin (1:1:1 w/w/w), melt together on a heat block at a low to medium setting (60–80 °C) and store at room temperature [24].
2. Polystyrene microbead (0.1 μm diameter), stored at 4 °C.
3. M9 buffer (3 g KH_2PO_4, 6 g Na_2HPO_4, 5 g NaCl, 1 ml 1 M $MgSO_4$, H_2O to 1 L). Sterilize by autoclaving.
4. 5% agarose.
5. Microscope slide.
6. Microscope cover glass.
7. Serotonin creatinine sulfate monohydrate. A small quantity of freshly made 50 mM serotonin solution (dissolved in M9) is needed for each experiment.
8. Optional: CherryTemp (microscope slide temperature controller, Cherry Biotech). Any temperature controller can be used.

2.3 MATLAB Codes and ImageJ Macros

1. All MATLAB codes and ImageJ macros needed for smFISH or MS2 data analysis can be found at: https://github.com/chlasic
2. smFISH_detection_analysis package (https://github.com/chlasic/smFISH_detection_analysis),
3. MS2_analysis package (https://github.com/chlasic/MS2_analysis).

3 Methods

3.1 smFISH in the C. elegans Germline

Samples are centrifuged at 2000 rpm for 30–60 s, unless otherwise stated.

3.1.1 smFISH Procedure

Probe Stock

1. Dissolve the dried probe mix (5 nmol) in 40 μL of RNase-free TE buffer (10 mM Tris–HCl, 1 mM EDTA, pH 8.0) to create a master probe stock of 125 μM (*see* **Note 3**).

2. Dilute the master stock 1:20 in RNAse-free TE buffer as a working stock. For the experiment, 1 μL of working stock is added to 49 μL of sample (1:50) for a final probe dilution of 1: 1000 and a final probe concentration of 125 nM (*see* **Note 4**).

Preparation for smFISH

1. Wipe down bench top, gloves, and pipettors with RNaseZap@ Wipes or spray. Rewipe gloves and pipettors if needed or exposed to RNases. Use RNase-free filtered tips and RNase-free tubes (*see* **Note 1**) and keep tip boxes closed when not in active use.
2. Prepare PBST (RNase-free 1× PBS + 0.1% Tween-20), permeabilization buffer (RNase-free 1× PBS + 0.1% Triton X-100), and RNase-free 70% ethanol, and 37% formaldehyde beforehand. These reagents can be premade and stored at 20 °C for at least a year.

Dissection of *C. elegans* Gonads

1. Prepare plate with 50–100 staged *C. elegans*. Collect worms by washing with PBST (can be nonsterile) using a P-1000 pipette and transfer the worm suspension to a 60-mm plastic petri dish cover or 100-mm glass petri dish (*see* **Note 5**). We typically use 4 mL worm suspension containing 50–100 worms in a in a 60 mm dish cover.
2. Add 0.25 mM levamisole (1:1000 from 0.25 M stock) directly to the worm suspension. For example, add 4 μL levamisole stock to a 4 mL worm suspension.
3. Using a scalpel, dissect worms by cutting behind the pharynx or before the rectum to release gonads. After levamisole treatment all dissection needs to be done within 20 min to prevent gonad deformation (*see* **Note 6**).
4. Collect dissected worms with pipette and transfer them to a new RNase-free 1.5 mL microcentrifuge tube or a new 15 mL conical tube (*see* **Note 1**). Avoid pipetting up intact (undissected) worms unless you plan to use them for your experiment.
5. Spin worm solution for 15 s at 2000 rpm. Remove supernatant and resuspend with 1 mL PBST.

Fixation and Permeabilization

1. Add formaldehyde (final concentration 3.7%. c.f., 1:10 dilution from 37% formaldehyde stock) directly to the dissected worms in PBST. For example, add 100 μL 37% formaldehyde to a 1 mL sample.
2. Incubate for 15–45 min at room temperature. We typically use 30 min. Rotate tubes (low rpm) or gently invert a few times every ~10 min.
3. Spin worm solution. Remove formaldehyde and dispose in the organic waste bottle. After this step, all reagents must be

RNase-free except for microscope slides and cover glass (*see* **Note 1**).

4. Add 1 mL permeabilization buffer (RNase-free 1× PBS + 0.1% Triton X-100) to the sample and incubate for 10 min at room temperature to permeabilize.
5. Wash with 1 mL PBST. Invert 5–6 times and spin at 2000 rpm for 30–60 s at room temperature. Repeat once for a total of two washes.
6. Resuspend the sample in 1 mL 70% EtOH and incubate at 4 °C for more than 4 h and less than 1 week.

smFISH Probe Hybridization (Day 2)

1. Bring hybridization buffer (HB) to room temperature before opening (*see* **Notes 2, 7,** and **8**). Thaw working probe stock, protecting from light. Probe dilutions vary depending on the probe set and need to be optimized before experiment (*see* **Notes 3** and **4**).
2. Spin fixed sample and remove ethanol.
3. Add 1 mL smFISH wash buffer, equilibrate 5 min at room temperature either in a tube rack or with gentle mixing.
4. While samples are equilibrating in wash buffer, mix HB and probe in a separate tube—prepare a new RNase-free 1.5 mL microcentrifuge tube (*see* **Note 1**) and add 49 μL HB + 1 μL probe dilution (*see* **Note 4**). If you multiplex probe sets, add 1 μL of each probe dilution (e.g., 1 μL of probe set 1 and 1 μL of probe set 2 in 48 μL of HB). If you have multiple samples (multiple tubes) prepare more HB solution (e.g., 392 μL HB + 8 μL probe dilution for eight samples). Vortex the solution and snap the tube to bring all solution to the bottom of the tube.
5. Spin down the fixed dissected worms and remove the wash buffer.
6. Add 50 μL of HB/probe solution in the tube containing sample. Gently flick the tube 5–6 times. Avoid making bubbles.
7. Wrap the tube with aluminum foil or place it in a sealed box to protect from light and incubate for more than 4 h to overnight at 37 °C on rotator (*see* **Note 9**).

Sample Washing and Mounting (Day 3)

1. All incubations should be done in the dark (*see* **Note 8**).
2. Add 1 mL wash buffer to sample, invert 5–6 times to mix. Spin and remove buffer.
3. Add 1 mL wash buffer to sample and incubate for 30 min on rotator at room temperature.
4. Spin down and remove the wash buffer.

5. Add 1 mL wash buffer +1 μg/mL DAPI and incubate for 30 min on rotator at room temperature. (*see* **Note 10**).
6. Wash 2× with 1 mL wash buffer (invert tube 5–10 times for each wash).
7. Mount samples on a microscope slide using ProLong Gold (PLG) Antifade Mountant (Life Technologies) (*see* **Note 11**).
8. Cover with 18 × 18 or 22 × 22 mm square cover glass. We use 10 μL or 12 μL PLG for the respective sizes (*see* **Note 12**).
9. Place the microscope slide on a flat surface in the dark, for example in a drawer at room temperature for 10–20 min, then gently push down the coverslip using a pipette tip to push out big air bubbles if present.
10. To cure the sample, leave the microscope slide in a dark place (in a box or a drawer) for at least 24 h (*see* **Note 13**).
11. Remove excess PLG on the slide by gently pushing down the edges of cover glass if needed. Seal the sample by putting nail polish around the edge of coverslip. Do not apply excess nail polish but make sure sample is sealed in all sides.
12. The slide can be stored in the dark at room temperature if it will be imaged within a week. Keep slides at 4 °C or, ideally, −20 °C for long-term storage (*see* **Note 14**).

3.1.2 smFISH Imaging

We have successfully imaged smFISH signals both with confocal and wide-field (compound) microscopes [21, 27]. Images taken by wide-field microscopy may need further image processing such as deconvolution. The deconvolution parameters (e.g., iterations and point spread function) must be carefully chosen to avoid over- or underimage processing which can create artificial signals (dots, granules, or tubules). Here, we report the image acquisition settings for both confocal and wide-field microscope systems.

Confocal Microscopy Settings

1. Microscope: Leica SP8 Confocal Laser Scanning microscope.
2. Objective lens: Leica HC PL APO CS2 63×/1.40 OIL.
3. Photon detector: Sensitive hybrid detector (HyD) for smFISH and photomultiplier (PMT) for DAPI.
4. Frequency: 400 Hz.
5. Zoom factor: 3×.
6. Pinhole: typically 95.5 μm (1 airy unit).
7. Laser power: between 1 and 5%. The power changes depending on the fluorophores conjugated to smFISH probes and the light source. For example, the *sygl-1* exon probe (CAL Fluor Red 610) was excited at 594 nm using 1.2% HeNe. Choose settings below saturation (255 for 8-bit images) so that the images can be quantitated.

8. Signal acquisition bandwidth: typically 100 nm. Start the acquisition window 5–10 nm from the end of the excitation laser range. For instance, the *sygl-1* exon probe conjugated to CAL flour 610 was excited with the 594 nm laser and the emission was acquired at 602–702 nm (*see* **Note 3**).
9. Gain: 40.
10. Acquisition settings: line average: 6–8 with 1–2 frame accumulation and unidirectional scanning.
11. *Z* step size: 0.3 μm.

Wide-Field Microscopy Settings

1. Microscope: Leica THUNDER Imager 3D Assay with RFID and LED8 (DMi8).
2. Objective lens: Leica HC PL APO 63×/1.40–0.60 OIL.
3. Filter set: DFT51011 quad cube.
4. LED Light intensity: 35–50%, typically 40%.
5. Exposure time: 150–300 ms, typically 250 ms.
6. *Z* step size: 0.3 μm.

3.1.3 smFISH Image Analysis

The original MATLAB code to process smFISH images and analyze the spatial patterns of RNA spots was presented in our previous work [21]. Here, we describe the latest version of the code (smFISH_detection_v2_7.m). We improved existing spot detection algorithms and implemented a few more tests to ensure the code detects transcripts and nuclei more accurately. It supports Leica image libraries (lif files), as in the previous versions, and has been updated to also handle TIFF format image files (*see* **Note 15**). The code consists of one main script (smFISH_detection.m) and several function codes in a folder called "Function."

1. Download all files and use the main script to run the code for smFISH image analysis on MATLAB. The code is available at: https://github.com/chlasic/smFISH_detection_analysis. New code will be posted when there are new updates.
2. User can choose one of two methods for selecting a folder containing image files with the extension "lif" or "tif" to be processed and another folder for saving the image processing results by using the variable "Method_of_choice" in the line 15 of the main script. Enter "1" if you want to use the user interface system, which will pop up a window for navigating to and selecting the folders. Enter "2" if you want to manually input the folder paths in lines 47 and 60 for the input and output folders, respectively.
3. Set the thresholds for detecting RNAs and nuclei (*see* **Notes 16** and **17**).

   (a) Threshold for mRNA detection using exon probes: "thresForExon" (line 20).

   (b) Threshold for ATS detection using intron probes: "thresForIntron" (line 21).

   (c) Threshold for nuclear detection: "thresForNuc" (line 22).

   The default is 1 and user can change the threshold from 0 to infinity. We found that most ~~of~~ detection works with thresholds between 0.3 and 1.8. The user must gauge the thresholds for the best detection results. The function "VisualRNAdetect" can be used to check the detection results. An example usage is "VisualRNAdetect(af, 1, 'ro', 50);". The user can change the second input argument "1" for indicating the *n*th RNA channel to visualize. The fourth argument ("50") is the brightness of the maximum projected image. Threshold levels for each channel are set using the same strategy. We try several thresholds for each channel for 5–10 images and choose a threshold that works best for most of the images within an experimental set. Noninteger values such as 1.25 can be used for the thresholds.

4. Indicate the type of channels in the image such as the exon or intron channel. The variable "ChOrder" in the line 29 of the main script records the channels the image contains. Use "1" for indicating the nucleus channel, "2" for the exon channel, and "3" for the intron channel. Use "5" if you want to skip processing the channel. For example, if you have three channels in your images in the order of (1) a nucleus channel, (2) an exon channel, and (3) an intron channel, use "ChOrder = [1 2 3]; ".

5. Set other parameters for accurate nucleus detection and mRNA counting (the number of mRNA per cell). The preset values work well with *C. elegans* germ cells in the distal gonad. The names of the variables storing these parameters, which are in lines 32–35 in the main script, are listed below. The variables "nrange" and "sensi" are for nucleus detection and "radius" and "vLim" are for mRNA counting in a cell. These parameters must be optimized for smFISH in systems other than the *C. elegans* germline (*see* **Note 18**).

   (a) nrange: defines the range of nuclear sizes in μm. The preset is [2.5 4.0] which means nuclei whose diameter is outside the range 2.5–4.0 μm will be excluded from the detection and analysis.

   (b) sensi: defines "circularity" for detecting nuclei. Circularity determines the sensitivity of the function "imfindcircles" embedded in the custom-made function "DetectNucleus" (line 87). The preset is 0.935 which is optimized for *C. elegans* germ cell nuclei.

(c) radius: defines the radius of a spherical region of interest (ROI) to count the number of mRNAs within the region. The center of ROI is placed on each nuclear center to estimate the number of mRNAs in vicinity of each nucleus. The preset is 2.5 (μm) which is roughly the radius of *C. elegans* germ cells in the distal gonad.

(d) vLim: defines the maximum radius of the 3D Voronoi cell, which estimates the boundary of each cell based on the 3D location of all nuclei. The number of mRNA is counted in each Voronoi cell and recorded as the total number of mRNA in each cell. The preset is 3 (μm) which is a typical radius of *C. elegans* germ cells located next to the rachis in the distal gonad.

6. All information pertaining to detected RNA and nuclear spots along with the image information are stored in the variable "af." The specific information saved in each column of "af" is annotated in the main script (lines 593–649). This information can be used for further analyses as shown in our previous work [21, 22].

3.2 MS2 RNA Live Imaging

3.2.1 Sample Preparation for Imaging

1. A gel-pad microscope slide is required for live *C. elegans* imaging. The method to prepare the gel-pad slide was modified from [24, 36] (*see* **Note 19**). To make the gel-pad slide, melt the 5% agarose/M9 (w/v) solution using a microwave or a heat block and put 300–500 μL solution onto the middle of a microscope slide using P-1000 pipette. Cut the tip of the pipette tip to make a larger opening for better flow of the agarose solution if needed. Immediately put another microscope slide on top of the microscope slide with agarose drop. Remove the top or bottom slide after a few seconds. The 5% agarose solution can be premade and reheated when the experiment is conducted.
2. Add 0.5–2 μL polystyrene microbead suspension onto the middle of the gel-pad (*see* **Note 20**).
3. Immediately add 0.5–2 μL of freshly made 50 mM serotonin solution (dissolved in M9) to bring its final concentration to 25 mM. Use the same volume of serotonin solution as the microbeads (*see* **Note 21**).
4. Move the gel-pad slide to the dissecting scope and pick 10–15 well-fed worms using a worm pick and OP50. Gently move worms in the serotonin solution on the gel-pad. Do not gouge the gel-pad with a worm pick. Use the surface tension of the serotonin solution to detach worms from the worm pick. Additional OP50 can be added around the edges of serotonin solution drop on the gel-pad.

5. Gently place the cover glass on the gel-pad. Do not move the cover glass once it sits on the gel-pad as moving it can kink, twist, or sever the worms.
6. Check the worms on the microscope to confirm they are immobilized. If worms still move, prepare a new gel-pad slide and repeat the procedure with a smaller volume of microbeads and serotonin. Make sure the pharynx of the worm on the microscope slide is pumping at a normal rate (*see* **Note 22**).
7. Apply VALAP (*see* Subheading 2) around edges of the cover glass to completely seal the slide. Failing to do so will result in a dried, wrinkled gel-pad.
8. The slide can be attached to a temperature controlling system such as CherryTemp and mounted on the microscope stage with a chip holder to keep a constant temperature throughout the live imaging process or to shift temperature during the experiment (*see* **Note 23**).

3.2.2 MS2 Imaging

Live imaging was performed immediately after making the gel-pad microscope slide. Both confocal and wide-field microscopes worked well for *sygl-1* RNA live imaging [24] (unpublished for the wide-field system). Here, we describe our image acquisition settings for both imaging systems (*see* **Note 24**).

Confocal Microscopy Settings

1. Microscope: Leica SP8 Confocal Laser Scanner.
2. Objective lens: Leica HC PL APO CS2 63×/1.40 OIL.
3. Photon detector: Sensitive hybrid detector (HyD).
4. Frequency: 900–1000 Hz.
5. Zoom factor: 2.5×.
6. Image size: 512 × 512 or 1024 × 512.
7. Pinhole: 95.5–105.1 μm (1–1.5 airy unit).
8. Laser power: 0.3–0.4% Argon laser (488 nm) for GFP and 0.6–1% HeNe laser (594 nm) for mCherry.
9. The signal acquisition bandwidth: 50 nm. Start the acquisition window 5–10 nm from the end of the excitation laser range. For example, GFP was excited with the 488 nm laser and the emission was acquired at 496–546 nm. No line/frame averaging or accumulation was used.
10. Gain: 40.
11. *Z* step size: 0.4 μm.
12. Optional: Autofocusing was enabled for every other time point with 594-nm Argon laser to keep the gonad that is imaged in the similar position on the *z*-axis.

Wide-Field Microscopy Settings

1. Microscope: Leica THUNDER Imager 3D Assay with RFID and LED8 (DMi8).
2. Objective lens: Leica HC PL APO 63×/1.40–0.60 OIL.
3. Filter set: DFT51011 quad cube.
4. LED Light intensity: 10–30%, typically 20%.
5. Exposure time: 50–100 ms, typically 50 ms.
6. *Z* step size: 0.4 μm.

3.2.3 MS2 Image Analysis

Even if the worm on a gel-pad slide is completely immobilized, the gonad inside the worm being imaged can move or twitch time to time due to the movement of the intestine, muscles, or embryos. This results in shift, drifting or fluctuation of gonads in the MS2 movies. Theses gonads can be aligned in two steps using our customized, automated ImageJ macros, which are available at https://github.com/chlasic/MS2_analysis.

1. Open ImageJ or Fiji and open the first macro "MultiStackReg_BatchProcess1of2.ijm." We use the shortcut "["to open a script window to open the macro (*see* **Note 25**).
2. Click "Run" on the bottom of the window to start the macro. It lets users choose two directories, one for images to be processed and the other for saving processed images. Then, it will process all images in the folder and save the results in the designated folder location.
3. Open the second macro "MultiStackReg_BatchProcess2of2.ijm" and click "Run." It will also ask two directories as above and will process all images (*see* **Note 25**).
4. The user can open the final images to check if all *z* planes are aligned throughout the time lapse.
5. If the automated alignment does not work properly, the user can manually align one time point using "TranslateStack.ijm" or multiple time points using "TranslateStackAllAfterPos.ijm."
6. Split all gonadal images (lif or tif files) in two halves on the *z*-axis in ImageJ to minimize overlap of germ cell nuclei for further analysis. Use "duplicate" function to make two hyperstacks, each containing a half of the *z*-stack image.
7. *Z*-project each half using sum slices.
8. Draw a circular ROI (diameter of 1 μm) around each *sygl-1* MCP::GFP dot in the *z*-projected images to measure signal intensity, which estimates the number of nascent transcripts at the locus. Manually follow the individual dots and record their signal intensity throughout the MS2 movies (*see* **Note 26**). The same ROI should be used to measure background, which is drawn inside the gonad where there is no MS2 signal. The

recorded signals can be further analyzed such as spatiotemporal pattern analysis of transcriptional burst duration or frequency as in our previous publications [24].

4 Notes

1. Regular (non–RNase-free) reagents and consumables such as tubes and pipette tips can be used until the end of sample fixation with formaldehyde. Thereafter, all reagents must be RNase-free. We wipe the bench and pipettes with RNaseZap and use RNase-free microcentrifuge tubes, filtered nuclease-free pipette tips. Unopened packages of non–RNase-free microcentrifuge tubes can also be used throughout the entire protocol if they are kept separately and handled cleanly. Use PPE at all times to protect the sample from RNase contamination.
2. Dextran sulfate, E. coli tRNA, and nuclease-free BSA can be excluded from the hybridization buffer. They help reduce overall background of smFISH images. Leave them out when making the hybridization buffer if smFISH signal is too weak even at the highest probe concentration.
3. We designed the smFISH probes using "Stellaris probe designer" (https://www.biosearchtech.com/support/tools/design-software/stellaris-probe-designer) and ordered from Biosearch Technologies (https://www.biosearchtech.com/). Available fluorescence dyes are listed here (https://www.biosearchtech.com/support/education/stellaris-rna-fish/dyes-and-modifications-for-stellaris). Carefully choose the dyes that are compatible with the microscope system used for the experiment. Choose the dyes that have the least overlapping spectra when multiplexing the probes to eliminate bleed-through between channels. We recommend empirically testing emission collection windows for bleed-through on confocal microscopes. For example, we found that exciting with a 561 nm laser and collecting wavelengths greater than 588 nm could detect CAL Fluor Red 610 signal when multiplexed, so we limited our Quasar 570 emission collection to 564–588 nm. For multiplexing probes, we used two among Quasar 570, CAL Fluor Red 610 and Quasar 670 for the Leica confocal microscope, and TAMRA and Quasar 670 for the Leica Thunder wide-field microscope. We confirmed specificity of the probes by conducting BLAST/BLAT search with the *C. elegans* whole genome sequence (https://wormbase.org//tools/blast_blat). Only probes that match to the target sequence were used.

4. The master smFISH probe stock solution is at 125 μM (5 nmol probes in 40 μL TE buffer). We dilute the master stock 1:20 in TE buffer (6.25 μM) to make a personal working probe stock. We dilute this working probe stock 1:50 in hybridization buffer (125 nM) during the smFISH experiment. Try different dilutions of the working stock to optimize the smFISH signal. For example, try a 1:10 working stock (final concentration: 250 nM) or 1:5 (500 nM) dilution if smFISH signal is too dim. In case of high background, try 1:50 (50 nM) or 1:100 (25 nM).
5. During worm dissection, use 2 mL PBST to wash one 60-mm NGM plate. You will be able to pipette up about 1.5–1.7 mL with worm suspension from the plate. In case of using 100-mm glass dish for dissection, we use 10 mL worm suspension containing 50–100 worms.
6. For worm dissection, finishing worm dissection within 20 min ensures preservation of an intact gonad. After 30 min of levamisole treatment, you may notice deformation of gonad.
7. Periodically rewipe gloves and pipettors with RNaseZap.
8. During and after smFISH probe hybridization, minimize exposure of the sample to light. Shield tube in a cover (your hand, aluminum foil fold, cardboard box tent, etc.) while changing buffers or tips. We always wrap the tubes containing the sample and smFISH probes in foil. To remove liquid from samples, use a P-200 or P-1000 without exposing tube to too much incident light such as the base light of the dissecting scope.
9. For smFISH probe hybridization, you can leave samples for up to 72 h. Longer incubation may improve smFISH signal.
10. **Steps 3–4**, Subheading "Sample Washing and Mounting (Day 3)" can be skipped for imaging with the confocal microscopy. The additional washes help with background when using wide-field microscopy.
11. For mounting smFISH samples on a microscope slide, remove as much wash buffer as possible from fixed samples. Resuspend gonads in 10–20 μL PLG depending on # of dissected worms, then collect 5–10 μL using a pipette and drop 8–10 small droplets around the center of microscope slide. Microscope slides and coverslips can be baked as an extra precaution against RNase contamination, or simply dedicate a slide box to smFISH. After mounting sample, use a P10 pipette tip or a worm pick to carefully reposition gonads if needed.
12. After mounting sample on a microscope slide, do not press the cover glass hard, which can rupture samples. Do not try to remove all small bubbles in the slide which can crack the cover glass or break the samples. Adding some extra ProLong Gold

to the sample can help removing bubbles on the slide. This also dilutes the residual wash buffer which can improve the smFISH signal.

13. We follow the ProLong Gold manual's recommendation to cure 24–60 h, but the sample slides can be cured for as little as 2 h before imaging. Longer curing times (48–60 h) improve smFISH detection on our hands. The ProLong manual also includes recommendations for cure times and optimal mounting medium for different fluorophores (https://assets.thermofisher.com/TFS-Assets/LSG/manuals/MAN0002469_ProLong_SlowFade_Antifade_Mountants_PI.pdf).
14. The smFISH microscope slides are generally good for at least 6 months from mounting when they are completely sealed with nail polish.
15. The MATLAB code for image processing reads image files along with their metadata, detects objects that fall in a particular category (e.g., the tissue outline, RNAs and nuclei), and records the information associated with objects (e.g., 3D coordinates, integrated intensities, and shape).
16. For image processing with the MATLAB code, a lower threshold allows detection of dimmer spots, which usually have lower signal-to-noise ratio, on microscope images.
17. Choosing the right threshold in the MATLAB code is critical to accurate RNA or nucleus detection but it can be time consuming. A good practice is to pick a portion of images (5–10 images) and use them to optimize the thresholds. The random number generator "randi" function in MATLAB can be used for unbiased image choices. The selected threshold typically works well for most of images with the same image acquisition settings.
18. The MATLAB code and the workflow described here can be used for detecting transcripts of interest in other tissues such as intestine, seam cells, and neurons in *C. elegans* as well as other model systems such as *S. cerevisiae*, *Neurospora*, and tissue culture (unpublished, confirmed by personal conversations). The parameters "nrange," "sensi," "radius," and "vLim" should be empirically determined for accurate detection and analysis with systems other than the *C. elegans* germline. For "sensi," a higher value will make the detection more stringent and exclude nuclear objects with relatively irregular circular shapes. A lower value will allow detection of even less-circular objects. Three methods are used to count mRNAs around each nucleus: (A) spherical ROIs with a user-set "radius," (B) 3D Voronoi cells without any limit, which can include the rachis in the distal gonad, and (C) 3D Voronoi cells with a limit for cell radius of "vLim". The mRNA counts from each of the three

methods are stored in the variables "nuc" and "af" (see the annotation in lines 325–328 of the main script) so that the user can choose an mRNA estimate that best represents the number of mRNA per cell for the type and shape of the cells used for smFISH. Method A is particularly useful for complete syncytial systems with no individual cell boundaries as it will consider which nucleus (or nuclei) the mRNAs belong to based only on their proximity to the nucleus (or nuclei) within the user-defined "radius". If two or more ROIs overlap, the number of mRNAs within the overlapped region will be divided by the number of overlapping ROIs and equally assigned to these ROIs because there is an equal chance these mRNAs belong to those ROIs in syncytium. In contrast, methods B and C are useful for systems with separate cells. They draw theoretical cell outlines around each nucleus and record the number of mRNAs within each Voronoi cell. Method B is typically for tissue with no spaces or gaps between individual cells and C is for tissue with gaps which can be excluded from mRNA counting by setting "vLim". We used 3D Voronoi cells with a limit to estimate the number of mRNAs per cell [21, 22].

19. The thickness of gel-pad is important for worm immobilization and long-term live imaging. 1.5–2 mm (four layers of time tape) thick gel-pad gave us the best result, keeping normal pharyngeal movement (3–5/s) for at least several hours. Thinner gel-pad can help immobilize worms but the pharynx tends to stop after a few hours of imaging (1–3 h). It was harder to keep worms immobilized with a gel-pad thicker than 2 mm.
20. To immobilize the worm, we have also tried 0.05-μm polystyrene microbeads which gave us a similar result for worms at L4 stage or older. We speculate the smaller microbead may better immobilize smaller worms (L1–L3).
21. After adding microbeads and serotonin, worms should be immediately and quickly loaded on the gel-pad with minimum exposure to any direct light such as the base light for the dissecting microscope. The solution on the gel-pad dries up quickly, which makes harder to load worms on it.
22. Healthy worms on the gel-pad slide have a pharyngeal pumping rate of 3–5/s and an egg laying rate of 2–5/h, and a mitotic index of about 6.25, consistent with worms on nematode growth (NGM) plates.
23. The consistent temperature helps keep worms alive for extended time. We used CherryTemp but any temperature controlling system can be used.
24. The multipoint imaging can be used for monitoring several worms on the same slide over time. Make sure to use autofocusing to minimize image drifting on the *z* axis.

25. The image processing macro "MultiStackReg_BatchProcess1of2.ijm" aligns gonadal images along the *z*-axis at each time-point. "MultiStackReg_BatchProcess2of2.ijm" uses the processed images from the first macro to further align them throughout the MS2 movie.
26. Most MCP dots did not move dramatically inside the nucleus in the *C. elegans* germ cells (95%), which made it easier to trace transcriptional bursting [24]. Other systems might exhibit more dynamic chromosomal movement.

Acknowledgements

CHL was supported by the American Heart Association (18POST34030263). TRL was supported by the National Science Foundation Graduate Research Fellowship Program (Grant No. DGE1256259). JK is grateful to NIH R01 GM134119 for support.

References

1. Siebel C, Lendahl U (2017) Notch signaling in development, tissue homeostasis, and disease. Physiol Rev 97:1235–1294. https://doi.org/10.1152/physrev.00005.2017
2. Kopan R, Ilagan MX (2009) The canonical notch signaling pathway: unfolding the activation mechanism. Cell 137:216–233. https://doi.org/10.1016/j.cell.2009.03.045
3. Gazave E, Lapébie P, Richards GS, Brunet F, Ereskovsky AV, Degnan BM, Borchiellini C, Vervoort M, Renard E (2009) Origin and evolution of the Notch signalling pathway: an overview from eukaryotic genomes. BMC Evol Biol 9:249. https://doi.org/10.1186/1471-2148-9-249
4. Imayoshi I, Shimojo H, Sakamoto M, Ohtsuka T, Kageyama R (2013) Genetic visualization of Notch signaling in mammalian neurogenesis. Cell Mol Life Sci 70:2045–2057. https://doi.org/10.1007/s00018-012-1151-x
5. Bolós V, Grego-Bessa J, de la Pompa JL (2007) Notch signaling in development and cancer. Endocr Rev 28:339–363. https://doi.org/10.1210/er.2006-0046
6. Aquila G, Pannella M, Morelli MB, Caliceti C, Fortini C, Rizzo P, Ferrari R (2013) The role of Notch pathway in cardiovascular diseases. Glob Cardiol Sci Pract 2013:364–371. https://doi.org/10.5339/gscp.2013.44
7. MacGrogan D, Nus M, de la Pompa JL (2010) Chapter eleven - notch signaling in cardiac development and disease. In: Kopan R (ed) Current topics in developmental biology. Academic Press, New York, pp 333–365
8. de la Pompa JL, Epstein JA (2012) Coordinating tissue interactions: Notch signaling in cardiac development and disease. Dev Cell 22:244–254. https://doi.org/10.1016/j.devcel.2012.01.014
9. High FA, Epstein JA (2008) The multifaceted role of Notch in cardiac development and disease. Nat Rev Genet 9:49–61. https://doi.org/10.1038/nrg2279
10. Bray SJ (2006) Notch signalling: a simple pathway becomes complex. Nat Rev Mol Cell Biol 7:678–689. https://doi.org/10.1038/nrm2009
11. Kimble J, Crittenden SL (2007) Controls of germline stem cells, entry into meiosis, and the sperm/oocyte decision in Caenorhabditis elegans. Annu Rev Cell Dev Biol 23:405–433. https://doi.org/10.1146/annurev.cellbio.23.090506.123326
12. Sato C, Zhao G, Ilagan MX (2012) An overview of Notch signaling in adult tissue renewal and maintenance. Curr Alzheimer Res 9:227–240
13. Kershner AM, Shin H, Hansen TJ, Kimble J (2014) Discovery of two GLP-1/Notch target

genes that account for the role of GLP-1/Notch signaling in stem cell maintenance. Proc Natl Acad Sci U S A 111:3739–3744. https://doi.org/10.1073/pnas.1401861111
14. Shi Y, Shu B, Yang R, Xu Y, Xing B, Liu J, Chen L, Qi S, Liu X, Wang P, Tang J, Xie J (2015) Wnt and Notch signaling pathway involved in wound healing by targeting c-Myc and Hes1 separately. Stem Cell Res Ther 6:120. https://doi.org/10.1186/s13287-015-0103-4
15. Falo-Sanjuan J, Bray SJ (2020) Decoding the Notch signal. Develop Growth Differ 62:4–14. https://doi.org/10.1111/dgd.12644
16. Hoyle NP, Ish-Horowicz D (2013) Transcript processing and export kinetics are rate-limiting steps in expressing vertebrate segmentation clock genes. PNAS 110:E4316–E4324. https://doi.org/10.1073/pnas.1308811110
17. Housden BE, Fu AQ, Krejci A, Bernard F, Fischer B, Tavaré S, Russell S, Bray SJ (2013) Transcriptional dynamics elicited by a short pulse of notch activation involves feed-forward regulation by E(spl)/Hes genes. PLoS Genet 9:e1003162. https://doi.org/10.1371/journal.pgen.1003162
18. Yoon K, Gaiano N (2005) Notch signaling in the mammalian central nervous system: insights from mouse mutants. Nat Neurosci 8:709–715. https://doi.org/10.1038/nn1475
19. Krebs LT, Xue Y, Norton CR, Shutter JR, Maguire M, Sundberg JP, Gallahan D, Closson V, Kitajewski J, Callahan R, Smith GH, Stark KL, Gridley T (2000) Notch signaling is essential for vascular morphogenesis in mice. Genes Dev 14:1343–1352
20. Batista MR, Diniz P, Torres A, Murta D, Lopes-da-Costa L, Silva E (2020) Notch signaling in mouse blastocyst development and hatching. BMC Dev Biol 20:9. https://doi.org/10.1186/s12861-020-00216-2
21. Lee C, Sorensen EB, Lynch TR, Kimble J (2016) C. elegans GLP-1/Notch activates transcription in a probability gradient across the germline stem cell pool. eLife 5:e18370. https://doi.org/10.7554/eLife.18370
22. Crittenden SL, Lee C, Mohanty I, Battula S, Knobel K, Kimble J (2019) Sexual dimorphism of niche architecture and regulation of the Caenorhabditis elegans germline stem cell pool. MBoC 30:1757–1769. https://doi.org/10.1091/mbc.E19-03-0164
23. Lee C, Seidel H, Lynch T, Sorensen E, Crittenden S, Kimble J (2017) Single-molecule RNA Fluorescence in situ Hybridization (smFISH) in Caenorhabditis elegans. Bio-Protocol 7:e2357. https://doi.org/10.21769/BioProtoc.2357
24. Lee C, Shin H, Kimble J (2019) Dynamics of Notch-dependent transcriptional bursting in its native context. Dev Cell 50:426–435.e4. https://doi.org/10.1016/j.devcel.2019.07.001
25. Raj A, van den Bogaard P, Rifkin SA, van Oudenaarden A, Tyagi S (2008) Imaging individual mRNA molecules using multiple singly labeled probes. Nat Methods 5:877–879. https://doi.org/10.1038/nmeth.1253
26. Lee C, Roberts SE, Gladfelter AS (2016) Quantitative spatial analysis of transcripts in multinucleate cells using single-molecule FISH. Methods 98:124. https://doi.org/10.1016/j.ymeth.2015.12.007
27. Lee C, Zhang H, Baker AE, Occhipinti P, Borsuk ME, Gladfelter AS (2013) Protein aggregation behavior regulates cyclin transcript localization and cell-cycle control. Dev Cell 25:572–584. https://doi.org/10.1016/j.devcel.2013.05.007
28. Lee C, Occhipinti P, Gladfelter AS (2015) PolyQ-dependent RNA-protein assemblies control symmetry breaking. J Cell Biol 208:533–544. https://doi.org/10.1083/jcb.201407105
29. Cinquin O, Crittenden SL, Morgan DE, Kimble J (2010) Progression from a stem cell-like state to early differentiation in the C. elegans germ line. Proc Natl Acad Sci U S A 107:2048–2053. https://doi.org/10.1073/pnas.0912704107
30. Li F, Jianhua Z, Lee CH, Gina K, Fang L, Petersen AJ, Evi L, Anderson CL, Orland KM, Robertson GA, Eckhardt LL, January CT, Kamp TJ (2021) Long QT syndrome 2 PAS domain variant induces hERG1a/1b subunit imbalance in patient-specific iPSC-cardiomyocytes. Circ Arrhythm Electrophysiol 4:e009343. https://doi.org/10.1161/CIRCEP.120.009343
31. Bertrand E, Chartrand P, Schaefer M, Shenoy SM, Singer RH, Long RM (1998) Localization of ASH1 mRNA particles in living yeast. Mol Cell 2:437–445. https://doi.org/10.1016/S1097-2765(00)80143-4
32. Larson DR, Singer RH, Zenklusen D (2009) A single molecule view of gene expression. Trends Cell Biol 19:630–637. https://doi.org/10.1016/j.tcb.2009.08.008
33. Corrigan AM, Chubb JR (2014) Regulation of transcriptional bursting by a naturally oscillating signal. Curr Biol 24:205–211. https://doi.org/10.1016/j.cub.2013.12.011

34. Li C, Cesbron F, Oehler M, Brunner M, Höfer T (2018) Frequency modulation of transcriptional bursting enables sensitive and rapid gene regulation. Cells 6:409–423.e11. https://doi.org/10.1016/j.cels.2018.01.012
35. Kafri P, Hasenson SE, Kanter I, Sheinberger J, Kinor N, Yunger S, Shav-Tal Y (2016) Quantifying β-catenin subcellular dynamics and cyclin D1 mRNA transcription during Wnt signaling in single living cells. eLife 5:e16748. https://doi.org/10.7554/eLife.16748
36. Kim E, Sun L, Gabel CV, Fang-Yen C (2013) Long-term imaging of Caenorhabditis elegans using nanoparticle-mediated immobilization. PLoS One 8:e53419. https://doi.org/10.1371/journal.pone.0053419

Chapter 12

An Automatic Stage Identification MATLAB Tool to Reveal Notch Expression Pattern in Drosophila Egg Chambers

Lily Paculis, Qiuping Xu, Qian Xie, and Dongyu Jia

Abstract

Many highly conserved pathways control the development and determine cell fate in organisms. One of these pathways is the Notch signaling pathway that allows for local cell–cell communication. Researchers have found that the timing for when Notch signaling activates the target gene is important for maintaining normal gene expression. Any alterations in the downstream gene expression could cause issues with development or certain diseases. The *Drosophila* oogenesis is a widely used model in developmental biology for analyzing the Notch pathway. However, determining the stage of oogenesis is difficult and varies depending on individual analyzing it. Here, we provide a MATLAB tool to automatically identify the stage of a *Drosophila* egg chamber and reveal the Notch expression pattern.

Key words *Drosophila*, Notch signaling, Notch expression, Egg chamber stage, MATLAB

1 Introduction

Notch signaling is a highly conserved and regulated pathway that helps regulate the formation and destruction of many animal tissues. The pathway allows for local communication between cells in the organism to determine their cell fate. Thus, this pathway is essential for the development and survival of many biological organisms [1, 2]. Any disruption or alteration of the Notch pathway has been known to cause developmental disorders, such as cancers, and diseases, such as cardiovascular diseases [3].

Recent research suggests that the timing for when the Notch signaling starts expression within the target cells is important for preserving the correct expression of the downstream genes [4, 5]. *Drosophila* oogenesis has 14 stages, and it is a frequently used model in developmental biology. The development of the egg

Supplementary Information The online version contains supplementary material available at [https://doi.org/10.1007/978-1-0716-2201-8_12].

Dongyu Jia (ed.), *Notch Signaling Research: Methods and Protocols*,
Methods in Molecular Biology, vol. 2472, https://doi.org/10.1007/978-1-0716-2201-8_12,

chambers is regulated by Notch signaling that controls the egg chamber stage transitions [5]. The process of identifying an egg chamber stage is not consistent since this process is difficult due to errors in human perception and biased depending on who is analyzing [5, 6]. A standardized, reliable algorithm for determining egg chamber stage could prevent these issues and provide consistency for researchers.

The DAPI Image Feature Extraction toolbox for MATLAB is an algorithm that will output several image characteristics including the stage of the egg chamber through the input of any DAPI image. This toolbox provides characteristics of the egg chamber such as its cell size, cell ratio, cell orientation, oocyte size, follicle cell distribution, blob-like chromosomes, and centripetal cell migration. The script first analyzes the egg chamber area and ratio to determine other morphological features [7]. Our toolbox has been tested for its validity and reliability to consistently determine the stage of the *Drosophila* egg chamber based on its characteristics [8]. Using this tool can be crucial for determining the accurate stages when the Notch pathway is regulating expression of its downstream genes.

2 Materials

2.1 Reagents

1. Primary antibodies: mouse anti-Cut 1:15 (2B10; DSHB), anti-Hnt 1:15 (1G9; DSHB), mouse anti-Br-core 1:30 (25E9; DSHB).
2. Secondary antibodies: Select corresponding secondary antibodies Alexa Fluor 488, 546 or 633 at a dilution of 1:400 in accordance to the primary antibodies.
3. DAPI (10 μg/mL).
4. 1× PBS solution: 137 mM NaCl, 2.7 mM KCl, 10 mM Na_2HPO_4, 1.8 mM KH_2PO_4.
5. 1× PBT solution: 137 mM NaCl, 2.7 mM KCl, 10 mM Na_2HPO_4, 1.8 mM KH_2PO_4, 0.2% Triton X-100.
6. PBTG: 5% normal goat serum, 0.2% bovine serum albumin diluted in 1× PBT.
7. Mounting Solution: 1 g *n*-propyl gallate, 5 mL 10× PBS, 40 mL glycerol, 5 mL dH_2O.

2.2 Equipment

1. Dissecting forceps.
2. CO_2 tank and pad.
3. Embryo dish.
4. Centrifuge for Eppendorf tubes.
5. Nutator.

6. Flow cytometer.
7. Stereoscopic (dissecting) microscope.
8. Confocal microscope.

2.3 Fiji or ImageJ (See Note 1)

1. Download the Fiji/ImageJ installer for your current operating software from the ImageJ website (www.imagej.net).
2. Extract the contents of the zipped file folder.
3. Open the installer and update all URLs as recommended by Fiji before completing the installation.

2.4 MATLAB

1. Create a MathWorks account through the "Login" tab (www.mathworks.com).
2. Once logged in, click on the "My Account" tab under your name.
3. Click the arrow next to your license number and download the latest version of the MATLAB software.
4. Choose the installer for the type of operating system you are currently using on the computer.
5. Login to your MathWorks account on the MATLAB installer.
6. Install all the recommended products that are preselected on your installer, including MATLAB, Simulink, and the Image Processing Toolbox.

2.5 DAPI Image Feature Extraction MATLAB Toolbox

1. Download the DAPI image feature extraction toolbox for MATLAB (https://github.com/qx0731/Work_DAPI_image_feature_extraction) from GitHub (www.github.com) [7].
2. Extract the contents of the zipped file folder.

3 Methods

3.1 Immunostaining of Drosophila Ovaries

1. Dissect *Drosophila* ovaries with proper genotype (*see* **Note 2**).
2. Transfer the ovaries into 500 μL of 1×PBS into a 1.5 mL Eppendorf tube. Keep the tube on ice until enough ovaries are collected.
3. Remove the 1× PBS and add 0.5 mL of fix solution.
4. Place the tube on a rocking nutator for 10 min.
5. Remove and appropriately discard of the fix solution.
6. Wash the ovaries 3 × 15 min with 1 mL of 1× PBT.
7. Remove the last PBT wash and add 1 mL of PBTG to inhibit nonspecific binding. Place the ovaries on the rocking nutator again for 1 h at room temperature or overnight at 4 °C

8. Remove the PBTG and add 150 μL of the primary antibody (select the desired markers to test for the cell cycle switches). Incubate overnight at 4 °C on the rocking nutator (*see* **Notes 3–5**).
9. Remove the primary antibody. Return the primary antibody to a 1.5 mL Eppendorf tube so that it can be reused. Then, wash for 3 × 15 min with 1 mL of 1× PBT.
10. Add 150 μL of the secondary antibody and then incubate on the rocking nutator for 2 h at room temperature or overnight at 4 °C.
11. Remove the secondary antibody. Return the secondary antibody to a 1.5 mL Eppendorf tube so it can be reused. Then, wash for 2 × 15 min with 1 mL of PBT.
12. Add 150 μL of DAPI (10 μg/mL) for 10–15 min.
13. Remove the DAPI and wash for 1 × 10 min with 1 mL of 1× PBT.
14. Appropriately discard of the PBT and wash 2 × 10 min with 1 mL of 1× PBS.
15. Remove the PBS until 300 μL is left in the tube with the ovaries.
16. Pipet the ovaries up and down several times, freeing the egg chambers with the 200 μL pipette tip.
17. Spin the tube and remove as much of the 1× PBS as possible from the ovaries.
18. Add 120 μL of mounting solution (*see* **Note 6**).
19. Cut 0.3 mm of a 200 μL pipette tip and use this to transfer the mounting solution with the egg chambers to a microscopic glass slide.
20. Cover with a coverslip glass and seal the edges with nail polish (*see* **Notes** 7 and **8**).
21. Collect images with a confocal microscope (*see* **Note 9**).

3.2 Converting Stacked DAPI Images to Individual TIF Images with Fiji or Image J (See Supplementary Video 1)

1. Open your DAPI images inside the Fiji application or ImageJ application (*see* **Note 1**).
2. Click on the "Image" tab, then highlight the "Colors" tab and click "Split Channels."
3. Click on the blue channel and click on the "Image" tab again, highlight the "Type" tab, and click "RGB Color."
4. Repeat **step 2**.
5. Click on the "Freehand Selections" tab. Click on the blue channel and drag the mouse around the single egg chamber you want to analyze.

6. Click on the "Image" tab and click "Crop."
7. Click on the "File" tab, highlight "Saves As," and click "Tiff" to save your DAPI image as a .TIF file.

3.3 Moving Your TIF Images into the data_ready Folder

1. Find and navigate to the "Work_DAPI_image_feature_extraction-master" folder inside the DAPI Image Feature Extraction MATLAB Toolbox installed earlier.
2. Navigate to the "data_ready" folder and place all the individual TIF images you saved from Fiji into this folder.

3.4 Preparing and Using the DAPI Image Feature Extraction MATLAB Toolbox (See Supplementary Video 2)

1. Open the MATLAB application.
2. Open the "Home" tab and click the "Get Add-Ons" tab.
3. Search for the MinGW-w64 C/C++ Compiler and install the add-on into your MATLAB NOTE:· If you already have this add-on, you may skip **steps 2** and **3**.
4. In the "Current Folder" tab, navigate to the folder "Work_-DAPI_image_feature_extraction-master" and then the "InsidePolyFolder".
5. Type in and execute the following command.

   >> mex insidepoly_dblengine.c
   >> mex insidepoly_sglengine.c
6. Navigate back to the "Work_DAPI_image_feature_extraction-master" folder and then execute the following command (*see* **Note 10**):

   >> [area ratio oocyte_size distance]=feature_extraction ('filename.tif');
7. Replace "princomp" with "pca" in line 12 and line 187 in the editor if the command comes back with an error message
8. The script will produce several figures that pop up in separate windows to provide certain features of the egg chamber visually. Other features of the egg chamber will be reported under the executed command. Repeat the command from **step 6** for each file you want to analyze.

3.5 Notch Expression Pattern Analysis

1. Using the Fiji/ImageJ application, the images will be split according to the channels with the DAPI channel being the blue channel, the Br/Cut/Hnt channel being the green channel with Alexa Fluor 488 secondary antibodies, and the PH3 being the red channel with Alexa Fluor 546 secondary antibodies.
2. Look at the separate Br/Cut/Hnt and PH3 channels to determine the presence and strength of the primary antibodies and PH3 present in the DAPI images.
3. Follow the previous steps in Subheading 3.4 to determine the stage of the egg chamber using the cropped image of the DAPI

image made in the previous procedure for Subheading 3.2. The same protocol will provide the area and ratio of the egg chambers as well.

4. Determine the presence and strength of the 5-blob phenotype of the nursing cells inside the egg chambers by analyzing the figure that displays the nurse cells. Use the generated figures and stage determination to analyze the Notch expression pattern for stage in which the upregulation/downregulation of Hnt, Cut, and Br occurs.

4 Notes

1. The installation procedures mentioned in this chapter are for the Windows operating software. Some installation steps may be different if installing these programs on different operating softwares. The Fiji software must be installed and used if you saved your DAPI photos as .CZI files. The ImageJ software can be directly used if you saved your DAPI photos as .LSM files. We recommend that you save the DAPI Images as .LSM files. The procedures for using Fiji and ImageJ are the same.
2. Detailed dissection steps are described in our previous protocol publication [8].
3. For the antibody staining, the primary antibodies can be mixed and costained with the corresponding secondary antibodies at different fluorescence channels as long as the primary antibodies are not from the same species. We recommend using the BrdU incorporation for better quality. EdU incorporation can be used as well, but it might result in poorer quality.
4. Since some primary antibodies cannot be obtained easy in large quantities due to their high cost, we recommend reusing antibodies. The higher the quality of the antibodies, the more times they can be reused.
5. The quality of antibodies might vary from batch to batch.
6. We use a 1000 μL pipette tip to transfer three drops of the mounting solution into the tube since the mounting solution is difficult to precisely transfer into a 120 μL tube due to its stickiness.
7. The thickness of the microscopic cover glass to use depending on the confocal imaging system used.
8. You must seal the edges of the cover glass to avoid the egg chambers from flowing inside the mounting solution. The nail polish used to seal the cover glass must be transparent.

9. We use a Zeiss 710 confocal microscope in our lab, any confocal microscopes with correct figure dimension information should work.
10. Replace the filename with the name of whatever file you want the script to analyze. The toolbox provides four examples to test the command with and ensure everything is set up correctly.

Acknowledgments

This work was supported by the Start-Up funding, Faculty Research Committee (FRC) Award, College of Science and Mathematics Research Grant Award, and Undergraduate Research Assistant Funding from Georgia Southern University and the Georgia Southern Biology Undergraduate Research Grant. MATLAB® and Simulink® are trademarks of the MathWorks Inc. These products are not sponsored nor endorsed by the authors. MathWorks does not authorize the accuracy of the tool used in this chapter.

References

1. Takebe N, Nguyen D, Yang SX (2014) Targeting Notch signaling pathway in cancer: clinical development advances and challenges. Pharmacol Ther 141(2):140–149. https://doi.org/10.1016/j.pharmthera.2013.09.005
2. Borggrefe T, Oswald F (2009) The Notch signaling pathway: transcriptional regulation at Notch target genes. Cell Mol Life Sci 66(10): 1631–1646. https://doi.org/10.1007/s00018-009-8668-7
3. Kopan R, Ilagan MXG (2009) The canonical Notch signaling pathway: unfolding the activation mechanism. Cell 137(2):216–233. https://doi.org/10.1016/j.cell.2009.03.045
4. Kopan R (2012) Notch signaling. Cold Spring Harb Perspect Biol 4(10):a011213–a011213. https://doi.org/10.1101/cshperspect.a011213
5. Jia D, Tamori Y, Pyrowolakis G, Deng W-M (2014) Regulation of broad by the Notch pathway affects timing of follicle cell development. Dev Biol 392(1):52–61. https://doi.org/10.1016/j.ydbio.2014.04.024
6. Hudson AM, Cooley L (2014) Methods for studying oogenesis. Methods 68(1):207–217. https://doi.org/10.1016/j.ymeth.2014.01.005
7. Jia D, Xu Q, Xie Q, Mio W, Deng W-M (2016) Automatic stage identification of *Drosophila* egg chamber based on DAPI images. Sci Rep 6(1): 18850. https://doi.org/10.1038/srep18850
8. Rowe M, Paculis L, Tapia F, Xu Q, Xie Q, Liu M, Jevitt A, Jia D (2020) Analysis of the temporal patterning of Notch downstream targets during *Drosophila melanogaster* egg chamber development. Sci Rep 10(1):7370. https://doi.org/10.1038/s41598-020-64247-2

Chapter 13

Employing the CRISPR Technology for Studying Notch Signaling in the Male Gonad of *Drosophila melanogaster*

Cordula Schulz

Abstract

The Notch (N) signaling pathway plays versatile roles in development and disease of many model organisms (Andersson et al., Development 138, 3593–3612, 2011; Hori et al., J Cell Sci, 126, 2135–2140, 2013). However, studying a role for N in adult tissue of *Drosophila melanogaster* is challenging, because the gene is located on the X-chromosome. To reduce the expression of N specifically in either the germline or the somatic cells of the adult gonad, we used the CRISPR technology in combination with the UAS/Gal4 expression system. After generation of the flies, gonads were investigated for germline survival. Here, we outline our detailed protocol.

Key words *Drosophila*, Notch signaling, Male gonad, Apoptosis, Immunofluorescence

1 Introduction

In *Drosophila melanogaster*, a plethora of tools were engineered to study development and adult tissue function, including the use of temperature sensitive alleles, mosaic analysis, and tissue-specific expression of target genes via the UAS/Gal4 system [3, 4]. In the UAS/Gal4 system, the Yeast transcription factor, Gal4, is expressed under the control of an endogenous, tissue- or cell type–specific promotor. In the male gonad of *Drosophila*, clusters of developing germline cells, derived from germline stem cells, are surrounded by somatic cyst cells (Fig. 1a, b), and signaling between these two cell types is essential for spermatogenesis [5, 6]. To study these cell types, distinct Gal4-lines have been established that express Gal4 either from a germline promotor, *nanos*-Gal4 (*nos*-Gal4), or from soma promotors, such as *traffic jam*-Gal4 (*tj*-Gal4) [7, 8]. The target genes, on the other hand, are expressed under control of

Supplementary Information The online version contains supplementary material available at [https://doi.org/10.1007/978-1-0716-2201-8_13].

Dongyu Jia (ed.), *Notch Signaling Research: Methods and Protocols*,
Methods in Molecular Biology, vol. 2472, https://doi.org/10.1007/978-1-0716-2201-8_13,

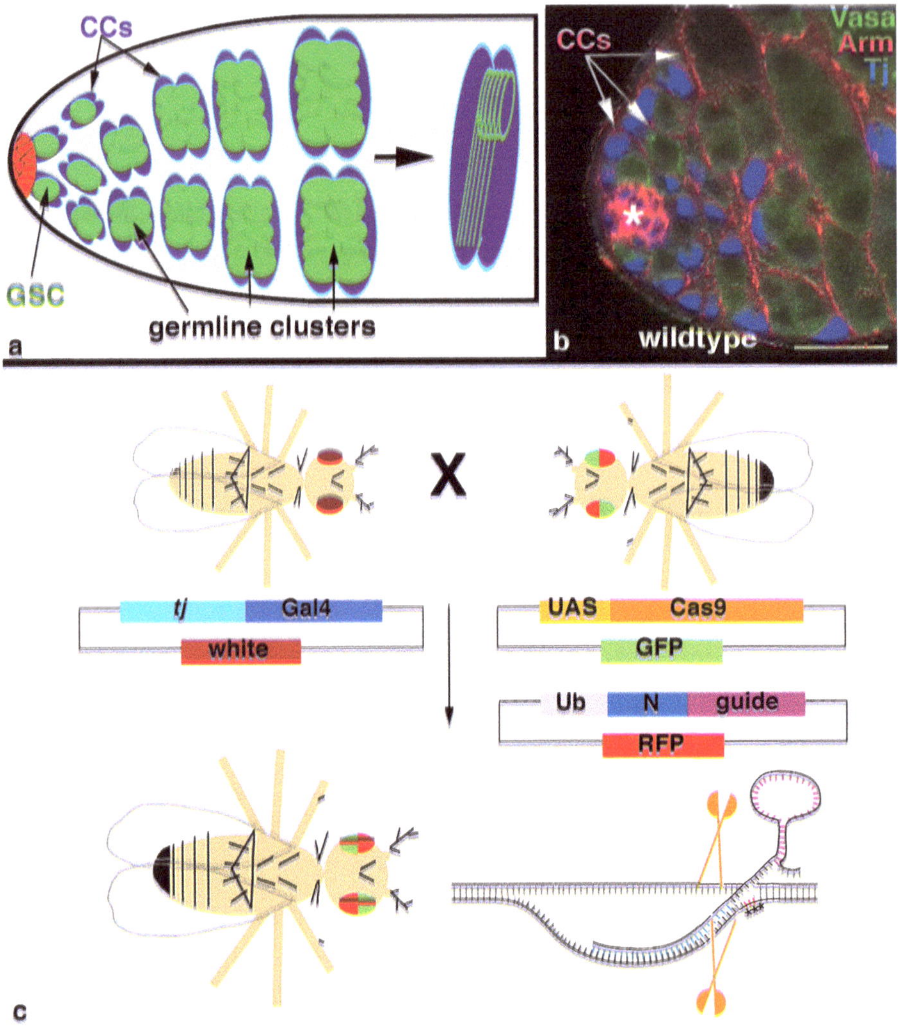

Fig. 1 CRISPR/Cas9 mediated conditional mutagenesis in the male gonad. (**a**) Cartoon illustrating spermatogenesis in *Drosophila*. Germline stem cells (GSCs) at the tip of the gonad produce daughter cells that proliferate into clusters of germline cells and eventually differentiate into sperm. Cyst cells (CCs) surround the germline cells. (**b**) The apical tip of a wildtype male gonad showing the germline cells in green and the surrounding cyst cells (arrows) in red with blue nuclei; stainings as indicated, asterix marks somatic hub cells, scale bar: 50 μm. (**c**) Top: Orange/red eyed virgins carrying the *tj*-Gal4 transactivator (w-/w-, *tj*-Gal4/*tj*-Gal4) were crossed to males carrying the UAS-Cas9 construct and the N guide-RNA (w/Y; Ub-N-guide/+; UAS-Cas/+). Bottom: Their progeny that carry all three constructs have orange/red eyes and express RFP and GFP in their eyes. Within the cyst cells of the gonad, Cas9 (indicated by orange scissors) recognizes the genomic-DNA-guide-RNA association and cuts the DNA prior to the PAM motif (NGG, in red and marked by asterisks)

the Yeast Upstream Activating Sequences (UAS). Binding of Gal4 to UAS activates transcription of the target genes specifically in the cell type in which the endogenous promotor is normally active [9, 10].

Previous studies using RNA Interference (RNA-i) and mosaic analysis with loss-of-function alleles indicated that *delta* (*dl*) is required within the germline cells for their survival [11]. However, none of the available UAS-RNA-i-lines for its receptor, N, caused an effect when expressed via *nos*-Gal4 or *tj*-Gal4. Thus, we turned toward the Clustered Regular Interspaced Short Palindromic Repeat (CRISPR) technology [12]. This system requires a guide-RNA that contains the target gene sequence and a palindromic sequence. For the system to work, the genomic target gene sequence has to be two to six nucleotides upstream of a photospacer adjacent motif (PAM). When the guide RNA associates with genomic DNA, this organization is recognized by the CRISPR associated protein 9 (Cas9). Cas9 will cut the genomic DNA three to four nucleotides upstream of PAM and introduce mutations. In *Drosophila*, the guide-RNA is normally expressed from a ubiquitous promotor, for example U6B or CR7T, while the Cas9 is under control of UAS (Fig. 1c) [13, 14]. Using a previously generated construct (Ub-N-guide), we were able to reproduce the mutant phenotype caused by loss of *dl* from the germline by knocking down N from the cyst cells [11].

2 Materials

All solutions and fly media should be prepared with ultrapure water. Unless otherwise noted, solutions need to be autoclaved and stored at room temperature. Fly media are stored at 4 °C and warmed up to room temperature prior to use.

2.1 Fly Work

1. 18 °C, (25 °C), and 29 °C incubators.
2. Dissection microscope equipped a with blue LED light, filters for red and green fluorescent proteins (RFP and GFP), and a CO_2-pad (*see* **Note 1**).
3. Fly bottles and vials with standard fly food (*see* **Note 2**).
4. Egg lay bottles, tape, and apple juice/agar plates: 15.94 g agar, 365 ml apple juice, 365 ml H_2O (*see* **Note 3**).
5. Fruit-flies carrying the guide-RNA, the UAS-Cas, and the Gal4-transactivators (*see* **Note 4**).
6. Instant food (Carolina Biological Supply Company) or filter paper.

2.2 Immunofluorescence Procedure

1. Fluorescent microscope.
2. Rocking platform.
3. Dissection dish and forceps.
4. Slides and coverslips.

5. Nail polish.
6. 37% Formaldehyde and designated chemical waste bottle
7. Triton X-100.
8. 10×TIB: 1.83 M KCl, 0.47 M NaCl, 100 mM Tris, pH 6.9
9. 10×PBS: 1.37 M NaCl, 27 mM KCl, 100 mM Na_2HPO_4, 18 mM KH_2PO_4, pH 7.4
10. TUNEL kit: In Situ Cell Death Detection Kit, TMR, red (Sigma-Aldrich, 12156792910; *see* **Note 5**).
11. 37 °C incubator or heat block
12. Primary antibodies: rat anti-Vasa (Developmental Studies Hybridoma Bank #anti-Vasa), guinea pig anti-Tj (D. Godt, U Toronto).
13. Secondary antibodies: donkey anti-rat, goat anti-guniea pig coupled with Alexa Fluor 488 (Invitrogen #A-11006, A-11073).
14. Mounting medium: SlowFade Gold antifade reagent with DAPI (Invitrogen #S36942, *see* **Note 6**).

3 Methods

3.1 Fly Work

Unless otherwise noted, grow and maintain flies on a standard cornmeal–molasses diet at room temperature, or in a 25 °C incubator. Set up single animal crosses in vials. For multi animal crosses (>40 females) and to amplify flies, place them on egg lays, replace the apple juice/agar plates every 24–48 h, and transfer the apple juice/agar with progeny into food bottles where they can develop into adults (you can get 4–5 bottles out of 1 cohort of flies, *see* **Notes 7–9**). Flies are anesthetized with CO_2 and sorted under a dissection scope. Orange/red eyes and other morphological markers can be observed with a regular light source, while fluorescent eyes can be seen with LED light and optic filters.

3.1.1 Establishment of Ub-N-Guide Stocks

1. Allow animals injected with Ub-N-guide to develop into adult flies, collect them as *virgins*, and cross them to *white* (*w*) mutant, *virgin* flies in single animal crosses. (*see* **Notes 10** and **11**).
2. Score progeny from each vial for RFP in the eye. Those lines with RFP-positive flies are the lines that you want to keep (*see* **Note 12**).
3. From the positive lines, cross *single*, virgin RFP-positive progeny (ideally males) to *w* mutant virgin flies to amplify them. If you don't have enough positive lines (10 is ideal), you can use more than one progeny with RFP from the same line in

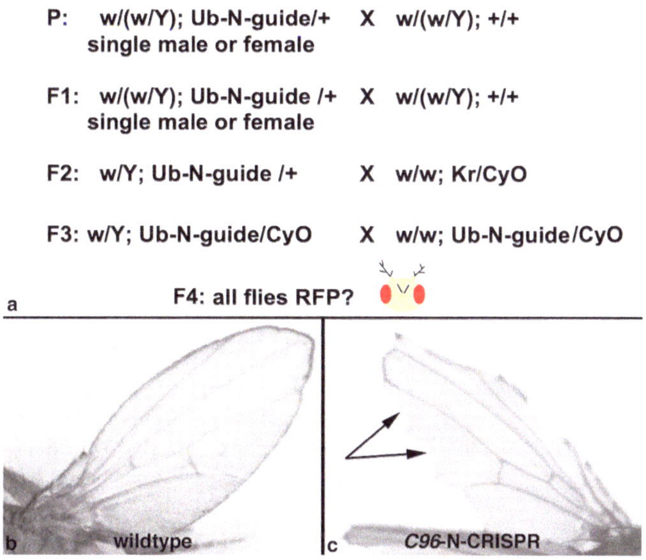

Fig. 2 Establishing and testing the Ub-N-guide-construct. (**a**) Crossing scheme for the establishment of the Ub-N-guide stocks. *Kr:* Krueppel (have small, rough eyes and segmentation defects), *CyO:* Curly of Oyster Balancer (curled wings), *w:* white eyes. (**b**) A wildtype wing. (**c**) Wing of a fly expressing the Ub-N-guide construct in the wing margin using the *C96*-Gal4 transactivator, arrows point to notched margins

separate single animal crosses, but keep track of their parental origin (for example, label your crosses 1a, 1b, etc., *see* **Note 11**)

4. From each vial in **step 3**, cross multiple RFP-positive, male progeny to balancer, virgin females in vials to determine on which chromosome the transgene is inserted and to establish fly stocks. Let the progeny develop and collect virgins with RFP eyes and the balancer you are using. Mate them to each other and score the progeny in the next generation. Figure 2a shows the crossing scheme with an example of how to established a stock for second chromosomal insertions.

3.1.2 Generation of Ub-N-Guide; UAS-Cas9 Flies

1. Cross males that carry the Ub-N-guide to virgin females that carry the UAS-Cas9 insert on another chromosome either at room temperature or in a 25 °C incubator (*see* **Note 13**).
2. Score for male progeny expressing RFP and GFP in the eye. These are used for the next cross. If you intend to use these flies only for one set of experiments, there is no need to establish a Ub-N-guide; UAS-Cas9 fly stock (*see* **Note 14**).

3.1.3 Testing the Efficiency of the Ub-N-Guide in Wings

1. Cross virgin females carrying the wing transactivator, *C96*-Gal4, to Ub-N-guide; UAS-Cas9 males (*see* **Note 13**), and grow the progeny in a 29 °C incubator [15].
2. Score hatched males for the expression of RFP, GFP, and for orange/red eyes. If the Ub-N-guide works well, animals with all three eye markers will display wing margin defects (compare Fig. 2b, c; *see* **Note 15**).

3.1.4 Generation of CRISPR Experimental and Control Flies for Investigating the Role of N in the Male Gonad

1. Cross virgin females either carrying *tj*-Gal4 on the second chromosome or *nos*-Gal4 on the third chromosome to Ub-N-guide; UAS-Cas9 males and place the cross and the progeny in a 18 °C incubator (*see* **Note 13**). Temporal control prevents Cas9 expression from the Gal4-transactivators during development as Gal4 has little to no activity at low temperatures (*see* **Note 16**).
2. After hatching, collect the progeny. Males with RFP, GFP, and orange/red eyes are the experimental flies and their siblings are the controls. To induce high levels of Gal4-activity, shift the males to 29 °C for 2 weeks (*see* **Note 16**). Then follow the immunofluorescence procedure described below.

3.2 Immunofluorescence Procedure

Unless otherwise noted, all steps are performed at room temperature and all tissue washes and incubations are performed on a rocking platform.

3.2.1 Preparation of Buffer Solutions

1. 10×TIB (500 ml): weigh out 68.2 g KCl, 13.75 g NaCl, and 6 g tris base, place salts into a 1000 ml bottle and add H_2O to 500 ml, close bottle and shake, measure pH. It should be at 6.9.
2. 1×TIB (1000 ml): in a 1000 ml measuring cylinder, measure out 900 ml of H_2O. In a 100 ml measuring cylinder, measure out 100 ml of 10×TIB and add to the H_2O, cover with Parafilm and mix solution by inverting the cylinder several times, pour into appropriate bottles (*see* **Note 17**).
3. 10×PBS (500 ml): weigh out 40 g NaCl, 1 g KCl, 7.2 g Na_2HPO_4, and 1.2 g KH_2PO_4, pour salts into a 1000 ml bottle and add H_2O to 400 ml, close and shake the bottle, adjust pH to 7.4 using 50% HCl and/or 10×NaOH as needed, add H_2O to 500 ml, close bottle and mix.
4. 10% Triton X-100 (100 ml): mix 10 ml of Triton X-100 with 90 ml H_2O in a 200 ml bottle until Triton has dissolved. After autoclaving, wrap the bottle with aluminum foil to prevent exposure to light.
5. 1× PBT (1000 ml): measure out 890 ml H_2O in a 1000 ml measuring cylinder, measure out 100 ml 10×PBS in a 100 ml measuring cylinder and add to H_2O, measure out 10 ml of 10%

Triton X-100 with a 10 ml pipette and add to the solution, cover measuring cylinder with parafilm and mix solution by inverting the cylinder several times, pour into appropriate bottles (*see* **Note 17**).

6. PBT + 3% BSA (50 ml): weigh out 1.5 g BSA in a sterile 50 ml tube, add PBT up to the 50 ml mark, place on a rocking platform at room temperature until all BSA is dissolved, if necessary, add PBT to the 50 ml mark, do not autoclave, can be stored at 4 °C for up to 1 week.

3.2.2 Dissection and Fixation

1. For each genotype, label a 1.5 ml tube and add 1.3 ml 1×TIB.
2. Set up the dissection station: place dissection dish with 1×TIB under dissection scope, place bucket with ice on the side of the station where your dominant hand is, place paper towels for discarding the carcasses (fly bodies) and forceps on each side of the microscope. You can also place a petri dish with tap water on the side of the station where your nondominant hand is and use this for discarding the carcasses. Place an open petri dish on the ice, and place the first labeled, open 1.5 ml tube corresponding to the genotype to be dissected on ice.
3. Put flies to sleep on CO_2 (*see* **Note 1**). Once the flies hold still, place them inside the petri dish on ice. Close the petri dish with its lid until you are sure the flies will not wake up and/or move around (*see* **Note 18**).
4. Grab a single male with the forceps in your dominant hand. Push the fly to the bottom of the dish. Make sure the fly is fully emerged in TIB and that your hands are resting on the bench. Then grab the fly with the other forceps at the beginning of the abdomen. Use the forceps in your dominant hand to pull out the gonads. Discard the carcass, and remove and discard all remaining tissue (cuticle, gut) except for the reproductive tract. While doing so, hold the reproductive tract with the forceps at the vas deferens, the white, triangular tissue that is connected to the end of the abdomen (Supplementary Movie 1). Still using the forceps in your dominant hand immediately transfer the gonads into the 1.5 ml tube. Keep collecting the tissue for no more than 30 min. (You may have to go through several rounds of this in order to collect enough tissue for one staining.)
5. Place the tube with the tissue in a rack at room temperature. Let the gonads settle to the bottom of the tube (*this step is necessary prior to the removal of the solution for any of the following steps or you will lose the tissue!!!*). Use a 1 ml pipette to remove the TIB from the tube—but not the gonads. It helps to leave some of the solution in the tube, for example you could always pipette away the solution until you reach the 100 μl

mark of the tube. Add 1 ml of PBT. Close lid and invert the tube 2–3 times. Once the gonads have settled remove PBT. Put on gloves. Add PBT so that the tissue is in 1 ml of PBT. Add 140 μl of 37% Formaldehyde. Close lid firmly.

6. Fix gonads for 30 min.
7. Rinse gonads with 1 ml PBT: Once gonads have settled remove the Fix-solution (*goes into chemical waste bottle!!*). Then add 1 ml fresh PBT. Invert the tube 2–3 times.
8. Repeat the rinse two more times with 1 ml PBT.
9. Take off gloves. Place tube on rocking platform until all samples are fixed and washing in PBT. If applicable, combine tissue of the same genotype. If necessary, gonads can be stored in PBT at 4 °C for up to 14 days.

3.2.3 TUNEL

1. For each tube of gonads you intend to stain, make a 500 μl active TUNEL solution by combining and mixing the contents of one tube A and one tube B of the kit.
2. After gonads settled, remove as much of the PBT as possible (you can use a 1 ml pipette to remove most of the PBT, followed by a smaller pipette to remove the remainder of the solution without pipetting out the tissue). Add the active TUNEL solution to each tube, and incubate at 37 °C for 1 h. This step can be done without rocking but invert the tubes every few minutes. Once a fluorescent tag, such as the TMR red or tagged secondary antibodies, is added the tubes should be kept in the dark. This can be done by placing a small cardboard box over them.
3. Place tubes back to room temperature, let them settle in a rack, and rinse three times with 1 ml PBT.

3.2.4 Antibody Staining

1. Wash tissue 3 × 20 min in 1 ml PBT.
2. Remove PBT and add 1 ml PBT + 3% BSA. Incubate for 30 min.
3. Prepare primary antibody solution in PBT + 3% BSA. Total volume per tube is 500 μl but as you leave 100 μl of solution on the gonads you need to calculate 400 μl solution for each tube (*see* **Note 19**).
4. Remove PBT + 3% BSA and add antibody solution. Incubate over night at 4 °C.
5. Place tubes back to room temperature, let gonads settle, and rinse three times with 1 ml PBT.
6. Wash tissue at least 3 × 20 min in 1 ml PBT.
7. Prepare secondary antibody solution in PBT + 3% BSA. Again: the total volume per tube is 500 μl but as you leave 100 μl of

solution on the gonads you need to calculate 400 μl solution for each tube (*see* **Note 19**).

8. Remove PBT + 3% BSA and add secondary antibody solution. Incubate for 2 h.
9. Rinse tissue three times with 1 ml PBT.
10. Wash tissue 3 × 20 min in 1 ml PBT.

3.2.5 Embedding and Imaging

1. Let the tissue settle and remove about half of the PBT. Place a slide on the bench. Cut off the tip of a 1 ml pipette tip. Set the volume to 1 ml. Pipette PBT with the tissue into the tip but hold the pressure of the pipette with your thumb once all the solution is inside the tip. Hold the pipette straight downward and let the tissue settle to the bottom of the tip where a drop will form. If necessary carefully release some pressure from the pipette so that the drop does not fall of too early. Once all tissue is inside the drop carefully place the drop with the tissue onto the slide.
2. Place slide under the dissection scope and remove any fibers that my have gotten in your sample.
3. Pile the tissue in one spot and use a Kimwipe to remove all buffer.
4. Place two drops of embedding medium on top of the tissue pile.
5. Use your forceps to carefully distribute embedding medium and the tissue on the slide such that it fits under a 20 × 30 mm coverslip. Add a coverslip.
6. Label the slide and nail polish around the sides of the coverslip.
7. Store in the dark.
8. Staining can be viewed and images taken with a fluorescent microscope equipped with proper filters and a camera (Fig. 3, *see* **Note 20**).

4 Notes

1. If you do not have a fluorescent set-up on your dissection scope, you can either purchase LED lights and filters from a microscope company, or you can build your own set-up using blue LED clamp lights, and plastic or glass optic filters [16]. If you do not have a CO_2 pad, you can use ether or Fly Nap (Carolina Biological Supply Company) to anesthetize the flies. Alternatively, you can immobilize them by cold but be aware that they remobilize the moment they warm up, so you have to work in a cold room or build a cold pad.

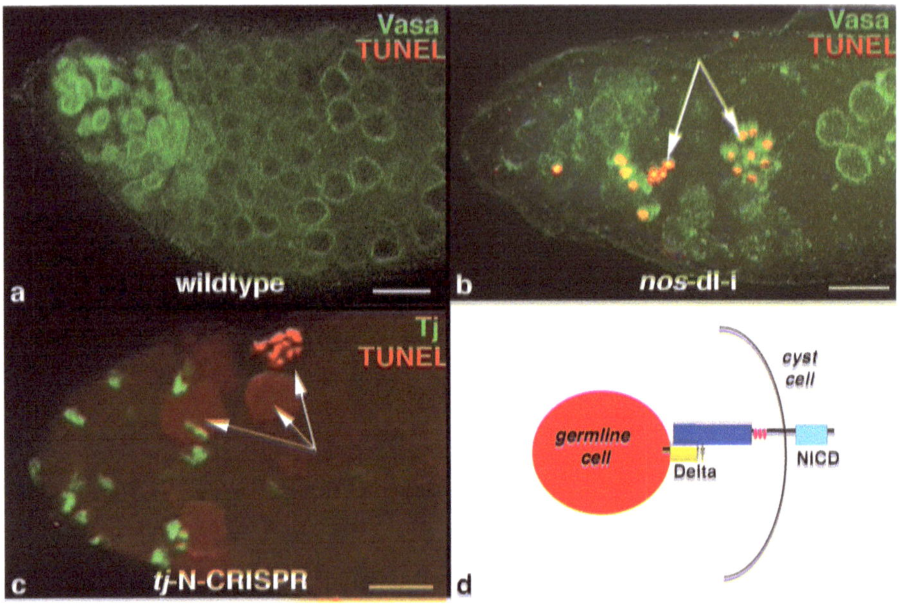

Fig. 3 N signaling prevents apoptosis. (**a–c**) Apical regions of male gonads from (**a**) wildtype and (**b**) flies expressing RNA-i against *dl* in the germline showing germline cells in green and (**c**) flies expressing CRISPR against N in the cyst cells showing cyst cells in green. Dying germline cell clusters in red (arrows), stainings as indicated, scale bars: 50 μm. (**d**) Cartoon illustration that germline cells signal via the transmembrane ligand Dl to the transmembrane receptor N on the cyst cells

2. Containers with or without fly food and closures can be purchased from a company, such as Genesee Scientific, from a facility at your research institution, or you can order supplies and prepare the food yourself. Example for fly food recipes can be found at bdsc.indicana.edu.
3. For egg lays, you can use empty plastic fly food bottles as containers (poke small holes on the sides for oxygen supply) and 35 mm cell culture dishes as plates (Corning 430165). To make apple juice/agar plates mix the ingredients in a beaker and bring to a boil in the microwave, watch it as *agar tends to boil over*. You may have to whisk and boil the solution several times until the agar is completely dissolved. Lay out plates on a bench (they should not touch each other or the solution will spill). Once the solution has cooled down to a point where it has a solidified top layer, remove the top layer with a spatula or spoon. Pour solution into small beaker and from there into the plates (pour enough to create a concave surface). After they solidified, apple juice/agar plates can be stored in a closed plastic container at 4 °C and dried off with a paper towel prior to use.
4. *nos*-Gal4 (BL#4937), *C96*-Gal4 (BL#43343), and *tj*-Gal4 (K#104055) are available from the Bloomington and Kyoto stock centers. The UAS-Cas9 flies were a gift from Ting Xie, yet

other UAS-Cas9 stocks are available from the stock centers. If knocking down a target gene in each cell type of the gonad, it is a advisable to use the same fly stocks for the experiments, except for the different Gal4-lines. If the goal is to knock down a gene of interest only in the germline cells, alternative fly lines can be obtained from the stock centers or the fly community. For example, fly stocks in which Cas9 is directly driven under a germline promoter, such as *vasa* (*vas*-Cas9), are available and can be directly crossed to flies carrying the guide-RNA. If flies with the guide-RNA are not available, they need to be created. We obtained the Ub-N-guide construct from Guanjun Gao [13]. To get the constructs inserted into flies, a commercial service can be used, such as The Best Gene or Rainbow.

5. Other in Situ Cell Death Detection Kits, such as the Fluorescein coupled kit (Roche/Sigma-Aldrich 11684795910), also work well in the male *Drosophila* gonad.
6. Other mounting media, such as Vectashield with DAPI (Vectashield, H-1200), are equally good for embedding *Drosophila* gonads.
7. Apple juice agar plates should be supplemented with yeast paste (promotes egg laying and provides Vitamin A). For this, dissolve dry yeast in water to create a smooth paste. Then place a bean-sized amount of it *onto the agar side* of the apple juice/agar plate (the agar side is soft and the plastic side is hard). After transferring the flies into the egg lay bottles (via a funnel), close the egg lays by placing the apple juice/agar plate onto the opening of the egg lay bottle such that the agar and yeast paste are inside. Then, use tape to tighten and seal the egg lay.
8. To avoid contamination when transferring the progeny into fly bottles, use a clean wooden stick or spatula for each plate.
9. If fly vials or bottles become too moist, remove the apple juice/agar (if applicable) with a clean spoon and add some instant food, or push a filter paper into the food.
10. Females must be collected as virgins because they can store sperm from a previous mating in their reproductive system. If you are new to flies, then look up the fly database and the literature for basic information on fly work [17].
11. The injected flies potentially have transformed gametes. As transformation is random, insertion of the DNA can happen within one cluster of developing germline cells, or in two or more clusters, or even in two independent events within one cluster. As the sibling progeny expressing RFP in the eye could be genetically distinct in each of the lines you want to perform single animal crosses. It is recommended to place single animal crosses in glass vials as, in our experience, flies develop/survive

less well in plastic vials. To get the most progeny out of a cross you can cross males to several virgin females; however, after 2–3 rounds of mating the males run out of accessory gland proteins that are essential for fertility and need to be reproduced [18]. Therefore, males should not be placed with too many virgin females at the same time. Ideally you want to add 2 fresh female virgins to the vial for each male every 2–3 days. When using single virgin females for a cross, place them with 2–3 males. If the female does not produce enough progeny for the vial to take, you can add several mated *w* mutant females and sort out the correct progeny based on the orange/red eyes. Once progeny develop, the food gets sticky. Therefore, the adult flies should be transferred to a fresh vial in which they can continue to reproduce. Label the vials accordingly.

12. If you used a commercial service for injection, they may provide you with the RFP-positive lines so that you can skip **steps 1** and **2**. You could also have them establish fly lines, in which case, you can skip the whole procedure described in Subheading 3.1.1.
13. Unless one of the transgenes is located on the X-chromosome, the direction of the cross is not important. However, it may be faster to amplify flies and collect virgins from the UAS-Cas and Gal4-lines than from the previous cross.
14. If you intend to establish a Ub-N-guide; UAS-Cas9 fly stock; you first need to cross the flies into a genetic background that allows you to follow each chromosome based on a visible marker, and to balance the flies to prevent recombination. For example, you could cross the Ub-N-guide and the UAS-Cas9 flies each to w/w; Kr/CyO; ftz,e/TM6 Hum, e flies. You want to set up these crosses with at least 50 virgin females, because the desired progeny are rare. In the next generation, you harvest w/Y; UB-N-guide/CyO; +/TM6, Hum, e males (RFP eyes, curled wings, extra bristles on the humerus) and w/w; +/CyO; UAS-Cas9/TM6, Hum, e virgin females (GFP eyes, curled wings, extra bristles on the humerus) and cross them to each other. Cross the virgin progeny with RFP and GFP in their eyes, curled wings, and extra bristles on the humerus to each other to establish the fly stock (w/w(Y); Ub-N-guide/CyO; UAS-Cas9/TM6, Hum, e).
15. Wings from wildtype and experimental flies can be harvested for molecular studies, if needed [19].
16. If the low temperature of 18 °C is not sufficient to prevent Cas9 expression during development, then a Yeast Gal80 construct (UAS-GAL80^{ts}, available from the Bloomington Drosophila Stock Center) can be added to the flies that binds to Gal4 at low temperatures and prevents it from binding to

UAS [20]. Should the effect of the Gal4 at 29 °C be too strong (cause lethality or other decremental effects) then the UAS/Ga4 flies can be shifted to lower temperatures, for example to 25 °C or room temperature [21].

17. Consider to aliquot 50 ml of 1×TIB into 100 ml bottles, and 100 ml of 1×PBT into 200 ml bottles prior to autoclaving. This size aliquots are big enough for one round of dissection and staining.
18. The petri dish can be closed at any time during the procedure to reimmobilize flies.
19. To stain the germline cells, use 350 μl PBT + 3% BSA and 50 μl rat anti-Vasa for each tube. To stain the cyst cells, use 400 μl PBT + 3% BSA and 1 μl guinea pig anti-Tj for each tube (as Tj is used at 1:5000 you first have to make a 1:10 stock dilution). Invitrogen secondary antibodies work as 1:1000. For each tube, use 400 μl PBT + 3% BSA, and 0.5 μl secondary antibody.
20. The fluorescent microscope has to be equipped with a UV source, and filters for the excitation and emission waves of Alexa 488 and 568.

Acknowledgments

The author thanks Robert Ng for technical assistance and Heath Aston, Nithin Reddy, Trang Bui, Austin Waits, and Raahim Islam for proofreading of the manuscript. This work was funded by NSF grants #0841419 and #1355009.

References

1. Andersson ER, Sandberg R, Lendahl U (2011) Notch signaling: simplicity in design, versatility in function. Development 138:3593–3612
2. Hori K, Sen A, Artavanis-Tsakonas S (2013) Notch signaling at a glance. J Cell Sci 126: 2135–2140
3. Suzuki DT (1970) Temperature-sensitive mutations in Drosophila melanogaster. Science 170:695–706
4. del Valle RA, Didiano D, Desplan C (2011) Power tools for gene expression and clonal analysis in Drosophila. Nat Methods 9:47–55
5. Hardy RW, Tokuyasu KT, Lindsley DL, Garavito M (1979) The germinal proliferation center in the testis of Drosophila melanogaster. J Ultrastruct Res 69:180–190
6. Zoller R, Schulz C (2012) The Drosophila cyst stem cell lineage: partners behind the scene? Spermatogenesis 2:145–157
7. Van Doren M, Williamson AL, Lehmann R (1998) Regulation of zygotic gene expression in Drosophila primordial germ cells. Curr Biol 8:243–246
8. Li MA, Alls JD, Avancini RM, Koo K, Godt D (2003) The large Maf factor Traffic Jam controls gonad morphogenesis in Drosophila. Nat Cell Biol 5:994–1000
9. Brand AH, Perrimon N (1993) Targeted gene expression as a means of altering cell fates and generating dominant phenotypes. Development 118:401–415
10. Phelps CB, Brand AH (1998) Ectopic gene expression in Drosophila using GAL4 system. Methods 14:367–379
11. Ng CL, Qian Y, Schulz C (2019) Notch and Delta are required for survival of the germline stem cell lineage in testes of Drosophila melanogaster. PLoS One 14:e0222471

12. Pickar-Oliver A, Gersbach C (2019) The next generation of CRISPR-Cas technologies and applications. Nat Rev Mol Cell Biol 20: 490–507
13. Xue Z, Wu M, We K, Ren M, Long L, Zhang X, Gao G (2014) CRISPR/Cas9 mediates efficient conditional mutagenesis in Drosophila. G3 (Bethesda) 4:2167–2173
14. Gratz SJ, Rubinstein CD, Harrison MM, Wildonger J, O'Connor-Giles KM (2015) CRISPR-Cas9 genome editing in Drosophila. Curr Protoc Mol Biol 111:31.2.1–31.2.20. https://doi.org/10.1002/0471142727.mb3102s111
15. Gustafson K, Boulianne GL (1996) Distinct expression patterns detected within individual tissues by the GAL4 enhancer trap technique. Genome 39:174–182
16. Angulo J, Astin C P, Bauer O, Blash K J, Bowen N M, Chukwudinma N J, Dinofrio A S, Faletti D O, Ghulam A M, Gusinde-Duffy C M, Horace K J, Ingram A M, Isaack K E, Jeong G, Kiser R J, Kobylanski J S, Long M R, Manning G A, Morales J M, Nguyen K H, Pham R T, Phillips M H, Reel T W, Seo J E, Vo H D, Wukuson A M, Yeary K A, Zheng G Y, Lukowitz W (2020) Targeted mutagenesis of the *Arabidopsis* GROWTH-REGULATING FACTOR (GRF) gene family suggests competition of multiplexed sgRNAs for Cas9 apoprotein. BioRxiv. https://doi.org/10.1101/2020.08.16.253203
17. The Flybase Consortium (2003) The FlyBase database of the Drosophila genome projects and community literature. Nucleic Acids Res 31:172–175
18. Sirot LK, Buehner NA, Fiumera AC, Wolfner MF (2009) Seminal fluid protein depletion and replenishment in the fruit fly, Drosophila melanogaster: an ELISA-based method for tracking individual ejaculates. Behav Ecol Sociobiol 63:1505–1513
19. Carvalho GB, Ja WW, Benzer S (2009) Non-lethal PCR genotyping of single Drosophila. BioTechniques 46:312–314
20. Nogi Y, Shimada H, Matsuzaki Y, Hashimoto H, Fukasawa T (1984) Regulation of expression of the galactose gene cluster in Saccharomyces cerevisiae. II. The isolation and dosage effect of the regulatory gene GAL80. Mol Gen Genet 195:29–34
21. Duffy JB (2002) GAL4 system in Drosophila: a fly geneticist's Swiss army knife. Genesis 34: 1–15

Chapter 14

Immunolocalization of Notch Signaling in Mouse Preimplantation Embryos

Elisabete Silva, Patrícia Diniz, Alexandre Trindade, Mariana Batista, Ana Torres, António Duarte, and Luís Lopes-da-Costa

Abstract

The Notch signaling pathway is an important determinant of cell diversity and identity in most developing embryonic tissues. The pathway components are expressed dynamically, and their function is critical for embryonic survival.

This protocol addresses the immunolocalization of Notch pathway components by confocal microscopy.

Key words Notch signaling, Mouse, Preimplantation embryo, Immunocytochemistry, Confocal microscopy, Notch intracellular domain (NICD)

1 Introduction

The Notch signaling pathway is highly conserved across metazoans. It is a molecular mechanism used throughout evolution to define cellular fates breaking symmetry in populations of uniform cells, through interactions between adjacent cells [1, 2]. In mammals, the Notch pathway family is composed of four transmembrane receptors (Notch1–4), three Delta-like ligands (Dll1,3–4) and two Jagged ligands (Jagged1,2). Notch receptor activation can lead to decreased ligand expression, by a process termed lateral inhibition, regulating the cell fate of several neighbor cells by binary fate choice. Alternatively, Notch receptor activation can lead to increased ligand expression, creating a wave of cells differentiating to a common fate, by lateral induction [2].

The Notch pathway is required throughout embryonic and fetal development for the proper differentiation of tissues, organs, and systems, with some components being strictly required for embryonic

Patrícia Diniz and Alexandre Trindade contributed equally with all other contributors.

Dongyu Jia (ed.), *Notch Signaling Research: Methods and Protocols*,
Methods in Molecular Biology, vol. 2472, https://doi.org/10.1007/978-1-0716-2201-8_14,

survival. Although the functions of this pathway in postimplantation embryogenesis are well documented [3], the role in preimplantation embryogenesis is still the subject of intense research. In fact, the relevance of Notch components in early embryo development [4, 5] and their carryover effects on later embryo–fetal development have only recently been demonstrated [6], which may have relevant applications in reproductive medicine and assisted reproductive techniques. The spatial localization of Notch components can be evaluated by locating: (1) the mRNA transcription pattern using in situ hybridization or single-cell RNA sequencing; (2) reporter proteins expressed under the control of Notch component promoter sequences, using transgenic animal models; and (3) the protein expression, using immunofluorescence. The protocol here presented (Fig. 1) describes the latter approach, which is technically less

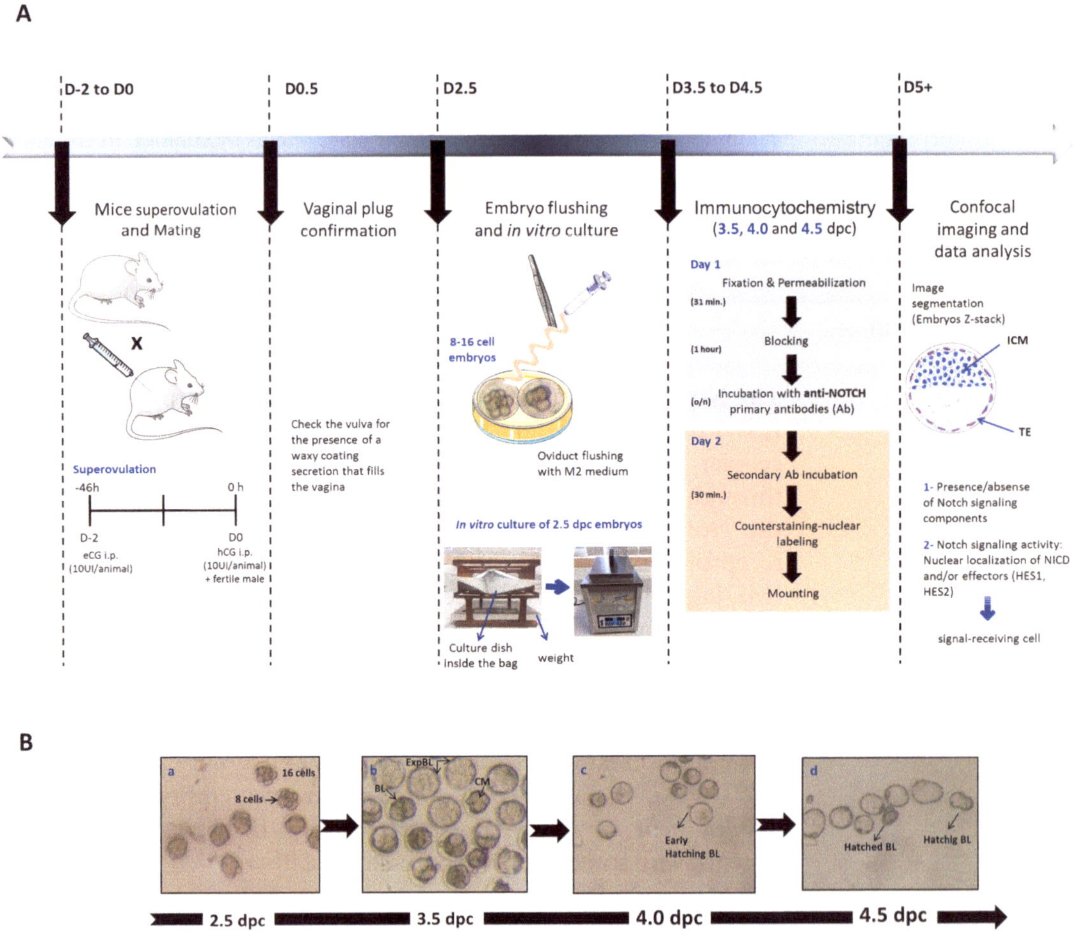

Fig. 1 Experimental procedure and protocol timeline. (**a**) Overview of the steps and duration of the protocol, from mice superovulation to analysis of Notch signaling in 3.5 dpc blastocysts. Experimental days (D) are counted from the moment of fertilization (D0). (**b**) Representative photographs illustrating the morphological staging of embryonic development. Embryos were in vivo collected at 2.5 dpc (**a**) and in vitro cultured until 4.5 dpc (**d**). This protocol illustrates the evaluation of Notch signaling in 3.5 dpc blastocysts. *CM* compact morula, *BL* blastocyst, *ExpBL* expanded blastocyst

challenging and provides the identification of the functional component of the pathway. Special attention will be given to the technical specificities that make it stand out from standard protocols, namely, due to the spherical nature of the preimplantation embryo and the presence of different cellular populations. It is also critical to distinguish between cells presenting full length receptors anchored in the membrane and cells where the Notch intracellular domain (NICD) or Notch effectors, from the Hes or Hey families [1, 2], can be located in the nucleus, to better interpret the activation dynamics of the pathway.

Although this protocol is specific for the mouse embryo, it can be adapted for other species, using appropriate antibodies.

2 Materials

2.1 Preimplantation Embryos

2.1.1 Mice Superovulation

1. Female mice (CD1), 2–3 months-old (Charles River, UK).
2. Male mice (CD1), 3 months and older (Charles River, UK).
3. Equine chorionic gonadotropin (eCG; Intergonan).
4. Human chorionic gonadotropin (hCG; Chorulon).
5. Syringes (1 mL) and needles (26 G; 5/8″).

2.1.2 Embryo Collection

1. Ketamine (Imalgene) and xylazine (Rompun).
2. Syringes (1 mL) and needles (26G; 5/8″).
3. Blunt tip needles (30G, 5/8″) and two-piece syringes (1 mL).
4. Stereoscopic binocular microscope.
5. Dissecting tools (fine-pointed scissors; micro forceps).
6. Sterile plastic petri dishes (35 mm, 90 mm).
7. Flushing (collection) and handling medium—M2 medium (with HEPES) (M-7167; Sigma).
8. CO_2 incubator at 37 °C and 5% CO_2 in air.
9. Stereoscopic binocular microscope, with upper stage illumination.
10. Aspirator tube assemblies for microcapillary pipettes (mouth pipettes).
11. Glass capillary (drawn Pasteur pipettes—plugged and flame-polished).
12. Warming plate.

2.2 Embryo In Vitro Culture

1. IVF Multidish 4 well (144,444, Nunc; Thermo Scientific).
2. Culture medium—KSOM (MR-101-DP, Millipore).
3. Mineral oil (ES-005-C, EmbryoMax®, Millipore).
4. Aspirator tube assemblies for microcapillary pipettes (mouth pipettes).

5. Glass capillary (drawn Pasteur pipettes—plugged and flame-polished).
6. Stereoscopic binocular microscope.
7. CO_2 Incubator at 37 °C and 5% CO_2 in air.
8. Foil/poly bags (10 × 20 cm) (Z183393, Sigma).
9. Water bath at 37 °C.

2.3 Immunocytochemistry

1. Cell culture dishes with and without inner rings (35 × 10 mm).
2. Phosphate-buffered saline (PBS).
3. Fixative solution: 4% paraformaldehyde (P6148, Sigma) in PBS.
4. Permeabilization solution: 0.5% Triton X-100 in PBS.
5. Blocking solution: 0.1% Tween 20 (EC-607, National Diagnostics) + 2.5% bovine serum albumin (BSA; A7906, Sigma) in PBS.
6. Primary Antibodies: anti-NOTCH1; anti-NOTCH2; anti-NOTCH3; anti-NOTCH4; anti-DLL1, anti-DLL4; anti-JAGGED1; anti-HES1; anti-HES2; Rabbit polyclonal IgG (Table 1).
7. Secondary antibodies: Alexa Fluor® 594 chicken anti-rabbit (Table 1).

Table 1
List of primary and secondary antibodies used in the immunocytochemistry procedure

	Species	Type	Dilution	Supplier (Reference)
Primary antibodies				
Anti-NOTCH1	Rabbit	Polyclonal	1:100	Abcam (ab8925)
Anti-NOTCH2	Rabbit	Polyclonal	1:100	Abcam (ab8926)
Anti-NOTCH3	Rabbit	Polyclonal	1:200	Abcam (ab23426)
Anti-NOTCH4	Rabbit	Polyclonal	1:50	Santacruz Biotechnology (sc5594)
Anti-DLL1	Rabbit	Polyclonal	1:100	Abcam (ab76655)
Anti-DLL4	Rabbit	Polyclonal	1:200	Abcam (ab7280)
Anti-JAGGED1	Rabbit	Polyclonal	1:50	Santacruz Biotechnology (sc8303)
Anti-HES1	Rabbit	Polyclonal	1:100	Abcam (ab71559)
Anti-HES2	Rabbit	Polyclonal	1:100	Abcam (ab134685)
Isotype control IgG	Rabbit	Polyclonal	1:50	Abcam (ab27478)
Secondary antibody				
Alexa Fluor 594 chicken anti-rabbit	Rabbit	Polyclonal	1:300	Life Technologies (A11012)

8. Hoechst 33258 (Sigma).
9. ProLong™ Gold Antifade reagent (P36930; Life Technologies).
10. Microscope slides.
11. Glass coverslips 24 × 24 mm.
12. Nail polish (transparent).

2.4 Image Acquisition

1. Confocal microscope.
2. Fiji software (National Institutes of Health).

3 Methods

3.1 Preimplantation Embryos

3.1.1 Mice Superovulation

Superovulation induces the release of a supraphysiologic number of oocytes, allowing for the reduction of donor animal numbers (3R's principle). The superovulatory response depends on several factors, including mouse strain, age, animal weight, and hormone dose [7]. Among these factors, for each strain, the weight of females and dose of eCG appear to be critical factors [7]. The protocol here presented (Fig. 1) was optimized for 2–3-months-old Crl:CD1 (ICR) female mice, with 20–30 g. Mice are kept on a 12-h light–dark cycle with ad libitum access to standard food and water in same sex and age groups (*see* **Note 1**)

1. Inject each female with 10 IU eCG intraperitoneally (i.p.) 6–7 h after the beginning of the mouse house light cycle (Fig. 1a). eCG stimulates follicle-stimulating hormone (FSH) activity in nonequid species.
2. Inject each female with 10 IU hCG (i.p.) 46 h after the eCG administration (Fig. 1a) (*see* **Note 2**). hCG stimulates luteinizing hormone (LH) activity and should be administered before the natural LH peak which occurs approximately 51 h after eCG administration. Ovulation should occur 10–13 h after the hCG administration.
3. House females overnight with stud mice (ideally one female per male). Keep a set of stud mice, slightly older than the females, with registered mating track record. Do not place more than 2 females with each male.
4. Check for the presence of a vaginal plug early the next morning (ideally up to 3 h after the beginning of the light cycle) with the help of a blunt atraumatic curved forceps to gently open the vulva, if necessary. Do this step early in the morning, as plugs are removed and/or dissolved over time, and can only be reliably detected in the first hours after copulation. Consider the morning of vaginal plug detection as 0.5 days postcoitum (dpc).

3.1.2 Embryo Collection and In Vitro Culture

Embryo Collection

Since embryo development in the mouse is relatively fast, timing of collection is important to guarantee that embryos are at the appropriate stage. As such, 8–16-cell stage embryos can be collected by oviduct flushing at 2.5 dpc (Fig. 1a).

1. Warm up 15 mL of M2 medium at 37 °C in an incubator (5% CO_2 in air). Incubate the KSOM medium and mineral oil in the same conditions. Collection medium has HEPES to maintain pH stability, and phenol red to visually control pH value. Correct pH with further incubations, if necessary.
2. Euthanize females by cervical dislocation under general anesthesia (150 mg kg^{-1} ketamine +10 mg kg^{-1} xylazine, i.p.). You may have to adjust this methodology to national and/or institutional regulations.
3. Perform a midline incision in the abdominal wall and expose the genital tract. Locate the ovary, dissect the ovarian fat and remove the adjacent oviduct with a small portion of ovary still attached. Separate the oviducts from the uterus by cutting the genital tract just cranially to the uterine-tubaric junction. Repeat the procedure for the contralateral oviduct (*see* **Note 3**).
4. Place each oviduct in a petri dish with a drop (25 μL) of prewarmed M2 medium.
5. Connect a sterile 30-gauge blunt tip needle to a 1 mL syringe filled with 0.5 mL of M2 medium.
6. Using the binocular stereoscopic microscope, with upper stage illumination, hold the infundibulum between the micro forceps and insert the tip of the needle (attached to the syringe) into the oviduct. Secure both the needle and the oviduct between the micro forceps and gently inject the M2 medium into the oviduct. The embryos are flushed out of the oviduct into the petri dish. Repeat the procedure for the other oviduct (*see* **Note 4**).
7. On a separate 50 μL droplet place both oviducts and gently, with the aid of two micro forceps, shred the oviducts to release any potential embryos that may have not been flushed. Let the embryos and tissue debris settle for 2–3 min before the next step.
8. Place the petri dish under the stereomicroscope with substage illumination and, using an aspirator tube attached to a glass capillary, collect the embryos from the medium droplets and immediately place them in a 35 mm petri dish with 2–3 mL fresh M2 medium.
9. At this step, embryo collection must be performed under a laminar flow cabinet, and media placed on a heated plate set for 37 °C to maintain optimal conditions for embryo maintenance outside the incubator.

10. From the previously collected embryo pool, select morphologically normal 8–16 cell embryos and place them in a petri dish with 2 mL of KSOM medium using an aspirator tube attached to a glass capillary.

In Vitro Culture of 2.5 dpc Embryos

1. Prepare a 4-well culture dish (IVF Multidish) with 500 μL KSOM per well. Keep it in the CO_2 incubator until needed.
2. Transfer the previously selected embryos to the culture dish in groups of 15–25 embryos per well. Cover each well with 400 μL of mineral oil.
3. Place approximately 1 mL of KSOM in the center of the culture dish and place the dish in a foil bag. Heat seal the bag and, with the help of a needle attached to the gas tank, inflate a mixture of 90% N_2 + 5% O_2 + 5% CO_2 humidified atmosphere (1 bar) into the bag (Fig. 2). The needle should perforate the bag at its corner in such a way it will be easy to heat seal the entry point once the modified atmosphere is inflated (*see* **Note 5**).
4. Incubate the culture dish in a water bath at 37 °C (Fig. 1a) (*see* **Note 6**).
5. After 24, 36, and 48 h of culture (corresponding to 3.5, 4.0, and 4.5 dpc, respectively), observe embryos in the culture wells under the stereomicroscope and classify the developmental stage using standard guidelines (Fig. 1b) [8]. During this procedure, if necessary, manipulate the embryos using an aspirator tube attached to a glass capillary and wash the tip on a separate petri dish with KSOM between wells to avoid introducing mineral oil droplets near the embryos, which can impair their visualization.

3.2 Immunocytochemistry

The immunocytochemistry protocol here presented includes the evaluation and preparation of 3.5 dpc blastocysts, fixation, permeabilization, blocking, primary and secondary antibody incubation, and nuclear staining (Fig. 1a). All procedures are performed under a stereoscopic microscope. Embryos are manipulated using an aspirator tube attached to a glass capillary, and a 35 mm petri dish with PBS must be available to wash glass capillaries whenever necessary.

1. Prepare a fresh solution of 4% paraformaldehyde (PFA) in PBS by adding 0.4 g of PFA to 10 mL of PBS and dissolving by heating in a >80 °C water bath in an airtight container. After the PFA is fully dissolved, let it cool to 4 °C.
2. Prepare a petri dish with inner rings by placing 100 μL drops of PBS in each ring (Fig. 2). Place the selected embryos in individual drops.
3. Wash the embryos in PBS by transferring them into a new PBS drop for 5 min. Repeat three times.

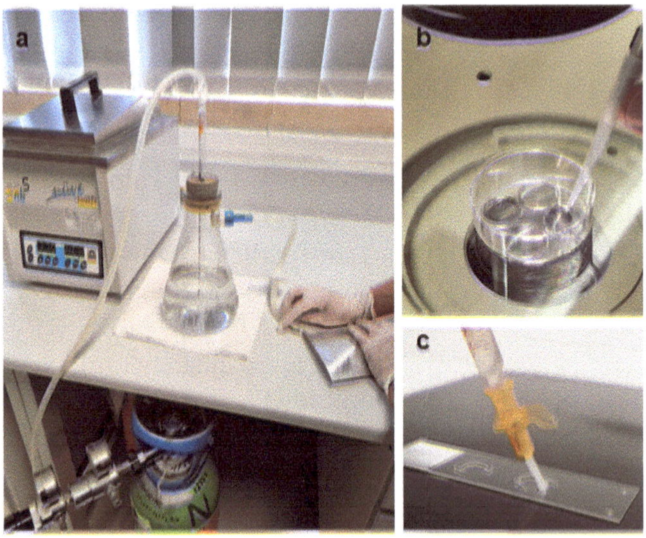

Fig. 2 Representative photos of specific steps of in vitro embryo culture and immunocytochemistry. (**a**) Infilling the foil bag with the humidified atmosphere; (**b**) petri dish with inner rings; (**c**) Preparation of microscope slides with chambers made with rings of nail polish

4. **Fixation:** prepare a petri dish (35 mm; w/o inner rings) with 2 mL of 4% PFA and transfer the embryos into it. Incubate for 30 min at 4 °C (*see* **Note 7**).
5. Wash the embryos in fresh PBS for 5 min.
6. **Permeabilization:** prepare a petri dish (35 mm; w/o inner rings) with 2 mL of 0.5% Triton X-100 in PBS. Place the embryos in this solution for 1 min and retrieve them quickly (*see* **Note 8**).
7. Wash in fresh PBS for 5 min.
8. **Blocking:** incubate the embryos in blocking solution for 1 h at room temperature in a humidified atmosphere (*see* **Note 9**). After blocking DO NOT WASH!
9. **Primary antibody incubation**: Prepare the appropriate dilutions of each antibody (Table 1) (*see* **Note 10**) in blocking solution. Place 100 μL drops of each antibody solution in a petri dish with inner rings. Transfer each group of embryos to their designated antibody drop. Incubate in a humidified atmosphere (*see* **Note 9**) overnight at 4 °C.
10. Wash the embryos in fresh PBS for 5 min (*see* **Note 11**). Repeat four times.
11. **Secondary antibody incubation:** Dilute Alexa Fluor® 594 chicken anti-rabbit secondary antibody in blocking solution (Table 1) (*see* **Note 12**). Prepare a petri dish with 100 μL of secondary antibody solution in each ring and transfer

embryos. Incubate for 30 min at room temperature, in a humidified atmosphere. From this moment on, decrease the stereomicroscope illumination to minimize photobleaching.

12. **Counterstaining-nuclear labeling:** wash the embryos once for 10 min in PBS, followed by incubation in 100 μL of Hoechst solution (0.02 mg mL^{-1} of Hoechst in PBS) for 10 min, and two final 10 min washes in PBS.
13. **Mounting:** prepare microscope slides by placing a nail polish ring (about 0.5 mm in diameter) on the slide (Fig. 2). The nail polish ring should be thick enough to build a chamber high enough to accommodate the tridimensionality of the embryo. Place a drop of ≤5 μL of mounting medium (ProLong™ Gold Antifade Mountant) in the centre of the ring.
14. Transfer the embryos into mounting medium with as little PBS as possible and let them settle apart from each other and sink to the bottom of the medium. Place the cover slide and seal the preparation with nail polish. After curing the sample for 24 h at room temperature in the dark on a flat, dry surface, store the slides at 4 °C until image acquisition, which should be performed as soon as possible, but can take up to 2 weeks without signal loss.

3.3 Confocal Imaging and Data Analysis

Here follows the example of a three-dimensional reconstruction of mouse blastocysts based on *z*-stacked images of whole-mount stained embryos using confocal microscopy and image analysis software (Fig. 3).

3.3.1 Confocal Imaging

1. Locate the embryo using low-magnification (100×) in an epi fluorescent blue filter and manually adjust focus until you detect cellular nuclei.
2. Increase the magnification, adjusting focus point as necessary to visualize the embryo, until you reach 1000× magnification.
3. Set up the image acquisition parameters using the microscope software: (a) define the top and bottom limits of the embryo *z*-stack; (b) set the thickness of each optical slice (0.5–6 μm depending on the objective) ensuring the quality of the 3D reconstruction, (c) adjust the intensity of the laser to ensure a good proportion of saturated pixels and dark pixels.

3.3.2 Data Analysis

The protocol here described allows for the detection of Notch pathway components in preimplantation embryos. Also, by evaluating the differential subcellular localization of Notch effectors and activated receptors (cleaved Notch intracellular domain, NICD), it is possible to depict the activation status of the pathway. By default, the nuclear localization of NICD indicates that the receptor was cleaved and translocated into the nucleus. This is a marker of Notch signaling activation, which contrasts to when NICD is found in the

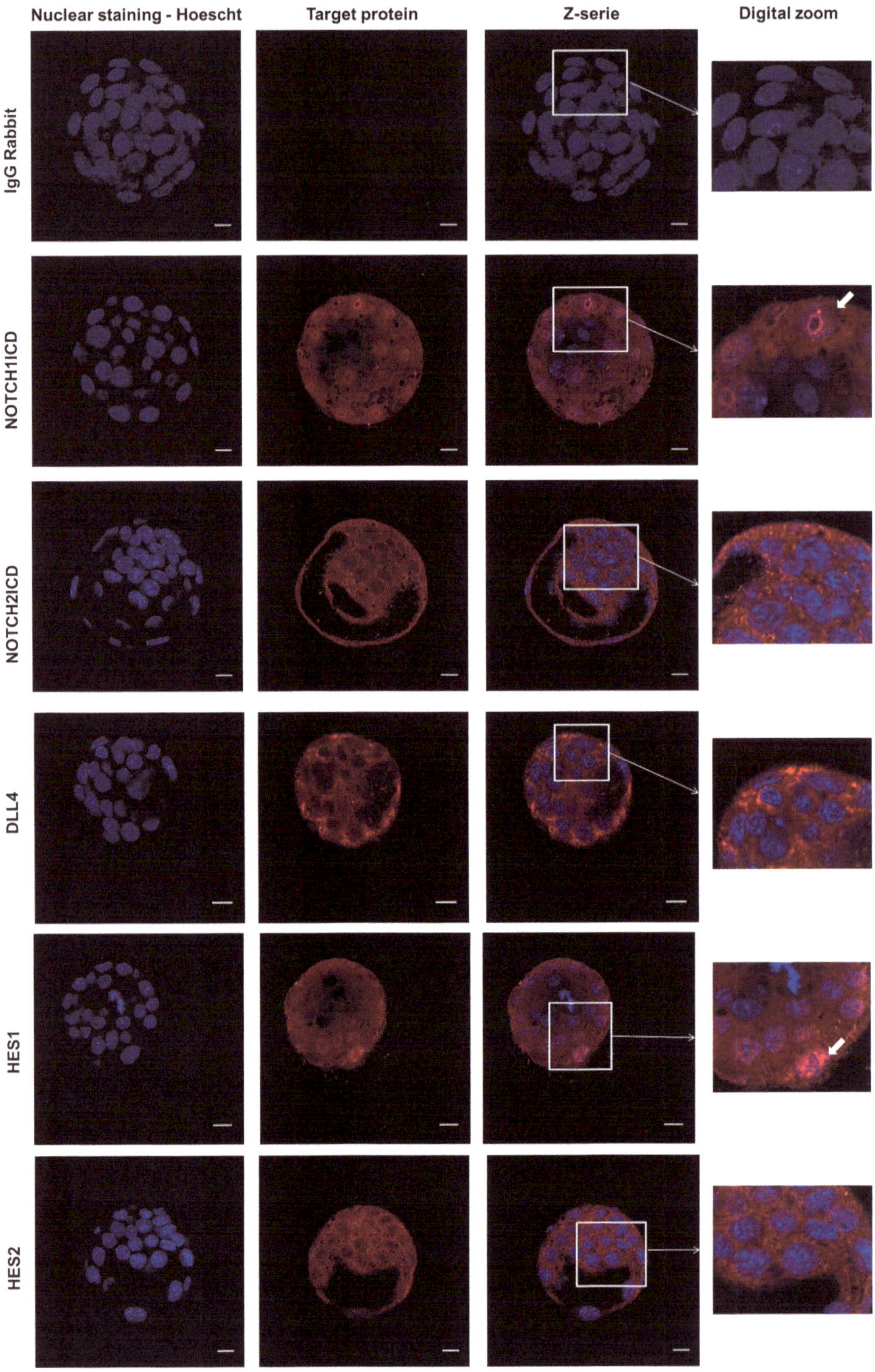

Fig. 3 Representative confocal single plane images of Notch signaling components in mouse 3.5 dpc blastocysts. Target proteins are stained red and nuclei are stained blue with Hoechst. The third column represents a digital zoom of positive staining (arrows). Scale bar 10 μm. Note that N1ICD and HES1 are present in the nucleus of TE cells (arrows) indicating that Notch signalling was activated through the NOTCH1 receptor. By contrast, N2ICD was exclusively found in the cytoplasm of both cell types. ICD = Notch intracellular domain

cytoplasm, indicating a nonactivated status. Recently, using this protocol, authors evidenced that the canonical Notch pathway is active in mouse blastocysts [4, 6]. As shown in Fig. 3, N1ICD displays a different subcellular localization in the trophectoderm (TE) and inner cell mass (ICM) cells, revealing a differential activation status of the pathway in these cell lines. Activation of Notch pathway at the blastocyst stage is further supported by the presence of Notch effectors, like the helix-loop-helix transcription factors of the Hes or Hey family [1, 2] in the nucleus of signal-receiving cells. Although not addressed here, the protocol may be used to colocalize Notch components with other signalling pathways/cell markers.

4 Notes

1. All animal manipulation and experimental procedures (including husbandry, breeding and euthanasia) are conducted according to the national and European Union legislation regarding the use of animals for experimental purposes (Directive 2010/63/UE) and under the license of the Institutional Animal Care and Use Committee.
2. Hormones are usually supplied and stored for long periods in the form of lyophilized powder. Their reconstitution must be performed under sterile conditions using the provided solvent and additional sterile saline if necessary, to a concentration of 50 IU mL^{-1}. After reconstitution, the solution can be stored in aliquots (usually 1 mL, enough for 5 mice) at −20 °C for up to a year. Allow the solution to fully thaw and just reach room temperature before injection.
3. While dissecting, it is important to leave a piece of ovary next to the oviduct to better locate the infundibulum. However, avoid bringing the ovarian fat pad, as it will cloud the medium with lipid droplets and impair embryo recovery.
4. During the flushing procedure, it is important to confirm that the oviduct ingurgitates, indicating that it is filling up with M2 medium. Otherwise, the tip of the needle may have exited the oviduct or the oviduct may have been perforated. In both cases, the flushing will not be successful.
5. To ensure that the desired atmosphere is achieved, the bag should be filled and emptied at least three times before sealing. Make sure the bag looks full and that the sealing was efficient. Alternatively, the incubation can be performed in a CO_2 incubator with variable oxygen control.
6. For incubation at 37 °C, the bag should be fully submerged to ensure temperature homogeneity and precautions should be

taken to ensure it maintains its upright positions while submerged. This can be achieved with a 50 mL tube rack in which the bag is placed in the upper opening and weights in the lower opening.

7. To make recovery of embryos easier, they should be placed in the centre of the dish and a circle can be drawn on the outside of the bottom of the dish to limit search area.
8. Due to the different surface tension of this detergent solution, embryos will tend to spread away from the capillary tip and precautions must be taken to ensure embryos are transferred with the lowest velocity and medium volume as possible to facilitate embryo recovery.
9. Blocking solution can be prepared and stored at −20 °C. Each aliquot must be thawed and allowed to reach room temperature before use. The humidified atmosphere can be achieved by incubating the dish with the Blocking solution in a chamber with a water deposit or a water-soaked paper tissue.
10. This protocol was optimized for the primary antibodies indicated in the Materials section. Using other antibodies will likely require further optimization of the protocol, namely, antibody concentrations.
11. As the protein concentration of the solution decreases, embryos will tend to stick to the plastic dish and capillary tip. If needed, the tip can be dipped in blocking solution between embryo transfers.
12. Other fluorochrome-coupled antibodies can be chosen in this step, with different spectral properties. This choice should be intertwined with the particularities of the confocal system being used and available laser lines.

Acknowledgments

We thank our present and past laboratory colleagues for helping perfect the techniques detailed in this chapter. Work in our laboratory is supported by FCT-Fundação para a Ciência e a Tecnologia, I.P. (Lisbon, Portugal) through grants. EXPL/CVTREP/2289/2013, UID/CVT/276/2013, UIDP/CVT/00276/2020 and UIDB/00276/2020. Patrícia Diniz and Alexandre Trindade contributed equally to this work.

References

1. Artavanis-Tsakonas S, Rand MD, Lake RJ (1999) Notch signaling: cell fate control and signal integration in development. Science 284: 770–776. https://doi.org/10.1126/science.284.5415.770
2. Henrique D, Schweisguth F (2019) Mechanisms of Notch signaling: a simple logic deployed in time and space. Development 146:dev172148. https://doi.org/10.1242/dev.172148

3. Reichrath J, Reichrath S (2020) Notch signaling and embryonic development: an ancient friend, revisited. Adv Exp Med Biol 1218:9–37. https://doi.org/10.1007/978-3-030-34436-8_2
4. Batista MR, Diniz P, Torres A et al (2020) Notch signaling in mouse blastocyst development and hatching. BMC Dev Biol 20(1):9. https://doi.org/10.1186/s12861-020-00216-2
5. Li S, Shi Y, Dang Y et al (2021) NOTCH signaling pathway is required for bovine early embryonic development. Biol Reprod. https://doi.org/10.1093/biolre/ioab056
6. Batista MR, Diniz P, Murta D et al (2021) Balanced Notch-Wnt signaling interplay is required for mouse embryo and fetal development. Reproduction 161(4):385–398. https://doi.org/10.1530/REP-20-0435
7. Luo C, Zuñiga J, Edison E et al (2011) Superovulation strategies for 6 commonly used mouse strains. J Am Assoc Lab Anim Sci 50(4):471–478
8. Nagy A, Gertsenstein M, Vintersten K et al (2003) Manipulating the mouse embryo: a laboratory manual, 3rd edn. Cold Spring Harbor Laboratory Press, New York

Chapter 15

Studying the NOTCH Signaling Pathway Activation in Kidney Biopsies

Laura Márquez-Expósito, Carolina Lavoz, Elena Cantero-Navarro, Raúl R. Rodrigues-Diez, Sergio Mezzano, and Marta Ruiz-Ortega

Abstract

The NOTCH signaling pathway is an evolutionarily conserved family of transmembrane receptors, ligands, and transcription factors. The NOTCH signaling is activated in many biological processes including nephrogenesis, tubulogenesis, and glomerulogenesis, as well as during pathological situations. Activation of Notch signaling is characterized by successive proteolytic cleavages triggered by the interaction between membrane-bound Notch receptors and ligands expressed on neighboring cells. In chronic kidney diseases, activation of the canonical NOTCH signaling pathway has been described. The following protocols will allow the direct assessment of Jagged-1/NOTCH signaling activation in biopsies of patients with chronic kidney diseases and in murine experimental models of renal damage.

Key words NOTCH, Jagged-1, Kidney biopsies, Immunohistochemistry, In situ hybridization

1 Introduction

The NOTCH signaling pathway is phylogenetically conserved and regulates multiple cellular processes, such as cell proliferation, differentiation, death; control of the immune system; and self-renewal of stem cells [1, 2]. This pathway is activated during embryonic development but silenced in the adult. However, in many progressive renal diseases, the components of the NOTCH signaling pathway are re-expressed in the kidney, as described in membranous nephropathy, IgA nephropathy, diabetic nephropathy, lupus nephritis, hypertensive nephrosclerosis, glomerulonephritis of minimal changes, or segmental focal glomerulosclerosis [3].

The components of NOTCH family include several canonical ligands, noncanonical ligands, receptors, and effector proteins. The canonical ligands are Jagged 1/2 and Delta-type 1/3/4. These ligands contain in its extracellular portion a Delta-Serrate-Lag1 (DSL) domain involved in receptor binding and EGF-like repeats

Dongyu Jia (ed.), *Notch Signaling Research: Methods and Protocols*,
Methods in Molecular Biology, vol. 2472, https://doi.org/10.1007/978-1-0716-2201-8_15,

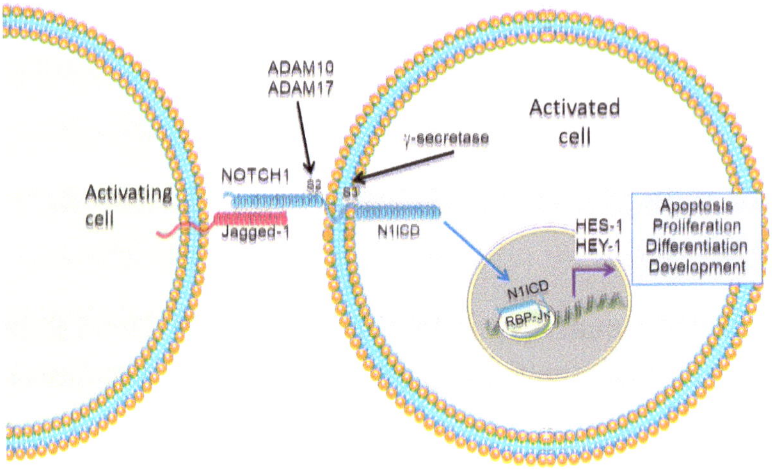

Fig. 1 NOTCH1 signaling pathway activation

that stabilize this binding [4, 5]. The noncanonical ligands are DLK1, DLK2, DNER, Jedi, F3/contactin1, OSM11, and EGFL7 [6, 7]. Among all these ligands the most relevant and studied is Jagged-1 [8]. NOTCH receptors are transmembrane type-III proteins belonging to the family of proteins similar to epidermal growth factor (EGF). There are four known NOTCH receptors: NOTCH 1/2/3/4 [9]. Finally, the effector proteins includes HES1/2/3/4/5/7 and the repressor proteins HERP or HEY 1/2/3 [9].

The activation of the NOTCH pathway has been involved in the progression of renal damage, both in experimental and human diseases [10, 11]. This activation starts when the ligand, for example Jagged-1, binds to the NOTCH receptor through DSL domain (Fig. 1). This binding alters the conformation of the NOTCH negative regulatory region, which leads to the release of the extracellular S2 domain due to two successive proteolytic cleavages, in a process mediated by the metalloproteinases/disintegrins ADAM-10 or ADAM-17 [10, 11]. These proteolytic processing generates an activated form of the receptor that remains attached to the membrane. Then, the γ-secretase enzyme catalyzes the second cleavage in the S3 domain [9, 11]. This final cleavage releases the intracellular domain of NOTCH (NICD), which migrates into the nucleus, where it is associated to RBP-Jκ, which in turn binds to DNA and activates the transcription of target gene effectors of the pathway, resulting in the regulation of several biological processes, such as differentiation, apoptosis and proliferation [9, 11].

Here, we described several methods to evaluate the activation of the Jagged-1/NOTCH signaling pathway in kidney biopsies, from human and mice samples, based on our previous published papers [12–16]. First, the evaluation of NOTCH activation can be observed by the nuclear localization of active NOTCH, measured using an antibody recognizing the active intracellular domain of NOTCH (NICD) by immunohistochemistry. In addition, the expression levels of the main canonical ligand Jagged-1 can be detected by immunohistochemistry. This protein presents a cytoplasmic expression pattern. Examples of these staining can be found in [12] for human and in [13–15] for murine kidney biopsies. Finally, NOTCH regulates the transcription of key genes, including *HES/HEY*. We described how changes in gene expression levels of *HES-1* can be evaluated by in situ hybridization. Examples of this technique can be found in [12, 16].

2 Materials

2.1 Reagents and Buffers

1. Avidin-Biotin Complex: Vectastain ABC kit peroxidase from Vector laboratories (Ref. PK-4000).
2. Antibody diluent solution: Ready-to-use diluent from DAKO (REF: S0809).
3. Biotin Blocking System from DAKO (Ref: X0590).
4. Biotin-stained probe cocktails for Jagged-1 and HES-1 (0.2 μg/mL) from R&D Systems).
5. Casein Solution from Vector Laboratories (Ref: SP-5020-250).
6. DAB (3.3-diaminobenzidine) substrate kit from Abcam (Ref.: ab64238).
7. DPX new, nonaqueous mounting medium for microscopy (contains *m*-xylene) from Merck (Ref. HX90969979).
8. EnVision FLEX Target Retrieval Solution, Low pH, (50×) from DAKO (Ref. K800521-2).
9. NBT/BCIP (nitro blue tetrazolium/5-bromo-4-chloro-3-indolyl phosphate) from R&D Systems.
10. PBS (10×): 80 g NaCl, 2 g KCl, 14.4 g Na_2HPO_4, 2.4 g KH_2PO_4 in 1 L distilled water.
11. Peroxidase-blocking solution from DAKO (Ref: S2023).
12. TBS buffer: Tris–HCl Buffer 50 mM pH 7.6.
13. TBST buffer: Tris-buffered saline (TBS) containing 2% Triton.
14. TPS buffer: Tris 0.05 M, Na_2HPO_4, 0.01 M, KH_2PO_4 30 mM, NaCl 0.1 M pH 7.8.

2.2 Equipment

1. PTlink system (Dako).
2. Lab's stove.
3. Coplin-like jars.
4. Light microscopy.

3 Methods

Carry out all procedures at room temperature unless otherwise specified. For immunohistochemical protocols, use 3 μm thick tissue sections transferred onto coated-glass slides suitable for immunohistochemistry. For in situ hybridization protocols, use tissue fixed in Bouin, embedded in paraffin wax, cut into 5 μm sections and mounted onto 3-aminopropyltriethoxy-silane–coated slides treated with DEPC.

3.1 Immunohistochemistry Protocol for Jagged-1 and N1ICD (Mouse)

1. Deparaffinize samples in a laboratory stove at 65 °C overnight (*see* **Note 1**).
2. Continue the deparaffinization 10–20 min in xylene (*see* **Note 2**).
3. Hydrate the tissue section from xylene through 100%, 96%, and 70% ethanol, 3–5 min each (*see* **Note 3**).
4. Immerse the tissue in distilled water.
5. Perform antigen retrieval to unmask the antigenic epitope using the EnVision FLEX Target Retrieval Solution Low pH at 1× and boil it at 95–100 °C for 20 min in the PTlink system (Dako) (*see* **Note 4**).
6. Block endogenous peroxidase activity by incubating sections in H_2O_2 peroxidase-blocking solution from DAKO twice, 10–15 min each. Wash with PBS 1× between incubations with gentle shaking for some seconds.
7. Eliminate nonspecific protein binding sites by incubating tissue sections in rabbit serum (prepared in PBS 1×/BSA 4%, dilution 1:10) for 40–60 min (*see* **Note 5**).
8. Rinse slides in PBS 1× for 5 min in constant shaking.
9. Incubate overnight at 4 °C with diluted primary antibody: anti-Jagged-1 goat (SC-6011, Santa Cruz Biotechnology) prepared in PBS 1×/BSA 4%, dilution 1:100 or anti-N1ICD (Ab8925, Abcam) dilution 1:300 prepared in antibody diluent solution.
10. Rinse slides in PBS 1× for 5 min in constant shaking.
11. Incubate 60 min at room temperature with the α-goat biotinylated secondary antibody (#31732, Invitrogen) diluted 1: 200, prepared in PBS 1×/4% BSA) in the case of Jagged-1 primary antibody, or with α-rabbit biotinylated secondary antibody (AP132B, Millipore) diluted 1:200 in antibody diluent solution in the case of the N1ICD primary antibody.
12. Rinse slides in PBS 1× for 5 min in constant shaking.
13. Incubate for 30 min in a humidified chamber with AB complex (9μLA + 9μLB) in 1 mL PBS 1× freshly made and incubate in constant agitation 30 min – 1 h before use protected from light.

14. Rinse slides in PBS 1× for 5 min in constant shaking.
15. Apply DAB solution (freshly made just before use: 20 μL chromogen +1 mL Buffer substrate) to the sections on the slides to reveal the color of antibody staining. Allow the color development for less than 5 min until the desired color intensity is reached (*see* **Note 6**).
16. Wash slides in PBS 1× for some minutes.
17. Rinse slides in distilled water before hematoxylin counterstain.
18. Counterstain slides by immersing slides in Carazzi's hematoxylin for 1–2 min. Rinse well the slides in running tap water for 10 min. Renew the tap water in the first steps (*see* **Note 7**).
19. Dehydrate the tissue section through increase percentage of ethanol (70%, 96%, 100%) 3 min each approximately (*see* **Note 8**).
20. Dehydrate completely the tissue in xylene for some minutes.
21. Coverslip using DPX (*see* **Note 9**).
22. Observe the color of the antibody staining in the tissue sections under a microscopy.

3.2 Immunohistochemistry Protocol for Jagged-1 (Human)

1. Deparaffinize samples in a laboratory stove at 65 °C overnight (*see* **Note 1**).
2. Continue the deparaffinization in xylene twice, 15 min each (*see* **Note 2**).
3. Hydrate the tissue section from xylene through 100% ethanol 5 min, 100% methanol 15 min, 96% ethanol 5 min, and 70% 5 min (*see* **Note 3**).
4. Immerse the tissue in distilled water for 5 min.
5. Block endogenous peroxidase activity by incubating sections in 3% H_2O_2 solution in methanol for 15 min. Wash with distilled water for 1 min and then with TPS for 5 min at room temperature.
6. Eliminate nonspecific protein binding sites by incubating tissue sections in rabbit serum (prepared in TPS–BSA 1%, 1:10) for 60 min at 22 °C.
7. Rinse slides in TPS for 5 min in constant shaking.
8. Incubate overnight at 4 °C with diluted primary antibody: anti-Jagged goat (SC-6011) prepared in TPS–BSA 1% dilution 1: 30.
9. Rinse slides in TPS three times for 5 min in constant shaking.
10. Incubate for 30 min at 22 °C with secondary α-goat biotinylated antibody (prepared in TPS–BSA 1%, dilution 1:300).
11. Rinse slides in TPS three times for 5 min in constant shaking.

12. Incubate for 30 min at 22 °C with streptavidin conjugate-HRP, prepared in TPS–BSA 1% dilution 1:1000.
13. Rinse slides in TPS three times for 5 min in constant shaking.
14. Apply DAB solution (freshly made just before use: 20 μL chromogen +1 mL Buffer substrate) to the sections on the slides to reveal the color of antibody staining. Allow the color development for less than 5 min until the desired color intensity is reached (*see* **Note 6**).
15. Wash slides in TPS for some minutes.
16. Rinse slides in distilled water before hematoxylin counterstain.
17. Counterstain slides by immersing slides in Carazzi's hematoxylin for 30 s, followed by sodium borate 1% for 30 s.
18. Dehydrate the tissue section through increase percentage of ethanol (70%, 96%, 100%) 2 min each approximately (*see* **Note 8**).
19. Dehydrate completely the tissue in xylene for 2 min twice.
20. Coverslip using Canada balsam.
21. Observe the color of the antibody staining in the tissue sections under a microscopy.

3.3 In Situ Hybridization Protocol for Jagged-1 and HES-1

1. Deparaffinize samples in xylene twice, 10 min each.
2. Hydrate the tissue section from xylene through 100%, 96%, and 70% ethanol, 5 min each.
3. Immerse the tissue in DEPC-treated water for 1 min (*see* **Note 10**).
4. Incubate the slides 10 min at 60 °C with 2× SSC 30% formamide.
5. Wash slides in DEPC-treated water for 1 min.
6. Block endogenous alkaline phosphatase activity by incubating sections in levamisole 5 mM for 30 min at room temperature. Wash with Tris–HCl 50 mM pH 7.6 for 5 min.
7. Digest tissue sections with proteinase K (5 μg/mL) in Tris–HCl 50 mM pH 7.6 for 20 min at 37 °C.
8. Block endogenous biotin by incubating sections in avidin and biotin for 10 min at room temperature.
9. Wash slides in PBS 1× for 1 min.
10. Fix slides in paraformaldehyde 0.4% in PBS 1× for 20 min at 4 °C.
11. Wash slides in DEPC-treated water for 1 min.
12. Incubate the slides with a prehybridization solution.

13. Incubate overnight at 37 °C with 100 μL of Biotinylated probes cocktails for each cytokine (0.2 μg/mL; R&D Systems) in a humidified chamber.

    Probes used for Jagged-1 detection (Sigma):

    Antisense oligonucleotide: 5′- CCTGACAGTATTATT GAAAAGGCT-3′.

    Antisense oligonucleotide: 5′-GTACGGCTGGCAAGGCTTG TACTG-3′.

    Antisense oligonucleotide: 5′- CACGCCTGCCTCTCT GATCCCTGT-3′.

    Probes used for HES-1 detection (Sigma):

    Antisense oligonucleotide: 5′-CTTCTCTCCTTGGTCCTGG AACAG-3′.

    Antisense oligonucleotide: 5′- AGCTCGCGGCATTC CAAGCTGGAG-3′.

    Antisense oligonucleotide: 5′- CTGCGCTGAGCACAGACC CAAGTG-3′
14. Incubate the slides in 2× SSC 30% formamide twice, then 1× SSC 30% formamide twice and finally 0.5× SSC 30% formamide, 5 min each at 37 °C.
15. Wash slides in TBST 1× for 15 min in constant shaking at room temperature.
16. Incubate the slides in Streptavidin–alkaline phosphatase conjugate, prepared in Tris–HCl 50 mM pH 7.6 (diluted 1:300), for 30 min in constant shaking at room temperature.
17. Rinse slides in TBST 1× three times for 5 min.
18. Wash slides in DEPC-treated water for 1 min.
19. Apply NBT/BCIP(R&D Systems) as the enzyme substrate to the sections on the slides, for 15 min at 37 °C protected from light, to reveal the color of staining (*see* **Note 11**).
20. Dehydrate the tissue section through increase percentage of ethanol (70%, 96%, 100%) 5 min each.
21. Dehydrate the tissue section in Xilol/Fenol for 1 min.
22. Dehydrate completely the tissue in xylene for 10 min twice.
23. Coverslip using DPX (*see* **Note 9**).

3.4 Technique Controls

Steps to confirm the specificity of the reaction:

1. Demonstrate the disappearance of hybridization signal when RNase (100 μg/mL; Sigma) is added in 0.05 mol/L Tris after the digestion with proteinase K.
2. Probe oligo DT.
3. Incubate the tissue without the antisense probe.

4 Notes

1. Deparaffination in the oven can be performed from 1 h to overnight, depending on the paraffine on the sample. If this is skipped, it is preferable to leave samples more time on xylene (from 20 to 30 min).
2. Modify depending on the previous step. If ON deparaffination in oven was done, 10 min of xylene should be appropriate. If 1 h or less was done, it is preferably to increase the time of the samples in xylene from 20 to 30 min.
3. Hydration process can be done from some seconds to some minutes. However, if quickly steps are to be done, allow the samples to absorb well each alcohol with gentle shaking and then change to the next alcohol. Be careful of not detaching the samples from the slides. When doing each step within some minutes, gentle shaking is not necessary.
4. The PTlink system allows the user to preheat the buffer solution to 65°. It is preferable to do this beforehand.
5. For the N1ICD antibody, block nonspecific protein binding sites by incubating tissue sections with casein solution (diluted 1× in PBS 1×) for 60–120 min. The N1ICD antibody presents high background, so it is highly recommendable to increase the blocking time to 2 h. In the Jagged-1 murine protocol, 1 h should be appropriate.
6. Be careful with the degree of staining to not to get too much background. N1ICD should be incubated with DAB no more than 1 min (or until the positive nuclei are visible). To achieve this, use a microscope while developing with DAB to get the desired positive staining.
7. In the case of the counterstain, it is highly recommendable optimizing the time of incubation with the hematoxylin in your own laboratory. Nevertheless, in 1 min of incubation there should be positive stained nuclei. In addition, it is up to the user the degree of hematoxylin incubation in the N1ICD protocol, since too much counterstain in this case can impede the view of the positive nuclei from the N1ICD staining.
8. Times of each step can be reduced if gentle shaking is applied to the samples in each alcohol, with careful of not detaching the tissue from the slides. Until mounting with the coverslip and the specified mounting solution, samples can be maintained in xylene a moderate time.
9. The mounted slides can be stored at room temperature permanently. It is highly recommended to clean the next day after mounting the slides with H_2Od or EtOH 70° before observing them into the microscope to not to have rests of the mounting solution.

10. The entire procedure must be performed in RNase-free water (DEPC-treated water), included the cutting tissue, and preparation alcohols to hydrate and dehydrate and buffers for the washes.
11. Check every 5 min for color development under a microscope.

Acknowledgments

This work was supported by grants from the Instituto de Salud Carlos III (ISCIII) and Fondos FEDER European Union (PI17/00119, PI20/00140, DTS20/00083), Red de Investigación Renal REDINREN: RD16/0009/003 and RICORS2040 (RD21/0005/0002) to M.R-O, Sociedad Española de Nefrología, "NOVELREN-CM: Enfermedad renal crónica: nuevas Estrategias para la prevención, Diagnóstico y tratamiento"; B2017/BMD-3751 to M.R-O and PEJD-2019-PRE/BMD-15423 to L.M-E), "Convocatoria Dinamización Europa Investigación 2019" MINECO (EIN2019-103294 to M.R-O); Innovation programme under the Marie Skłodowska-Curie grant of the European Union's Horizon 2020 (IMProve-PD ID: 812699) to M.R-O.

References

1. Andersson ER, Sandberg R, Lendahl U (2011) Notch signaling: simplicity in design, versatility in function. Development 138:3593–3612. https://doi.org/10.1242/dev.063610
2. Siebel C, Lendahl U (2017) Notch signaling in development, tissue homeostasis, and disease. Physiol Rev 97:1235–1294. https://doi.org/10.1152/physrev.00005.2017
3. Murea M, Park JK, Sharma S et al (2010) Expression of notch pathway proteins correlates with albuminuria, glomerulosclerosis, and renal function. Kidney Int 78:514–522. https://doi.org/10.1038/ki.2010.172
4. Bray SJ (2006) Notch signalling: a simple pathway becomes complex. Nat Rev Mol Cell Biol 7:678–689. https://doi.org/10.1038/nrm20095
5. Kopan R, Ilagan MXG (2009) The canonical notch signaling pathway: unfolding the activation mechanism. Cell 137:216–233. https://doi.org/10.1016/j.cell.2009.03.045
6. D'Souza B, Meloty-Kapella L, Weinmaster G (2010) Canonical and non-canonical notch ligands. Curr Top Dev Biol 92:73–129. https://doi.org/10.1016/S0070-2153(10)92003-6
7. Nueda ML, González-Gómez MJ, Rodríguez-Cano MM et al (2018) DLK proteins modulate NOTCH signaling to influence a brown or white 3T3-L1 adipocyte fate. Sci Rep 8:16923. https://doi.org/10.1038/s41598-018-35252-3
8. Cordle J, Johnson S, Zi Yan Tay J et al (2008) A conserved face of the Jagged/Serrate DSL domain is involved in Notch trans-activation and cis-inhibition. Nat Struct Mol Biol 15:849–857. https://doi.org/10.1038/nsmb.1457
9. McIntyre B, Asahara T, Alev C (2020) Overview of basic mechanisms of notch signaling in development and disease. In: Advances in experimental medicine and biology. Springer, New York, pp 9–27. https://doi.org/10.1007/978-3-030-36422-9_2
10. Fleming RJ (2020) Ligand-induced cis-inhibition of notch signaling: the role of an extracellular region of serrate. In: Advances in experimental medicine and biology. Springer, New York, pp 29–49. https://doi.org/10.1007/978-3-030-36422-9_3
11. Marquez-Exposito L, Cantero-Navarro E, Lavoz C et al (2018) Análisis de la vía Notch como una posible diana terapéutica en la patología renal. Nefrologia 38:466–475. https://doi.org/10.1016/j.nefro.2017.11.027

12. Lavoz C, Droguett A, Burgos ME et al (2014) Estudio traslacional de la vía Notch en nefropatía hipertensiva. Nefrologia 34:369–376. https://doi.org/10.3265/Nefrologia.pre2014.Jan.12436
13. Lavoz C, Poveda J, Marquez-Exposito L et al (2018) Gremlin activates the Notch pathway linked to renal inflammation. Clin Sci 132:1097–1115. https://doi.org/10.1042/CS20171553
14. Lavoz C, Rodrigues-Diez R, Benito-Martin A et al (2012) Angiotensin II contributes to renal fibrosis independently of notch pathway activation. PLoS One 7:e40490. https://doi.org/10.1371/journal.pone.0040490
15. Marquez-Exposito L, Rodrigues-Diez RR, Rayego-Mateos S et al (2021) Deletion of delta-like 1 homologue accelerates renal inflammation by modulating the Th17 immune response. FASEB J 35:e21213. https://doi.org/10.1096/fj.201903131R
16. Mezzano SA, Droguett MA, Burgos ME et al (2000) Overexpression of chemokines, fibrogenic cytokines, and myofibroblasts in human membranous nephropathy. Kidney Int 57:147–158. https://doi.org/10.1046/j.1523-1755.2000.00830.x

Chapter 16

Exosomes as Carriers for Notch Molecules

Guya Diletta Marconi, Francesca Diomede, Oriana Trubiani, Cristina Porcheri, and Thimios A. Mitsiadis

Abstract

Exosomes are extracellular vesicles involved in cell-to-cell communication as well as extrusion of biological material. Using dental pulp stem cells culture as a model, we hereby describe a method for the packaging of Delta-like 4 (DLL4), a representative Notch ligand, into newly generated exosomes. We then provide methods of analysis to confirm the presence of Notch proteins and transcripts internalization and transport via exosomes.

Key words Exosomes, Extracellular vesicles, Notch pathway, Secretome, Dental pulp stem cells

1 Introduction

Extracellular vesicles (EVs) are lipid bilayer-bound vesicles produced by cells and have been implicated in intercellular communication. The membrane vesicles released by eukaryotic cells are exosomes, microparticles, macrovesicles and apoptotic bodies, that can be preserved as a dynamic vesicular compartment. The paracrine or autocrine biological outcomes of EVs play an important role in tissue metabolism (Fig. 1).

In detail, EVs and soluble secretory factors, also entitled "secretome," include several proteins, nucleic acids, lipids and a collection of soluble cytokines. Exosomes are present in both healthy and pathological conditions. EVs can be found in most of the biological fluids, such as urine, serum, plasma, lymph, or cerebrospinal fluid. Due to their size (50–1000 nm), EVs can circulate in the entire body [1].

EVs are delivery carriers that transport molecules to target cells and release their content upon internalization. Basically, they offer a mechanism of cell-to-cell communication and of intercellular interchange of cell factors, such as nucleic acids, cytokines, lipids, and

Dongyu Jia (ed.), *Notch Signaling Research: Methods and Protocols*,
Methods in Molecular Biology, vol. 2472, https://doi.org/10.1007/978-1-0716-2201-8_16,

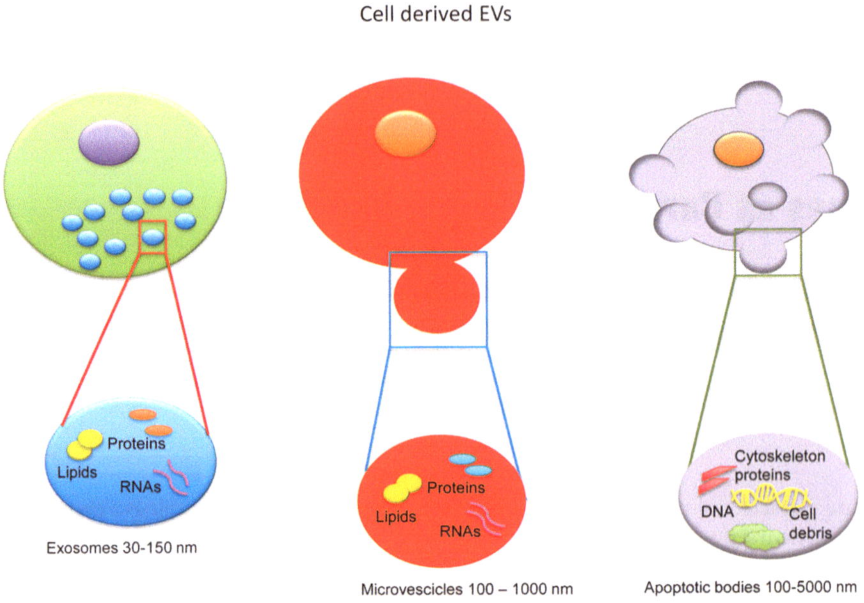

Fig. 1 Different types of extracellular vesicles (EVs) can be used as carriers for macromolecules and bioactive substances. EVs include exosomes (30–150 nm), microvesicles (100–1000 nm) and apoptotic bodies (100–5000 nm), and represent key players in cell-to-cell communication in normal physiology and in pathological conditions

proteins. They operate as signals in homeostatic events and in pathological conditions [2].

In mammals, four types of Notch receptors and five types of ligands (Delta-like 1, 3, and 4 and Jagged1 and 2) exist [3, 4]. The canonical Notch signaling is activated upon receptor–ligand interaction between neighboring cells, followed by a sequential series of cleavages of the receptor protein [5]. This results in the detachment of the intracellular domain of the Notch receptor and its translocation to the nucleus, where it regulates the transcription of downstream target genes [6]. Plasma membrane-derived vesicles can contain active components of the Notch pathway, indicating a novel, nonconventional activity for the Notch signaling [7].

The Notch receptors and ligands can be packaged into endothelial cell-derived exosomes, transported to other endothelial cells and integrated into their cell membrane, resulting in modulation of the Notch signaling. Hence, exosomes can potentially be used as possible carriers of a transmembrane molecules for Notch modulation [8].

Here, we describe a method for the packaging of the Notch ligand DLL4 into exosomes produced by human primary dental pulp cells and its detection via western blot analysis. As exosomes might also be RNA carriers, we additionally provide a method for mRNA extraction and detection of Notch components via RT-PCR.

2 Materials

2.1 Cell Culture

1. TheraPEAK™Chemically Defined Mesenchymal Stem Cell Growth Medium (MSCGM-CD™) (LONZAWalkersville Inc., Walkersville, MD, USA). MSCGM-CD™ (including MSCGM-CD™ basal medium and MSCGM-CD™ SingleQuots™).
2. TryPLE™ Select Enzyme (Gibco, Life Technologies, Eugene, OR, USA).
3. Dulbecco's phosphate buffered saline (0.00095 M) without Ca ++ and Mg++.

2.2 Western Blot for Protein Detection

1. RIPA Buffer, (50 mM Tris–HCl, 150 mM NaCl, 1.0% (v/v) NP-40, 0.5% (w/v) sodium deoxycholate, 1 mM EDTA, and 0.1% (w/v) SDS).
2. Tris buffered saline (TBS), (20 mM Tris and 150 mM NaCl at 7.6 pH).
3. 1× SDS Sample Buffer.
4. Tris–Glycine SDS Running Buffer.
5. Tris–Glycine Transfer Buffer containing 20% methanol.
6. Blocking buffer: TBS with 5% BSA.
7. Sodium dodecyl sulfate–polyacrylamide gel electrophoresis.
8. Western blot chambers for transfer.
9. Nitrocellulose membranes (0.2 μm pore size10. α-Dll4 (Cell Signaling; 1:1000).
10. Secondary antibodies HRP-conjugated (DAKO, 1:5000).
11. Chemiluminescent detection kit.

2.3 RNA Isolation Reagents and Materials

1. miRNA Wash Solution 1[1,2] (Add 21 mL of 100% ethanol before use),
2. Wash Solution 2/3 (footnote 2) (Add 40 mL of 100% ethanol before use).
3. Collection tubes.
4. Filter cartridges.
5. Exosome Resuspension Buffer 4 °C.
6. 2× Denaturing Solution (footnote 1) (Add 375 μL of 2-mercaptoethanol before use)

[1] These reagents contain guanidinium thiocyanate. This is a potentially hazardous substance and should be used with appropriate caution. Contact with acids or bleach liberates toxic gases. DO NOT ADD acids or bleach to any liquid wastes containing this product.

[2] Store at room temp for up to 1 month. For longer term storage, store at 2–8 °C, but warm to room temperature before use.

7. Acid-phenol–chloroform[3] 4 °C.
8. Elution Solution 4 °C.

2.4 Additional Required Materials

1. 2-Mercaptoethanol (14.3 M).
2. 100% ethanol, ACS grade or better.
3. Phosphate buffered saline (PBS).
4. Microcentrifuge capable of at least 10,000 × *g*.
5. Heat block set to 95–100 °C.
6. (optional) Vacuum manifold: to pull solutions through the filter cartridges.
7. RNase-free 1.5 mL or 2.0 mL polypropylene microfuge tubes, adjustable pipettors, and RNase-free tips (use gloves).

3 Methods

3.1 Surgical Procedures

1. Patients sign written consent for clinical research and for the processing of personal data before surgical procedure.
2. Prepare patient for surgery by pretreatment for 1 week with professional dental hygiene and decontamination of the oral cavity with chlorhexidine 0.2%.
3. Carry out dental pulp biopsies from human teeth.
4. Exposed the dental pulp tissue using a cylindrical diamond bur mounted on high speed handpiece with water spry cooling.
5. Collect the dental pulp tissue with a sterile excavator [9] (*see* **Note 1**).

3.2 Laboratory Procedures to Isolate Human Dental Pulp Stem Cells (hDPSCs)

1. Put the specimens of dental pulp tissue in a sterile tube (15 mL) with DPBS.
2. Wash gently the tissue five times in DPBS (10 mL) supplemented with 5% of Gentamicin.
3. Place the tissue in Corning® BioCoat™ dish (Corning Inc., Corning, NY, USA) (diameter 10 mm).
4. Wash gently twice with DPBS supplemented with 5% of streptomycin (5 mL).
5. Dissociate mechanically the dental pulp tissue into small clumps using sterile scalpel.
6. Add appropriate volume (5 mL) of completed medium (MSCGM-CD™) to the dish.
7. Put the dish in the incubator at 37 °C and 5% CO_2.

[3] This reagent contains phenol, which is a poison and an irritant. Use gloves and other personal protection equipment when working with this reagent.

8. Change medium three times a week.
9. Wait 7–20 days of culture the spontaneous cell migration from the explants.
10. Detach the cells at 80% of confluence.
11. Remove medium.
12. Wash gently twice with DPBS (5 mL).
13. Add TryPLE™ Select to the dish (1 mL).
14. Put dish in the incubator at 37 °C and 5% CO_2 for 10 min.
15. Control cell morphological changes at light microscopy.
16. Transfer solution from dish to a sterile tube.
17. Centrifuge at 210 × *g* for 8 min at 25 °C.
18. Remove supernatant.
19. Resuspend the cell pellet in a minimal volume of temperature equilibrated MSCGM-CD™ (This step allows the selection of hDPSCs from the dental pulp tissue).
20. Remove a sample for counting.
21. Count the cells with a hemocytometer or cell counter and calculate the total number of cells.
22. Assess cell viability using Trypan Blue exclusion dye test.
23. Use the following equation to determine the total number of viable cells.

    Total # of viable cells = (Total cell count x percent viability)/ 100
24. Determine the total number of dishes to seed by using the following equation. The number of dishes needed depends on the cell yield and seeding density.

    Total # of dishes to inoculate = Total # of viable cells/ (Growth area x Rec: seeding density: 15–25 × 10³ cells/mL).
25. Use the following equation to calculate the volume of cell suspension to seed into your dishes. Determine the volume of MSCGM-CD™ to add to each flask so that the final culture volume is 0.2–0.4 mL/cm^2.

    Seeding volume = Total volume of diluted cell suspension/# of dishes as determined in ***step 2****.*
26. Prepare dishes by labeling each dish with the passage number, strain number, cell type, and date.
27. Add the MSCGM-CD™ at RT to the dish (5 mL).
28. Incubate at 37 °C and 5% CO_2.
29. After 3 days remove completely the medium.
30. Replace with an equal volume of MSCGM-CD™.
31. Cultures will be near confluence by day 5 or 6 and ready to subculture.

3.3 Exosomes Isolation from Conditioned Medium (CM) Derived from hDPSCs Through ExoQuick-TC Solution (See Fig. 2)

1. Culture hDPSCs in completed medium (MSCGM-CD™) for 48 h.
2. Collect the medium (10 mL) derived from hDPSCs (CM).
3. Centrifuge the CM at 3000 × *g* for 15 min to remove suspension cells and debris.
4. Extract exosomes using ExoQuick-TC commercial agglutinant (System Biosciences, Palo Alto, CA, USA).
5. Add 2 mL of ExoQuick-TC to 10 mL of CM.

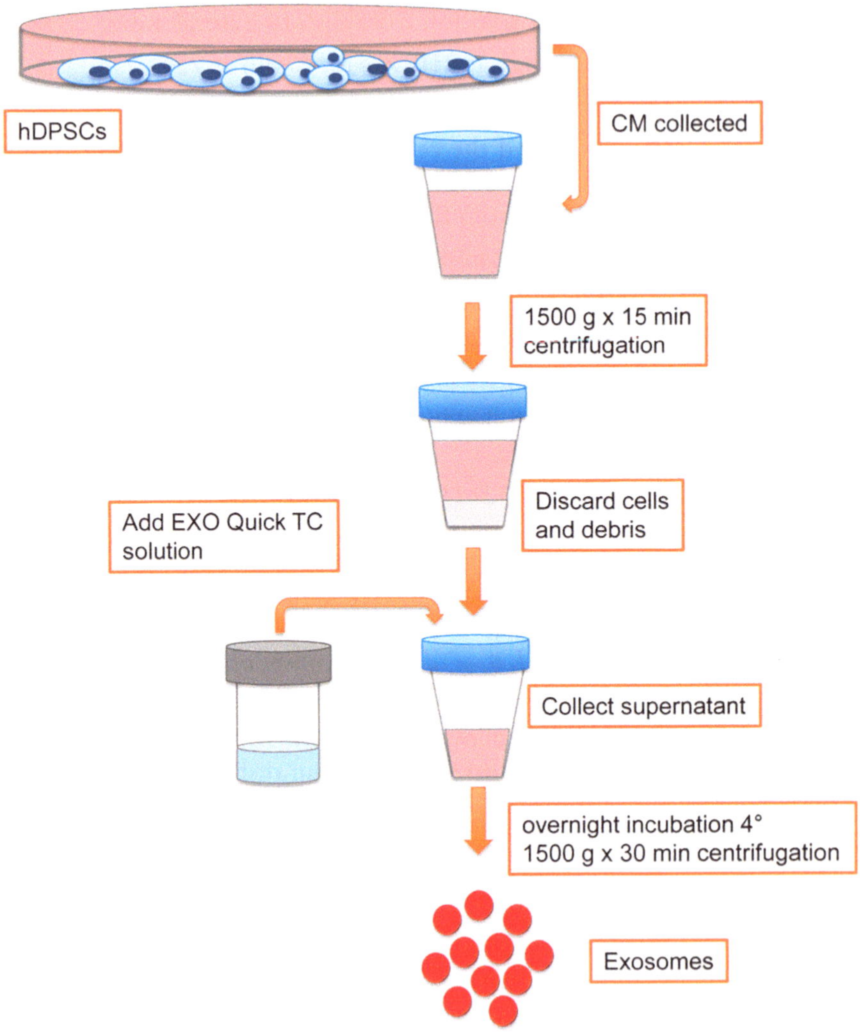

Fig. 2 Schematic representation of exosomes isolation derived from conditioned medium (CM) collected from hDPSCs culture through ExoQuick-TC solution. CM collected from hDPSCs was centrifuge at 1500 × *g* for 15 min to remove discard cells and debris. Then, Add ExoQuick-TC solution to CM. Incubate the mix overnight at 4 °C without rotation, one centrifugation step was performed at 1500 × *g* for 30 min to sediment the exosomes, in order to remove the supernatant

6. Incubate the mix overnight at 4 °C without rotation, one centrifugation step was performed at 1500 × *g* for 30 min to sediment the exosomes (Fig. 2).
7. Collect the pellets and resuspend in 200 μL PBS and split into two aliquots.
8. Precipitate and quantify exosomes using the whole-homogenate proteins [10, 11].

3.4 Exosomes Isolation from hDPSCs by Differential Ultracentrifugation (See Fig. 3)

1. Culture hDPSCs in completed medium (MSCGM-CD™) for 48 h.
2. Collect the medium (10 mL) derived from hDPSCs (CM).
3. Centrifuge the CM at 3000 × *g* for 10 min to remove discard cells and debris.
4. Filter the CM containing the exosomes through a 0.42 μm filter and centrifuge at 12,000 × *g* for 30 min at 4 °C to remove debris and apoptotic bodies.
5. Isolate exosomes from CM by twice ultracentrifugation at 100,000 × *g* for 90 min/each at 4 °C, with an interval wash with PBS, and the exosomal protein content (Fig. 3).
6. Measure exosomal protein content by Bradford method [12].

3.5 Delta-Like 4 Incorporation into Exosomes

1. Culture hDPSCs in completed medium (MSCGM-CD™) for 48 h before being harvested.
2. Plate hDPSCs on plates coated overnight with recombinant human DLL4 extracellular domain (R&D Systems) at a concentration of 1 μg/mL in 0.2% gelatin, to induce the production of the protein and subsequently the production of exosomes that contained the ligand [13] (*see* **Note 2**).
3. Collect the CM after 48 h.
4. Isolate Exosomes (*see* Subheading 3.3) containing the Notch ligand [14].
5. The resulting exosomes can be resuspended in PBS or lysed in RIPA buffer.
6. *Optional:* The exosomes can be labeled with the fluorescent membrane dye PKH67 (Sigma-Aldrich).
7. The exosomes lysed in RIPA buffer are analyzed by Western blotting for the detection of DLL4 and other exosome markers.

3.6 Concentration of Exosomes Derived from hDPSCs and Resuspension of Pellet

After pelleting exosomes derived from hDPSCs by precipitation through ExoQuick-TC solution (*see* Subheading 3.2) or ultracentrifugation (*see* Subheading 3.3) resuspend the exosome pellet according to the following directions:

1. Resuspend the exosome pellet with ice cold resuspension buffer or 1× PBS (*see* **Note 3**).

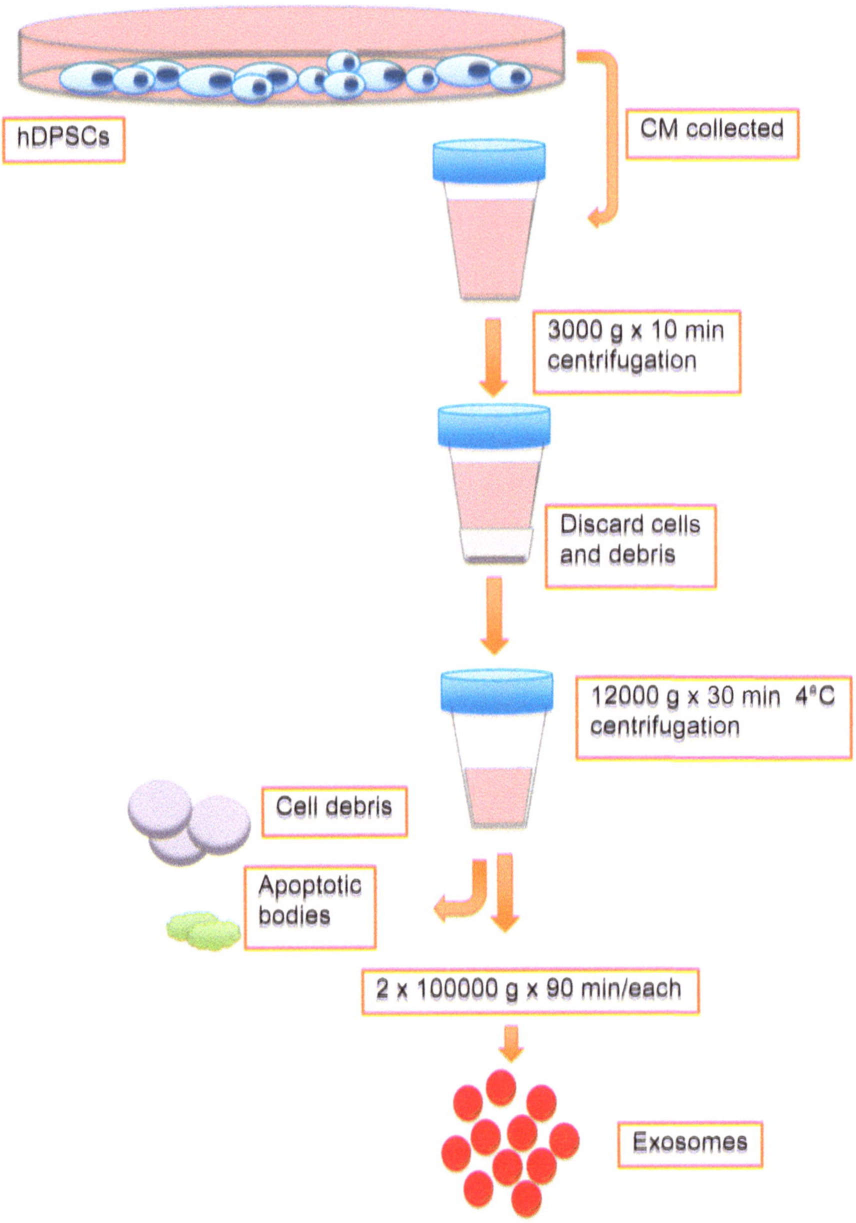

Fig. 3 Schematic representation of purification of exosomes by differential ultracentrifugation from conditioned medium (CM) collected from hDPSCs culture. CM collected from hDPSCs was centrifuge at 3000 × *g* for 10 min to remove discard cells and debris. Then, centrifuge at 12,000 × *g* for 30 min at 4 °C to remove debris and apoptotic bodies. Next, twice ultracentrifugation at 100,000 × *g* for 90 min/each at 4 °C to isolate exosomes

2. Incubate the sample for 5–10 min at room temperature to allow the pellet to dissolve.
3. Gently pipet up and down to thoroughly resuspend the sample.
4. Proceed to RNA isolation (*see* **Note 4**).

3.7 RNA Isolation from Exosomes Derived from hDPSCs

1. Add 1× PBS to the exosome sample in an RNase-free tube so that the total volume is 200 μL.
2. Add one volume of 2× Denaturing Solution and mix thoroughly.

   Important: Prewarm the 2× Denaturing Solution at 37 °C to dissolve precipitate if necessary.
3. Incubate the mixture on ice for 5 min.
4. Add one volume of acid-phenol–chloroform to each sample.
5. The volume of acid-phenol–chloroform should be equal to the total volume of the sample plus the 2× Denaturing Solution (e.g., if the initial sample lysate volume was 200 μL and it was mixed with 200 μL of 2× Denaturing Solution in **step 1**, add 400 μL acid-phenol–chloroform) (*see* **Note 5**).
6. Mix samples by vortexing for 30–60 s.
7. Centrifuge for 5 min at maximum speed (≥10,000 × *g*) at room temperature to separate the mixture into aqueous and organic phases. Repeat the centrifugation if the interphase is not compact.
8. Carefully remove the aqueous (upper) phase without disturbing the lower phase or the interphase, and transfer it to a fresh tube. Note the volume recovered.
9. Proceed to Isolate RNA—Purify the RNA (*see* **Note 6**).
10. Add 1.25 volumes 100% ethanol to the aqueous phase, and mix thoroughly (e.g., if 300 μL was recovered, add 375 μL ethanol).
11. For each sample, place a filter cartridge into one of the collection tubes (supplied in kit).
12. Pipet 700 μL of the lysate–ethanol mixture onto the filter cartridge. For sample volumes >700 μL, apply the mixture in successive applications to the same filter.
13. Centrifuge at 10,000 × *g* for ~15 s, or until the mixture has passed through the filter. Alternatively, use a vacuum to pull the samples through the filter.
14. Discard the flow-through, and repeat until all of the lysate–ethanol mixture has been passed through the filter. Save the Collection Tube for the washing steps.
15. Add 700 μL miRNA Wash Solution 1 (working solution mixed with ethanol) to the Filter Cartridge.
16. Centrifuge at 10,000 × *g* for ~15 s or use a vacuum to pull the solution through the filter. Discard the flow-through from the Collection Tube, and replace the Filter Cartridge into the same Collection Tube.
17. Apply 500 μL Wash Solution 2/3 (working solution mixed with ethanol) and draw it through the filter as in the previous step.

18. Repeat with a second 500 μL of Wash Solution 2/3.
19. After discarding the flow-through from the last wash, replace the Filter Cartridge in the same Collection Tube and centrifuge the assembly at 10,000 × *g* for 1 min to remove residual fluid from the filter.
20. Transfer the Filter Cartridge into a fresh Collection Tube (supplied in kit).
21. Apply 50 μL of preheated (95 °C) Elution Solution or nuclease-free water to the center of the filter.
22. Centrifuge for ~30 s to recover the RNA.
23. Repeat the elution once more with an addition aliquot of 50 μL of Elution Solution or nuclease-free water.
24. Collect the eluate (which contains the RNA) and place on ice for immediate use, or store it at ≤−20 °C [15].

3.8 Analysis of mRNA Expression by qRT-PCR

1. mRNA is retrotranscribed using the iScript cDNA synthesis Kit by Bio-Rad.
2. RT-PCR in biological replicates is performed using human-specific primers for Notch-transcripts with the following thermocycling conditions.

   95 °C for 10 min.

   40 cycles of 95 °C for 15 s.

   60 °C for 60 s
3. Primers for qRT-PCR analysis are displayed in Table 1.

Table 1
Primers sequences for the detection of Notch family members via qRT-PCR

Target	Forward primer 5′-3′	Reverse primer 5′-3′
Notch1	GAG GCG TGG CAG ACT ATG C	CTT GTA CTC CGT CAG CGT GA
Notch2	ACA GTT GTG TCT GCT CAC CAG GAT	GCG GA A A CC ATT CA C A CC GTT GA T
Notch3	TGG CGA CCT CAC TTA CGA CT	CAC TGG CAG TTA TAG GTG TTG AC
Notch4	GAT GGG CTG GAC ACC TAC AC	CAC ACG CAG TGA AAG CTA CCA
Jagged1	GTC CAT GCA GAA CGT GAA CG	GCG GGA CTG ATA CTC CTT GA
Jagged2	TGG GCG GCA ACT CCT TCT A	GCC TCC ACG ATG AGG GTA AA
DLL1	GAT TCT CCT GAT GAC CTC GCA	GTG TGC AGT AGT TCA GGT CCT
DLL3	CCG GAT GCA CTC AAC AAC CT	CCG AGC GTA GAT GGA AGG AG
DLL4	TGG GTC AGA ACT GGT TAT TGG A	GTC ATT GCG CTT CTT GCA CAG
Hes1	TCA ACA CGA CAC CGG ATA AAC	GCC GCG AGC TAT CTT TCT TCA
Hes5	GCC CAT CTT CTG CCA AGT GT	GAG CCC CGG CAC TAC AAA TA
β-actin	GTC TCC TCT GAC TTC AAC AGC G	ACC ACC CTG TTG CTG TAG CCA A

4. Expression levels is calculated by the comparative ΔΔCt method ($2^{-\Delta\Delta Ct}$ formula), after being normalized to the Ct-value of the β-*Actin* housekeeping gene.

4 Notes

1. Patient inclusion criteria: Age range: 18–35 years, nonsmoking, oral and general healthy conditions.
2. The same principle can be applied to enrich the content of exosome with any other Notch ligand and Notch receptors (DLL1, Jagged1, Jagged2, Notch1, Notch2, Notch3, and Notch4).
3. Resuspension volume depends on the exosome source, and desired volume and sample concentration.
4. If you are not immediately moving on to isolate RNA after resuspending the sample in the Exosome Resuspension Buffer or 1× PBS, store it at −20 °C until needed (Total Exosome RNA and Protein Isolation Kit, catalog #4478545; Thermo Fisher Scientific, Rockford, IL, United States) [16].
5. Be sure to withdraw the bottom phase containing acid-phenol–chloroform, not the aqueous buffer that lies on top of the mixture.
6. Final RNA isolation is performed on the aqueous phase from acid-phenol–chloroform extraction. Before starting, be sure to: Preheat the Elution buffer or nuclease-free water to 95 °C for use in eluting the RNA from the filter at the end of the procedure. Nuclease-free water can be used in place of the elution buffer, especially if concentrating the RNA with a centrifugal vacuum concentrator. 100% ethanol must be at room temperature. If the 100% ethanol is stored cold, warm it to room temperature before starting the RNA isolation.

Acknowledgments

CP and TAM were supported by the UZH (University of Zurich). CP is a recipient of the Olga Mayenfisch Stiftung grant. *The author would like to thank* Christian Thomas Meisel *for primers design.*

References

1. Trubiani O, Marconi GD, Pierdomenico SD, Piattelli A, Diomede F, Pizzicannella J (2019) Human oral stem cells, biomaterials and extracellular vesicles: a promising tool in bone tissue repair. Int J Mol Sci 20(20):4987. https://doi.org/10.3390/ijms20204987
2. Pulliero A et al (2019) Extracellular vesicles in biological fluids. A biomarker of exposure to cigarette smoke and treatment with chemopreventive drugs. J Prev Med Hyg 60(4):E327–E336. https://doi.org/10.15167/2421-4248/jpmh2019.60.4.1284

3. Artavanis-Tsakonas S (1999) Notch signaling: cell fate control and signal integration in development. Science 284(5415):770–776. https://doi.org/10.1126/science.284.5415.770
4. Koch U, Lehal R, Radtke F (2013) Stem cells living with a Notch. Development 140(4): 689–704. https://doi.org/10.1242/dev.080614
5. Lai EC (2004) Notch signaling: control of cell communication and cell fate. Development 131(5):965–973. https://doi.org/10.1242/dev.01074
6. Ohtsuka T, Ishibashi M, Gradwohl G, Nakanishi S, Guillemot F, Kageyama R (1999) Hes1 and Hes5 as notch effectors in mammalian neuronal differentiation. EMBO J 18(8):2196–2207. https://doi.org/10.1093/emboj/18.8.2196
7. Wang Q, Lu Q (2017) Plasma membrane-derived extracellular microvesicles mediate non-canonical intercellular NOTCH signaling. Nat Commun 8(1):709. https://doi.org/10.1038/s41467-017-00767-2
8. McGough IJ, Vincent J-P (2016) Exosomes in developmental signalling. Development 143(14):2482–2493. https://doi.org/10.1242/dev.126516
9. Diomede F et al (2019) A novel role of ascorbic acid in anti-inflammatory pathway and ROS generation in HEMA treated dental pulp stem cells. Materials (Basel) 13(1):130. https://doi.org/10.3390/ma13010130
10. Diomede F et al (2018) Three-dimensional printed PLA scaffold and human gingival stem cell-derived extracellular vesicles: a new tool for bone defect repair. Stem Cell Res Ther 9(1): 104. https://doi.org/10.1186/s13287-018-0850-0
11. Pizzicannella J et al (2019) Engineered extracellular vesicles from human periodontal-ligament stem cells increase VEGF/VEGFR2 expression during bone regeneration. Front Physiol 10:512. https://doi.org/10.3389/fphys.2019.00512
12. Alzahrani FA et al (2018) Potential effect of exosomes derived from cancer stem cells and MSCs on progression of DEN-induced HCC in rats. Stem Cells Int 2018:8058979. https://doi.org/10.1155/2018/8058979
13. Sheldon H et al (2010) New mechanism for Notch signaling to endothelium at a distance by Delta-like 4 incorporation into exosomes. Blood 116(13):2385–2394. https://doi.org/10.1182/blood-2009-08-239228
14. Théry C, Amigorena S, Raposo G, Clayton A (2006) Isolation and characterization of exosomes from cell culture supernatants and biological fluids. Curr Protoc Cell Biol Chapter 3, Unit 3.22. https://doi.org/10.1002/0471143030.cb0322s30
15. Silvestro S et al (2020) Extracellular vesicles derived from human gingival mesenchymal stem cells: a transcriptomic analysis. Genes (Basel) 11(2):118. https://doi.org/10.3390/genes11020118
16. Chiricosta L et al (2020) Extracellular vesicles of human periodontal ligament stem cells contain MicroRNAs associated to proto-oncogenes: implications in cytokinesis. Front Genet 11:582. https://doi.org/10.3389/fgene.2020.00582

Chapter 17

In Vivo and Ex Vivo Experimental Approach for Studying Functional Role of Notch in Pulmonary Vascular Disease

Pritesh P. Jain, Susumu Hosokawa, Aleksandra Babicheva, Tengteng Zhao, Jiyuan Chen, Patricia A. Thistlethwaite, Ayako Makino, and Jason X. -J. Yuan

Abstract

Pulmonary arterial hypertension (PAH) is a severe disease characterized by sustained vasoconstriction, concentric wall thickening and vascular remodeling leading to increased pulmonary vascular resistance, causing right heart failure and death. Acute alveolar hypoxia causes pulmonary vasoconstriction, while sustained hypoxia causes pulmonary hypertension (PH). Activation of Notch signaling is implicated in the development of PAH and chronic hypoxia induced PH via partially its enhancing effect on Ca^{2+} signaling in pulmonary arterial smooth muscle cells (PASMCs). Pharmacological experiments and genetic approach using animal models of experimental PH (e.g., chronic hypoxia-induced PH) have been routinely utilized to study pathogenic mechanisms of PAH/PH and identify novel therapeutic targets. In this chapter, we describe protocols to investigate the role of Notch by measuring pulmonary hemodynamics in vivo and pulmonary arterial pressure ex vivo in mouse models of experimental PH. Using these experimental protocols, one can study the role of Notch or Notch signaling pathway in the pathogenic mechanisms of pulmonary vascular disease and develop novel therapies by targeting Notch ligands and receptors.

Key words Notch signaling, Hypoxic pulmonary vasoconstriction, Hypoxia, Pulmonary hypertension

1 Introduction

Pulmonary arterial hypertension (PAH) is a severe disease characterized by sustained pulmonary vasoconstriction, concentric pulmonary vascular remodeling, and right ventricular hypertrophy which result in increased pulmonary vascular resistance and pulmonary arterial pressure in patients with PAH and animals with experimental PH. Acute alveolar hypoxia causes pulmonary vasoconstriction (HPV), while sustained hypoxia causes PH.

Notch signaling plays an important role in vascular development and Notch-3, a subtype of Notch receptor, is implicated in

Dongyu Jia (ed.), *Notch Signaling Research: Methods and Protocols*,
Methods in Molecular Biology, vol. 2472, https://doi.org/10.1007/978-1-0716-2201-8_17,

the development of PAH [1, 2]. Upregulated expression of Notch-3 from lung tissue of PAH patients has been reported when compared to control human subjects [2]. Furthermore, upregulation of Notch-3 expression has been demonstrated in animals with hypoxia-induced PH and monocrotaline (MCT)-induced PH [2]. Knockout of Notch-3 prevents the mice from developing hypoxia-induced PH [2]. We have previously shown that acute inhibition of Notch signaling in ex vivo isolated perfused/ventilated mouse lung attenuates hypoxic pulmonary vasoconstriction (HPV) [3]. Moreover, pharmacological targeting of Notch signaling pathway using γ-secretase inhibitor, *N*-[*N*-(3,5-difluorophenacetyl)-L-alanyl]-*S*-phenylglycine *t*-butyl ester (DAPT), in vivo partially reversed the development of hypoxia-induced PH in mice [3].

Isolated perfused/ventilated mouse lung is an important ex vivo model to study the role of ion channels and membrane receptors in HPV using specific pharmacological activators or blockers [4]. We have previously shown that, using 20% FBS in physiological salt solution (PSS), a consistent HPV response is observed with an amplitude of around 3–4 mmHg without the need of any pretone or priming substances [5]. It is now routinely used to study HPV due to lack of influence of central and peripheral nervous system, and systemic circulation but provides an intact lung for the experimentation [5–7]. Our data indicate that the design to utilize isolated perfused/ventilated lung from wild-type (WT) and knockout (KO) mice is useful for studying the mechanisms of HPV by comparing KO/transgenic and WT mice. Although the book chapter is focused on isolated perfused lung, it is worth noting that intact animal models, isolated pulmonary artery rings and pulmonary artery smooth muscle cells are also important preparations to study HPV [8–15].

An ideal animal model should exhibit key symptomatic features and histopathologic characteristics of human PH. Since PH is a complex heterogeneous disease, and until now, no single animal model can entirely recapitulate the features of human PAH [16]. An experimental animal model should present key features such as increased mean pulmonary arterial pressure (mPAP) or right ventricle systolic pressure (RVSP), a surrogate measure for pulmonary arterial systolic pressure, and increased pulmonary vascular resistance. The animal model should also replicate common histopathological features of human PH such as pulmonary vascular remodeling including large pulmonary artery stiffening, small artery medial and adventitial hypertrophy and arteriole muscularization [17]. In severe PH model, presence of plexiform lesions (or plexiform-like lesions), occlusive intimal lesions, perivascular inflammation and pruning of peripheral arteries is somehow important [18, 19].

Currently, there are several animal models of PH, although none can completely recapitulate the above mentioned features, at least some models can reflect clinical situation. Chronic hypoxia, Sugen/hypoxia, and monocrotaline are most routinely used classic experimental challenges to develop PH in animals [20]. Experimentally, chronic hypoxia-induced PH (HPH) is a routinely used in mice and rats due to its predictability reproducibility and simplicity [20, 21]. Although mice and rats both develop PH in response to alveolar hypoxia, the severity of phenotype changes variably [22, 23]. Mice develop a milder form of PH compared to rat and this could be attributed to the distinct gene expression profile in both species [24]. Nevertheless, the PH induced by chronic hypoxia is reversible upon return to normoxia in both mice and rats [25]. Thus, these animal models can serve as good starting point for investigation of innovative therapeutics for therapy of PH and could be used in conjunction with more severe models like Sugen/hypoxia- or monocrotaline (MCT)-induced PH. These animal models serve as a tool to provide valuable information about numerous pathways that contributes toward the pathogenesis of PH. Rigorous use of these animal models will permit continuous testing of novel hypothesis pertinent to pathology and treatment of PH.

Here, we describe the in vivo HPH mouse model confirmed by the measurement of hemodynamics parameters and ex vivo isolated perfused mouse lung technique to acutely measure pulmonary arterial pressure by challenging the lungs with alveolar hypoxia to study the functional role of Notch.

2 Materials

2.1 Animals

1. 20–25 g male C57/Bl6 mice.
2. Rodent chow.
3. Rodent cages.
4. Rodent bedding.
5. Pure drinking water.

2.2 Chronic Hypoxia-Induced Pulmonary Hypertension Mice Model

1. Normobaric white color propylene hypoxia chamber with a clear acrylic door 30″ W × 20″ D × 20′H.
2. Oxygen sensor.
3. CO_2 absorbent (Sodasorb®).

2.3 Right Heart Cathcterization

1. Plexiglass induction chamber.
2. Oxygen tank.
3. Isoflurane.

4. Isoflurane vaporizers (Fluosorbers) to scavenge excess isoflurane.
5. Heating pad.
6. Adhesive tape.
7. Mouse Pressure Volume Catheter 1F, 4E, 3.0 mm, 4.5 cm, PI (PVR 1030, Millar Instruments, Colorado, USA).
8. Surgical braided black silk suture 6-0 metric 0.7.
9. Cotton bud tips.
10. Surgical tools.
11. Microscope.
12. MPVS-Ultra single segment foundation system for Mice.
13. Data acquisition and analysis software.

2.4 Isolated Perfused Mouse Lung Model

1. Temperature controlled basic unit for isolated perfused mouse lung.
2. Ismatec Peristaltic pump.
3. Pressure transducer P75.
4. Rodent mini ventilator.
5. Hypoxia and normoxia gas tank.
6. Tracheal cannula, internal diameter (ID) 1.0 mm, outer diameter (OD) 1.3 mm, length (*L*) 20 mm.
7. Pulmonary Artery Cannula, Stainless Steel, ID 1.0 mm, OD 1.3/1.6 mm, *L* 28 mm.
8. Atrial Cannula for Mouse, ID 1.0 mm, OD 1.6 mm, *L* 24 mm.
9. Powerlab 8/30, Quad Bridge Amp, and LabChart.
10. The physiological salt solution (PSS) was composed of the following (mM): NaCl 120, KCl 4.3, $NaHCO_3$ 19, KH_2PO_4 1.1, glucose 10, $CaCl_2$ 1.8, and $MgCl_2$ 1.2 (pH 7.4). 40 mM K^+ (40K) PSS was prepared by replacing NaCl with equimolar KCl. 20% FBS was added to the PSS and now the solution is termed as perfusate.

3 Methods

3.1 Animals

1. Wild type (WT) male C57BL/6 mice were acclimatized to new environment in the vivarium of 12-h light–dark cycles at Room temperature (RT) and 40–50% humidity for a minimum of 48 h after receiving them from a commercial supplier, Jackson's laboratory. Notch 3 homozygous global knockout (*Notch 3*$^{-/-}$) mice were purchased from Jackson's laboratory (Stock No: 010547). All the animal experiments were approved by the IACUC of University of California, San Diego that complied with international guidelines.

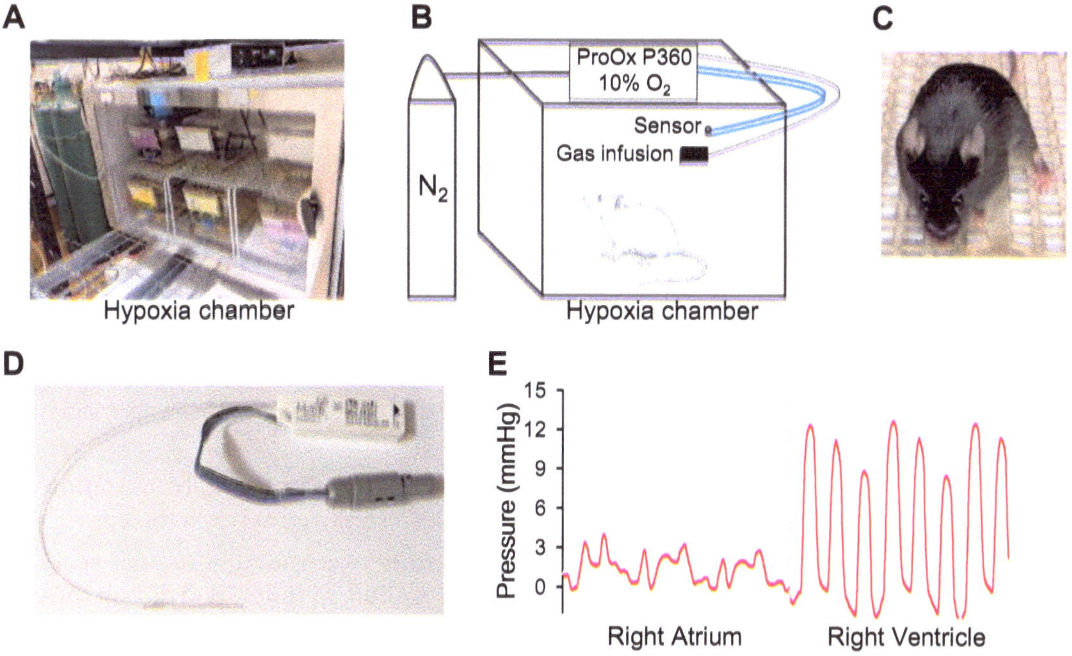

Fig. 1 Experimental set up for in vivo induction of chronic hypoxia-induced PH in mice. (**a**) Image of the BioSpherix hypoxia chamber we used in the experiment, which can accommodate six cages of mice. (**b**) schematic diagram of the normobaric hypoxia chamber (BioSpherix) which has an O_2 controller ProOx360 that detects alterations in O_2 concentrations, adjust, and maintain 10% O_2 inside of the chamber by purging N_2 gas. (**c**) A wild type C57Bl6 mouse. (**d**) A Miller pressure catheter for RHC in the study. (**e**) Representative recording of pressure when the catheter is in the right atrium and right ventricle

2. Mice were randomly assigned in four groups: Wild type (Normoxia and Hypoxia), *Notch 3*$^{-/-}$ (Normoxia and Hypoxia).

3.2 Induction of Pulmonary Hypertension

1. Mice were placed in normobaric hypoxic chamber for 4 weeks (at 10% O_2). An O_2 controller/sensor continuously monitored partial pressure of O_2 (PO_2). N_2 gas is continuously flushed in the chamber to maintain the O_2 concentration at 10% (Fig. 1a–d). The wild type normoxic mice (control group) were placed in the same room exposed to room air (21% O_2).
2. Food and water were provided ad libitum and bedding in cages were changed once per week. CO_2 absorbent was changed when pellets changed color from white to blue.
3. The mice were maintained at 12-h light/dark cycle at room temperature with humidity between 40 and 50% for 4 weeks.

3.3 Right Heart Catheterization

1. Place the tip of mouse 1.4F PV catheter in syringe filled with warm phosphate buffered saline (PBS) for at least 30 min. Turn on the MPVS system and data acquisition software (*see* **Note 1**).
2. Weigh and record the body weight of mice (*see* **Note 2**).

3. Anesthetize the mice using 2% isoflurane by placing them in anesthetic chamber. Verify the anesthesia by checking the absence of toe reflex. Shave the neck and chest of the mouse.
4. Place the mice in supine position with upper and lower limbs taped on the heating pad having mice head pointing toward the personnel.
5. Clean the surgical area with 70% ethanol and Betadine.
6. Work aptly and relaxed, the remainder of the right heart catheterization could be completed within 10 min, ideally 5–6 min (*see* **Note 3**).
7. Carefully cut the skin from mandible to sternum. Carefully separate the adipose tissue to expose the external right jugular vein.
8. Micro dissect the right jugular vein by carefully separating the connective tissue using clean sterilized surgical forceps.
9. Place two pieces of silk suture over the right jugular vein. Tie the suture closest to the mandible to shut blood flow in the vein and tape it. Take the second suture close to sternum and make a lose knot. Make a small incision in the right jugular vein between the sutures using fine micro scissors (*see* **Note 4**).
10. Gently hold the Millar pressure catheter (Fig. 1d), insert in the hole of right jugular vein, slightly tighten the suture and advance further in the right heart. While passing the catheter, keep observing the screen for pressure curves. Initially, the right atrium peaks (between 0 and 2 mmHg) will be displayed, advancing the catheter forward will display right ventricle peaks (between 0 and 20 mmHg) as shown in Fig. 1e. Record the pressure for 2 min. Select 1 s within the stable 2 min recording and plot the values for right ventricle systolic pressure (RVSP), estimated mean pulmonary arterial pressure (mPAP), right ventricular contractility (RV-±dP/dt) and heart rate (HR) for groups, Wild type (Normoxia and Hypoxia), *Notch*$^{-/-}$ (Normoxia and Hypoxia). As shown in Fig. 2a, b, exposure of WT mice to chronic hypoxia increased RVSP, mPAP, and RV-±dP/dt compared to normoxic WT mice. In *Notch3*$^{-/-}$ mice, nevertheless, RVSP, mPAP, and RV-±dP/dt during chronic hypoxia was significantly lower than the RVSP, mPAP, and RV-±dP/dt in WT mice. Global KO of Notch-3 alone had minimal effect on basal RVSP mPAP, and RV-±dP/dt, however deletion of Notch3 significantly inhibited the hypoxia-mediated increase in RVSP.
11. For Fulton index measurements, isolate the heart. Carefully dissect the right ventricle (RV) from left ventricle (LV) and septum (S) and weigh them individually. Right ventricle hypertrophy (RVH) is evaluated by dividing the weight of RV to the weight of LV + S. The ratio obtained is also known as Fulton

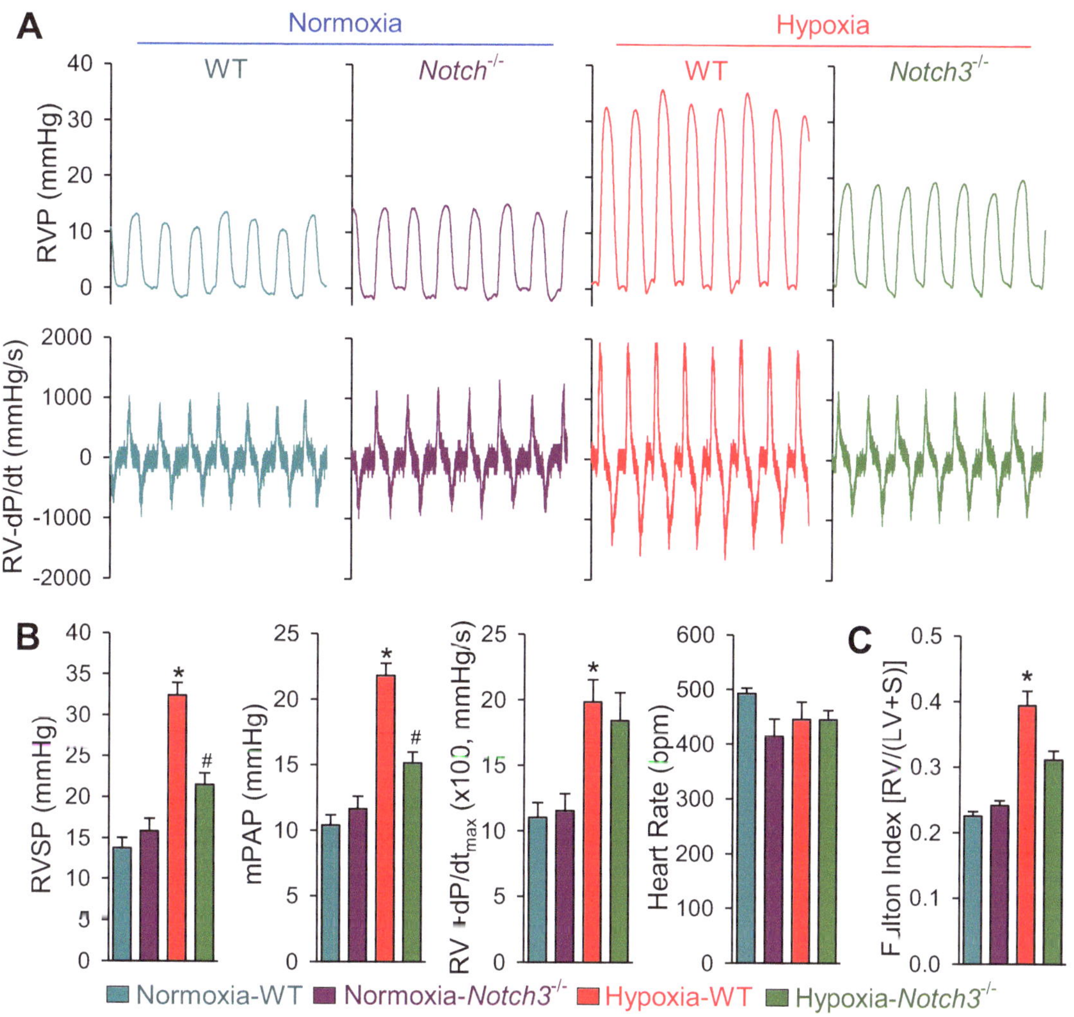

Fig. 2 Chronic hypoxia induced pulmonary hypertension is attenuated in *Notch3*$^{-/-}$ mice. (**a**) Representative records of right ventricular pressure (RVP) (upper panels) and right ventricle (RV) contractility (RV-±dP/dt) (lower panels) in wild type (WT) and Notch-3-KO (*Notch3*$^{-/-}$) mice under normoxic (Normoxia) and hypoxic (Hypoxia) conditions. (**b**) Summarized data (means) showing right ventricular systolic pressure (RVSP), estimated mean pulmonary arterial pressure (mPAP), RV-+dP/ta_x, and heart rate in normoxic WT ($n = 4$) and *Notch3*$^{-/-}$ ($n = 5$) mice and chronically hypoxic WT ($n = 7$) and *Notch3*$^{-/-}$ ($n = 7$) mice. $^*p < 0.05$ vs. Normoxia-WT; $^{\#}p < 0.05$ vs. Hypoxia-WT (ANOVA). (**c**) Summarized data (means) showing Fulton Index, the ratio of the RV weight to the left ventricle (LV) and septum (S) weight [RV/(LV + S)], in normoxic WT ($n = 7$) and *Notch3*$^{-/-}$ ($n = 7$) mice and chronically hypoxic WT ($n = 9$) and *Notch3*$^{-/-}$ ($n = 9$) mice. $^*p < 0.05$ vs. Normoxia-WT (ANOVA)

index that was increased after chronic hypoxia as shown in Fig. 2c. Global knockout of Notch3 alone had minimal effect on Fulton index however, deletion of Notch3 showed a trend in RV hypertrophy as measured by the Fulton index ratio. These data indicate that Notch3 is a requisite for the development of PH.

3.4 Isolated Perfused Mouse Lung

1. Switch on the water bath to circulate distilled water at 37 °C into the interior of isolated perfused lung system as shown in Fig. 3a. All the tubing's connection, the peristaltic pump, pressure transducer and cannula should be flushed with clean distilled water with no air bubbles.
2. The pressure transducer is zeroed by pressing the automatic zero button on PLUGSYS amplifier by opening the metallic valve toward the atmosphere. Once zeroed, the valve toward the atmosphere is closed and valve toward the cannula is opened.
3. Anesthetize the mice by intraperitoneally injecting 120 mg/kg pentobarbital sodium and ensure deep anesthesia by confirming no response to toe reflex. Prepare the mice for surgery by gently pinning the limbs to rubber bed on the isolated perfused system.
4. Clean the chest with Betadine and 70% ethanol and cut 1 cm skin at the neck using straight scissors.
5. Once the interior of neck is exposed, separate the glandular tissue and muscle, to identify the trachea. Carefully expose the trachea.
6. Use bent forceps under the trachea and place a lose suture knot under the trachea. Make a small incision using fine micro scissors and slide the tracheal cannula into the trachea. Tighten the suture knot to secure the tracheal cannula. Quickly begin the ventilation with room air using the mini ventilator (80 breaths/min with deep inspiration).
7. Spray the chest with Betadine and 70% ethanol; open the chest by median sternotomy. Carefully remove the pericardium, thymus and adipose tissue without damaging the lungs. Immediately, inject heparin (20 IU) which acts as anticoagulant (*see* **Note 5**).
8. Perform a right ventriculotomy by smoothly scooping a blunt tweezer from top of heart under left atrium/ventricle to proceed forward under the aorta and pulmonary artery (PA) without any resistance and make a lose knot. Prime the PA catheter by switching on the peristaltic pump at low speed. Make a small cut in right ventricle free wall, insert the stainless steel catheter into the main PA, and tighten the knot. Pulmonary artery pressure (PAP) is measured using pressure transducer, which is connected to PA catheter. The other end of the PA catheter is connected to Tygon tubing for perfusion (*see* **Note 6**) as shown in Fig. 3b.
9. Perform a left ventriculotomy by making a small incision in left ventricle and insert the stainless steel left atrium cannula. The other end of the left atrium cannula is connected to Tygon tubing for draining the perfusion (*see* **Note 7**) as shown in Fig. 3b.

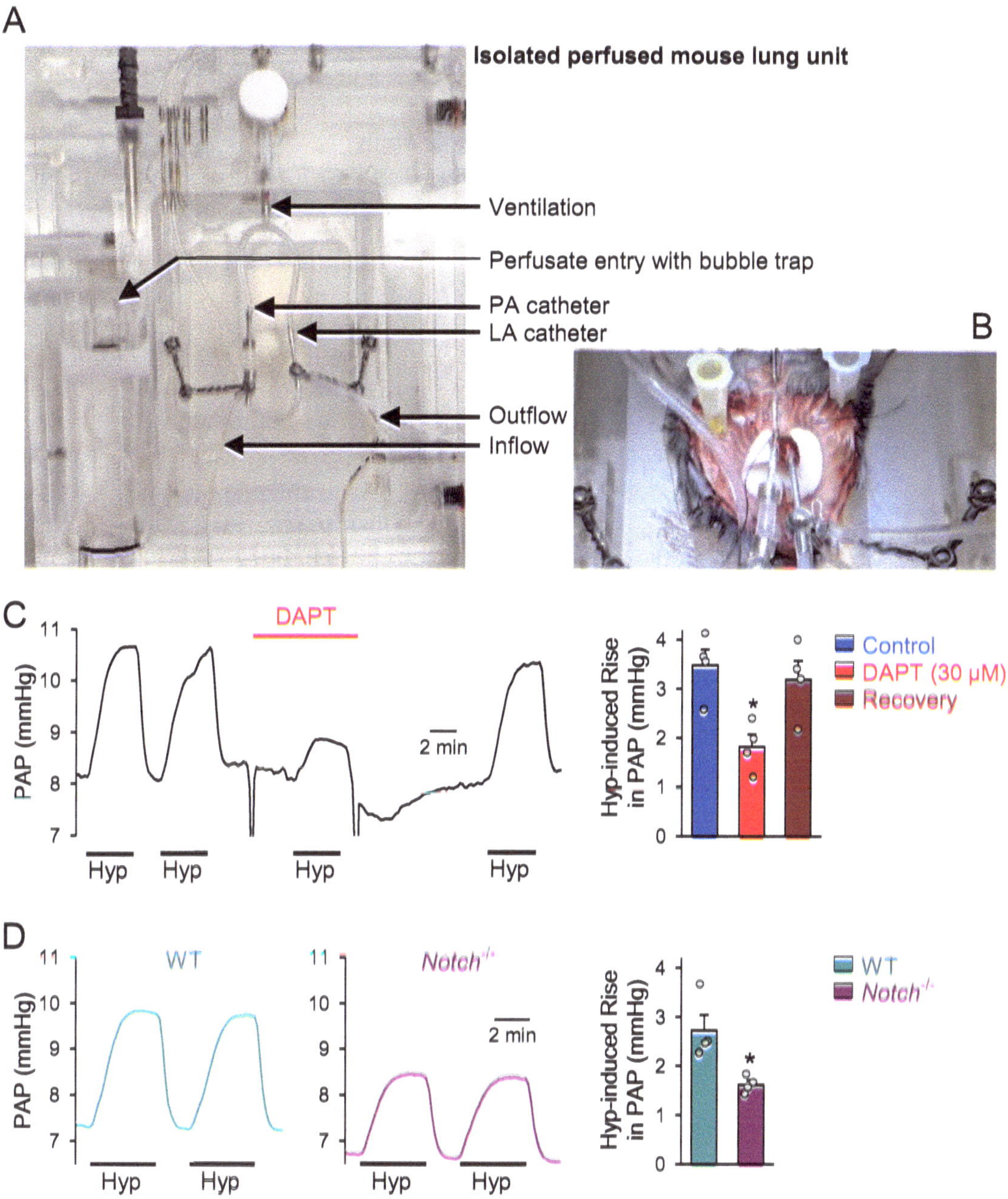

Fig. 3 Experimental setup for ex vivo measurement of pulmonary artery pressure (PAP) in isolated perfused/ventilated mouse lungs. (**a**) Schematic diagram of isolated perfused/ventilated lung setup. As depicted by black arrows, the warm physiologic salt solution (PSS, perfusate) is perfused by a peristaltic pump via bubble trap to pass through the pulmonary artery (PA) catheter into the pulmonary circulation, then the perfusate is drained via outflow left atrium (LA) catheter (LA). (**b**) Picture of isolated perfused/ventilated mouse lung turning white after intrapulmonary perfusion of perfusate. (**c**) Representative records (*left*) showing PAP before, during and after ventilation of hypoxic gas (Hyp, 1% O_2) in the absence and presence (DAPT) of intrapulmonary perfusion of 30 μM DAPT (a Notch inhibitor). Summarized data (means, $n = 4$; *right*) showing the hypoxia-induced increases in PAP before (Control), during (DAPT), and after (Recovery) intrapulmonary superfusion of DAPT. $^*p < 0.05$ vs. Control (ANOVA). (**d**) Representative records (*left*) showing PAP before, during and after ventilation of hypoxic gas (Hyp, 1% O_2) in isolated lungs from WT and *Notch3*$^{-/-}$ mice. Summarized data (means, $n = 4$; *right*) showing the hypoxia-induced increases in PAP in isolated lungs from WT and *Notch3*$^{-/-}$ mice. $^*p < 0.05$ vs. WT

10. Gradually increase the flow and then maintain the pulmonary perfusion by keeping the flow rate at 1 mL/min using pump. The outflow tubing will be slight pinkish and then as perfusion progresses, will turn clear and lungs will turn white in color replacing the blood as indicated in Fig. 3b. If no flow is seen from outflow tubing, readjust the left atrium (LA) cannula. Even after adjusting the LA cannula, if no outflow perfusate is seen, check for a tear in PA. Any damage to PA cannot be repaired and one should restart the experiment with fresh mice. To achieve a stable baseline, perfuse the lung three times with high K^+ (40 mM)-containing physiologic salt solution (PSS) (40K) before beginning the experiments.
11. After a stable baseline is achieved, stimulate lungs acutely with repeated hypoxic challenges by ventilating hypoxic gas (1% O_2 in N_2) into airway. Then perfuse the lung with γ-secretase inhibitor, DAPT (indirect inhibition of Notch signaling pathway) for 10 min and reventilate the lungs using hypoxic gas in the presence of DAPT. After alveolar hypoxia stimulus, wash out the DAPT to obtain a recovery as shown in Fig. 3c. Intrapulmonary perfusion of DAPT significantly attenuated the alveolar hypoxia-induced rise in PAP due to HPV (Fig. 3c).
12. Similarly, acute hypoxia-induced rise in PAP was measured in isolated perfused/ventilated lungs from WT and *Notch3*$^{-/-}$ mice using the same procedure as mentioned above. It was found that acute alveolar hypoxia-induced rise in PAP was significantly inhibited in the isolated perfused/ventilated lungs from *Notch3*$^{-/-}$ mice in comparison to the lungs from wild-type mice (Fig. 3d). These data indicate that Notch-3 or activation of Notch-3 signaling may be involved in mediating or modulating acute HPV.

4 Notes

1. It is important that the catheter is immersed in warm PBS for 30 min to allow stable recordings.
2. Handle the mice gently and quietly to prevent stress and minimize internal variation between the groups.
3. It is critical that surgical site should be kept moist for the catheter to easily glide in the vein and proceed inside the vein over this moisture.
4. It is very important to breathe and relax while conducting the procedure which will ensure that hands are steady to obtain consistent recordings.
5. Injecting heparin prevents clotting of blood in lung and permits a smooth and uninterrupted perfusion.

6. Using dull forceps minimizes the chances of puncturing the PA. Make sure that all the tubing's are free of air bubbles. Lung perfusion with distilled water or air bubbles will cause edema.
7. Both PA and LA catheters should be parallel to the plane of the isolated perfused lung system to avoid edema formation.

Acknowledgments

This work is supported in part by the grants from the National Heart, Lung and Blood Institute of the National Institutes of Health (R35 HL135807 and R01 HL146764). Aleksandra Babicheva was supported by the American Heart Association Postdoctoral Fellowship (20POST35210959).

References

1. Gridley T (2010) Notch signaling in the vasculature. Curr Top Dev Biol 92:277–309
2. Li X, Zhang X, Leathers R, Makino A, Huang C, Parsa P et al (2009) Notch3 signaling promotes the development of pulmonary arterial hypertension. Nat Med 15(11): 1289–1297
3. Smith KA, Voiriot G, Tang H, Fraidenburg DR, Song S, Yamamura H et al (2015) Notch activation of Ca^{2+} signaling in the development of hypoxic pulmonary vasoconstriction and pulmonary hypertension. Am J Respir Cell Mol Biol 53(3):355–367
4. Jain PP, Hosokawa S, Xiong M, Babicheva A, Zhao T, Rodriguez M et al (2020) Revisiting the mechanism of hypoxic pulmonary vasoconstriction using isolated perfused/ventilated mouse lung. Pulm Circ 10(4): 2045894020956592
5. Yoo HY, Zeifman A, Ko EA, Smith KA, Chen J, Machado RF et al (2013) Optimization of isolated perfused/ventilated mouse lung to study hypoxic pulmonary vasoconstriction. Pulm Circ 3(2):396–405
6. Keseru B, Barbosa-Sicard E, Schermuly RT, Tanaka H, Hammock BD, Weissmann N et al (2010) Hypoxia-induced pulmonary hypertension: comparison of soluble epoxide hydrolase deletion vs. inhibition. Cardiovasc Res 85(1): 232–240
7. Fuchs B, Rupp M, Ghofrani HA, Schermuly RT, Seeger W, Grimminger F et al (2011) Diacylglycerol regulates acute hypoxic pulmonary vasoconstriction via TRPC6. Respir Res 12:20
8. Karamsetty MR, Klinger JR, Hill NS (2002) Evidence for the role of p38 MAP kinase in hypoxia-induced pulmonary vasoconstriction. Am J Physiol Lung Cell Mol Physiol 283(4): L859–L866
9. Aaronson PI, Robertson TP, Knock GA, Becker S, Lewis TH, Snetkov V et al (2006) Hypoxic pulmonary vasoconstriction: mechanisms and controversies. J Physiol 570(pt 1): 53–58
10. Rathore R, Zheng YM, Li XQ, Wang QS, Liu QH, Ginnan R et al (2006) Mitochondrial ROS-PKCepsilon signaling axis is uniquely involved in hypoxic increase in [Ca2+]i in pulmonary artery smooth muscle cells. Biochem Biophys Res Commun 351(3):784–790
11. Wang Z, Jin N, Ganguli S, Swartz DR, Li L, Rhoades RA (2001) Rho-kinase activation is involved in hypoxia-induced pulmonary vasoconstriction. Am J Respir Cell Mol Biol 25(5): 628–635
12. Waypa GB, Schumacker PT (2002) O_2 sensing in hypoxic pulmonary vasoconstriction: the mitochondrial door re-opens. Respir Physiol Neurobiol 132(1):81–91
13. Zheng YM, Wang QS, Rathore R, Zhang WH, Mazurkiewicz JE, Sorrentino V et al (2005) Type-3 ryanodine receptors mediate hypoxia-, but not neurotransmitter-induced calcium release and contraction in pulmonary artery smooth muscle cells. J Gen Physiol 125(4): 427–440
14. Sylvester JT, Shimoda LA, Aaronson PI, Ward JP (2012) Hypoxic pulmonary vasoconstriction. Physiol Rev 92(1):367–520
15. Moudgil R, Michelakis ED, Archer SL (2005) Hypoxic pulmonary vasoconstriction. J Appl Physiol 98(1):390–403

16. Ryan J, Bloch K, Archer SL (2011) Rodent models of pulmonary hypertension: harmonisation with the world health organisation's categorisation of human PH. Int J Clin Pract Suppl 172:15–34
17. Tuder RM (2017) Pulmonary vascular remodeling in pulmonary hypertension. Cell Tissue Res 367(3):643–649
18. Stacher E, Graham BB, Hunt JM, Gandjeva A, Groshong SD, McLaughlin VV et al (2012) Modern age pathology of pulmonary arterial hypertension. Am J Respir Crit Care Med 186(3):261–272
19. Rabinovitch M, Guignabert C, Humbert M, Nicolls MR (2014) Inflammation and immunity in the pathogenesis of pulmonary arterial hypertension. Circ Res 115(1):165–175
20. Stenmark KR, Meyrick B, Galie N, Mooi WJ, McMurtry IF (2009) Animal models of pulmonary arterial hypertension: the hope for etiological discovery and pharmacological cure. Am J Physiol Lung Cell Mol Physiol 297(6): L1013–L1032
21. Stenmark KR, Fagan KA, Frid MG (2006) Hypoxia-induced pulmonary vascular remodeling: cellular and molecular mechanisms. Circ Res 99(7):675–691
22. Grover RF (1965) Pulmonary circulation in animals and man at high altitude. Ann N Y Acad Sci 127(1):632–639
23. Hoshikawa Y, Nana-Sinkam P, Moore MD, Sotto-Santiago S, Phang T, Keith RL et al (2003) Hypoxia induces different genes in the lungs of rats compared with mice. Physiol Genomics 12(3):209–219
24. Voelkel NF, Tuder RM, Wade K, Hoper M, Lepley RA, Goulet JL et al (1996) Inhibition of 5-lipoxygenase-activating protein (FLAP) reduces pulmonary vascular reactivity and pulmonary hypertension in hypoxic rats. J Clin Invest 97(11):2491–2498
25. Voelkel NF, Tuder RM (2000) Hypoxia-induced pulmonary vascular remodeling: a model for what human disease? J Clin Invest 106(6):733–738

Chapter 18

Metastasis Model to Test the Role of Notch Signaling in Prostate Cancer

Shiqin Liu, En-chi Hsu, Michelle Shen, Merve Aslan, and Tanya Stoyanova

Abstract

Distant metastasis is the main cause of death in prostate cancer patients. Notch signaling plays an important role in driving prostate cancer aggressiveness and metastasis. In this chapter, we describe a protocol to measure prostate cancer metastatic colonization, incidences of metastasis, accurately quantify the burden of metastasis, and test the role of NOTCH1 receptor on prostate cancer metastatic colonization and homing to distant sites. The metastasis model presented here is established by intracardiac injection of control human prostate cancer cells and NOTCH1 downregulated cells. The cells are engineered to express both red fluorescent protein (RFP) and luciferase. In this model, whole body bioluminescence imaging, high-resolution, and quantitative fluorescence imaging are utilized for quantitative assessment of metastatic colonization and metastasis burden. Further, histopathology analyses of diverse metastatic organs are performed. This model is a powerful and versatile tool to investigate the mechanisms underlying the function of NOTCH receptors in metastatic colonization in prostate cancer.

Key words Prostate cancer, Metastatic colonization, Intracardiac injection, Notch

1 Introduction

Prostate cancer remains the most commonly diagnosed non-skin cancer among men in the USA [1]. Clinically, significant improvement of median survival has been made for patients with metastatic prostate cancer, due to the development of biomarkers for early detection of high-risk prostate cancer as well as new therapeutic strategies for advanced prostate cancer [2–16]. NOTCH signaling plays a central role in regulation of cell proliferation and differentiation, stem cell maintenance, and cancer initiation and progression [17–24]. NOTCH1 can act as a tumor suppressor, such as in

Shiqin Liu and En-chi Hsu contributed equally with all other contributors.

Supplementary Information The online version contains supplementary material available at [https://doi.org/10.1007/978-1-0716-2201-8_18].

Dongyu Jia (ed.), *Notch Signaling Research: Methods and Protocols*,
Methods in Molecular Biology, vol. 2472, https://doi.org/10.1007/978-1-0716-2201-8_18,

cutaneous squamous cell carcinoma (CSCC) and small cell lung cancer (SCLC), and can also play an oncogenic role in different types of cancers including T-cell leukemia, breast cancer, colon adenocarcinoma, and others [25–29]. In prostate cancer, it has been shown that loss of NOTCH1 activity inhibits metastasis, while activation of NOTCH1 signaling induces luminal cell hyperproliferation and synergizes with multiple pathways such as MYC, Ras/Raf/MAPK, and PTEN/AKT to promote prostate cancer progression and metastasis [18, 20, 21, 30, 31]. Given the important functional role of NOTCH signaling in advanced prostate cancer, numerous metastatic models have been developed to investigate the mechanisms underlying NOTCH signaling and its role in prostate tumorigenesis [18, 21, 30, 32].

Downregulation or loss of NOTCH1 using small interfering RNA or CRISPR-Cas9 system significantly reduces cell migration, invasion, and metastatic colonization in prostate cancer [30, 32]. Moreover, high levels of NOTCH1 are reported in advanced castration-resistant prostate cancer [21]. The levels of NOTCH1 in metastatic cancer cells are also much higher when compared to the normal prostate tissues in the TRAMP (Transgenic adenocarcinoma of the prostate) model [33]. Additionally, targeting the NOTCH activity by gamma secretase inhibition (GSI) decreases the tumor growth of castration-resistant prostate cancer xenografts and enhances docetaxel-mediated tumor response of CRPC xenografts in bone [30, 34]. Additionally, inhibition of Notch combined with Hedgehog signaling pathways suppresses prostate cancer growth and reverses the acquisition of docetaxel resistance, suggesting that activation of NOTCH signaling promotes prostate cancer progression and leads to the development of drug resistance [35].

In this protocol, we describe a powerful and useful model to test the functional role of NOTCH receptors on prostate cancer metastatic colonization and homing to distant sites. We utilized a knockdown strategy to downregulate NOTCH1 in a previously described metastatic prostate cancer cell line model (LNCaP-Trop2-OV, Trop2 driven neuroendocrine prostate cancer, TD-NEPC) [36]. TD-NEPC cells were engineered to express both red fluorescent protein (RFP) and luciferase [36]. This model is established by intracardiac injection of TD-NEPC NOTCH1 knockdown cells and TD-NEPC control cells directly into left ventricle of the mouse heart. Downregulation of NOTCH1 in TD-NEPC cells dramatically decreases prostate cancer metastatic burden and incidences of metastasis compared to control cells accessed by bioluminescence imaging (BLI) and high-resolution and quantitative fluorescence imaging. The model allows for further characterization of the metastatic nodules in distant organs by histopathologic assessment.

2 Materials

2.1 Cell Lines

LNCaP (ATCC, Manassas, VA), and TD-NEPC [36] (*see* **Note 1**).

2.2 Animals

6–8-week-old NSG (NOD-SCID-IL2Rγ-null) mice (Jackson Laboratory, Stock No: 005557) (*see* **Note 2**).

2.3 Reagents

1. Roswell Park Memorial Institute (RPMI) 1640 (Gibco™, 31800022).
2. Fetal bovine serum (FBS) (Sigma-Aldrich, F0926).
3. Penicillin–streptomycin (Sigma-Aldrich, P4333).
4. GlutaMAX (Gibco™, 35050061).
5. Puromycin (MP Biomedicals, 0219453910).
6. 0.05% Trypsin (Sigma-Aldrich, T4174)
7. Phosphate-buffered saline (PBS) (VWR, SH30256).
8. Polybrene (Thermo Fisher Scientific, NC0663391).
9. 0.4% Trypan Blue Solution (Sigma-Aldrich, T8154)
10. Plasmid: Control scramble shRNA pLKO.1 (shCTL) lentiviral vector was a gift from David Sabatini ((CCTAAGGT TAAGTCGCCCTCGCTCGAGCGAGGGCGACTTAACCT TAGG, Addgene plasmid # 1864; http://n2t.net/addgene:1 864; RRID: Addgene_1864) [37]. shNOTCH1 #1: CCGGC GCTGCCTGGACAAGATCAATCTCGAGATT GATCTTGTCCAGGCAGCGTTTTTG (TRCN0000350253); shNOTCH1 #2: CCGGCCGGGA CATCACGGATCATATCTCGAGATATGATCCGT GATGTCCCGGTTTTTG (TRCN0000350330) (Millipore Sigma).
11. Ophthalmic ointment (Neo-Poly-Bac).
12. MediGel CPF (ClearH2O, 74-05-5022).
13. D-Luciferin (PerkinElmer, 122799-5).
14. 10% formalin (Thermo Scientific, 5701TS)
15. Clearify (Fisher Scientific, NC9837230).
16. Ethanol (Thermo Fisher Scientific, 04-355-451).
17. Hematoxylin (Sigma-Aldrich, GHS332).
18. Eosin (Fisher Scientific, E511-25).
19. DAPI Fluoromount-G (SouthernBiotech, 0100-20).
20. Cell scrapers (Fisher Scientific, 08-100-241).
21. BCA protein assay kit (Fisher Scientific, PI23227).
22. Bovine serum albumin (BSA) (Thermo Fisher Scientific, BP9706100).

23. RIPA buffer: 20 mM Tris–HCl pH 7.6, 150 mM NaCl, 1% NP-40, 1% sodium deoxycholate, and 0.1% SDS.
24. Protease and phosphatase inhibitor cocktails (Fisher Scientific, PI78443).
25. 4× SDS Protein Sample Buffer: 40% glycerol, 200 mM Tris/HCl pH 6.8, 8% SDS, 0.08% bromophenol blue, 5% beta-c.
26. Universal nuclease for cell lysis (Fisher Scientific, PI88701).
27. Running buffer: 0.025 M Tris, 0.192 M glycine, 0.1% SDS, pH 8.5.
28. Transferring buffer: 39 mM glycine, 48 mM Tris, 0.0375% SDS, 20% methanol.
29. Washing buffer: TBST (0.1% Tween 20, Fisher Scientific, BP337500).
30. Tris–Glycine mini gels, 8–16%, 15-well (Fisher Scientific, XP08165BOX).
31. Nitrocellulose membrane, 0.2 μm (Bio-Rad, 1620112).
32. Western blotting filter paper (Fisher Scientific, PI84783).
33. Anti-NOTCH1 antibody (Cell Signaling Technology, 3608S).
34. Anti-Actin antibody (Santa Cruz Biotechnology, sc-8432).
35. Goat anti-Rabbit secondary antibody, HRP (Fisher Scientific, PI31462).
36. Goat anti-Mouse secondary antibody, HRP (Fisher Scientific, PI31432).
37. ECL western blotting Substrate (Thermo Fisher Scientific, PI32106).

2.4 Equipment and Supplies

1. Automated cell counter.
2. Insulin syringes, 1/2 cc, 29G × 1/2 in (Exel International, 26028).
3. Digital caliper.
4. Shaver.
5. Lago optical imaging system.
6. Anesthesia system: 2.5% isoflurane with oxygen.
7. Animal surgery tools.
8. Tissue processor.
9. Microtome.
10. Leica stereomicroscope (Leica M205 FA).
11. Histology microscope (Leica DM4).
12. Tissue embedding station (Thermo Scientific).

3 Methods

3.1 Generating the Cell Lines

1. Co-infect LNCaP cells with a lentivirus expressing Trop2 oncogene and red fluorescent protein (RFP) and lentivirus expressing luciferase (LNCaP-Trop2-OV-RFP-Luc, TD-NEPC-RFP-Luc). Confirm RFP and bioluminescence signals at 72 h post infection.
2. Infect TD-NEPC-RFP-Luc cells with lentiviruses carrying control or NOTCH1 targeting short-hairpin RNAs (shCONTROL (shCTL), shNOTCH1 #1, and #2 respectively). A detailed protocol for the generation of lentiviral particles have been described as previously [38].
3. Seed 2 × 10^5 TD-NEPC-RFP-Luc cells per well in 6 well-plate and incubate the plates at 37 °C in a 5% CO_2 humidified incubator overnight.
4. Change the medium with medium containing polybrene (10 μg/ml) 2 h before the infection.
5. Add the shCTL, shNOTCH1 #1, and shNOTCH1 #2 viruses respectively in each well. Gently swirl the plates to mix. Change the medium 72 h post infection.
6. Culture the cells in medium with puromycin (0.5 μg/ml) for 12 days to select for infected cells.
7. Determine the NOTCH1 protein levels by western blot as previously described [21, 36]. Harvest the cells by cell scrapers, and centrifuge at 1000 × *g* for 3 min to collect the cell pellet.
8. Add proper amount of RIPA lysis buffer supplemented with protease-phosphatase inhibitor and universal nuclease reagent. Vortex every 10 min for three times and incubate on ice. Incubate the lysate at room temperature for 30 min.
9. Centrifuge in full speed for 15 min, and transfer supernatant to a new tube.
10. Quantify the protein concentration via BCA assay.
11. Load 40 μg of each sample into each well of 8–16% SDS-PAGE gel. Run the gel at constant voltage of 80–110 V until blue dye reaches the bottom of the gel.
12. Transfer the gel onto a 0.22 μm nitrocellulose membrane.
13. Block the membrane with 5% non-fat milk at room temperature for 1-h and incubate the membrane with anti-NOTCH1 primary antibody (1:1000) at 4 °C overnight.
14. After washing the membrane with TBS-T three times, incubate the membrane with HRP conjugated secondary antibody (1: 5000) at room temperature for 1-h.
15. Develop the membrane with ECL Western Blotting Substrate.

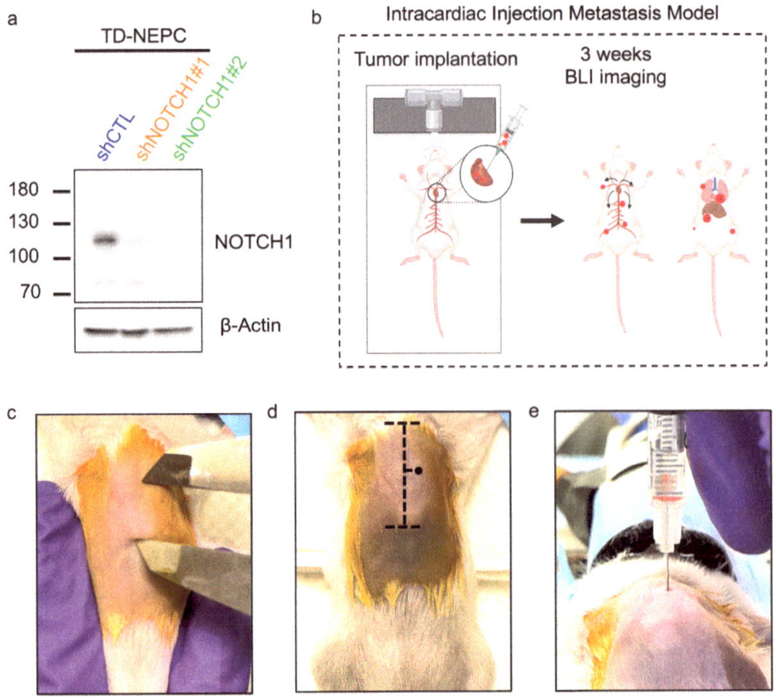

Fig. 1 Intracardiac Injection Technique. (**a**) Validation of knockdown of NOTCH1 in LNCaP Trop2-overexpression (LNCaP-Trop2-OV, TD-NEPC) cell lines by western blot. B-actin is chosen as a loading control. (**b**) Cartoon model of the intracardiac injection metastasis model. Mouse is placed in the supine position on the surgical platform under anesthesia. Tumor cells are injected directly into the left ventricle of the mouse heart. Whole body luminescence imaging (BLI) is obtained after 3 weeks. Diverse organs are harvested for histologic assessment. The cartoon was generated using BioRender (https:/biorender.com). (**c**-**d**) Locate the injection site. Calibrator is used to find the midpoint between the clavicle and the xiphoid process. Injection site is 2–3 mm leftward to the midpoint on the mouse and is indicated with the block dot. (**e**) Image of successful injection into left ventricle of the mouse heart

16. Validate the knockdown of NOTCH1 in TD-NEPC cells by western blot (Fig. 1a).

3.2 Preparing the Cell Lines

Count the cells by Cell Counter with 0.4% trypan blue dye. Resuspend 1×10^5 viable shCTL, shNOTCH1#1, and shNOTCH1#2 cells per mouse in 100 μl PBS before the injection. Keep the cells on the ice before the injection.

3.3 Intracardiac Injection

1. After anesthetizing the mouse, place the mouse on the surgery platform in the supine position, apply ophthalmic ointment, and insert mouse nose into the nose cone for continuous anesthesia.

2. Remove hair on the chest with shaver and disinfect the chest area with alcohol and iodine swabs sequentially.
3. Push up the upper chest of the mouse with a padding on the back for better exposure and visualization of the injection site.
4. Use the caliper to measure the midpoint between the clavicle and the xiphoid process.
5. From the midpoint, locate the tip of the caliper 2–3 mm leftward on the mouse as injection point.
6. Prepare the syringe while leaving some air between the inoculum and the plunger.
7. Slowly load 100 μl of PBS with 1×10^5 cells into the syringe.
8. Locate the injection point and insert the needle 6–8 mm deep perpendicularly. Wait to see a blood spurt flushes back into the syringe and very slowly inject the inoculum (Fig. 1b–d and Electronic Supplementary Movie 1) (*see* **Note 3**).
9. Slowly and vertically pull the needle out of the chest, and let the mouse fully recover from the injection and anesthesia on a heating pad.
10. Feed the mouse with gel food and monitor for 24 h.

3.4 Whole Body Bioluminescence Imaging (BLI)

1. Inject D-Luciferin (5 mg/kg) into the mice via intraperitoneal injection.
2. After 5 min, image the mouse using Lago optical imaging system (Fig. 2a) (*see* **Note 4**).

3.5 High-Resolution and Quantitative Fluorescence Imaging

1. Post euthanasia of the mice, harvest lungs, liver, kidneys, bones, and brain, as well as other RFP positive organs. Fix the organs with 10% formalin overnight at 4 °C (*see* **Note 5**).
2. Replace the 10% formalin with 70% Ethanol before the imaging.
3. Image the intact tissues on Leica stereomicroscope under brightfield and RFP filters. Take several images to cover the whole organ surfaces for quantification (Fig. 2b–d) (*see* **Note 6**).
4. After imaging, process the tissues by the Tissue Processor for histological analysis.

3.6 Pathological Assessment

1. After processing the tissues, paraffin embed the tissue samples by using the tissue embedding station.
2. Slice the tissues in 4-microns sections and place them on slides for hematoxylin and eosin (H&E) and immunofluorescence staining. The staining has been performed as previously described [30, 36] (Fig. 3).

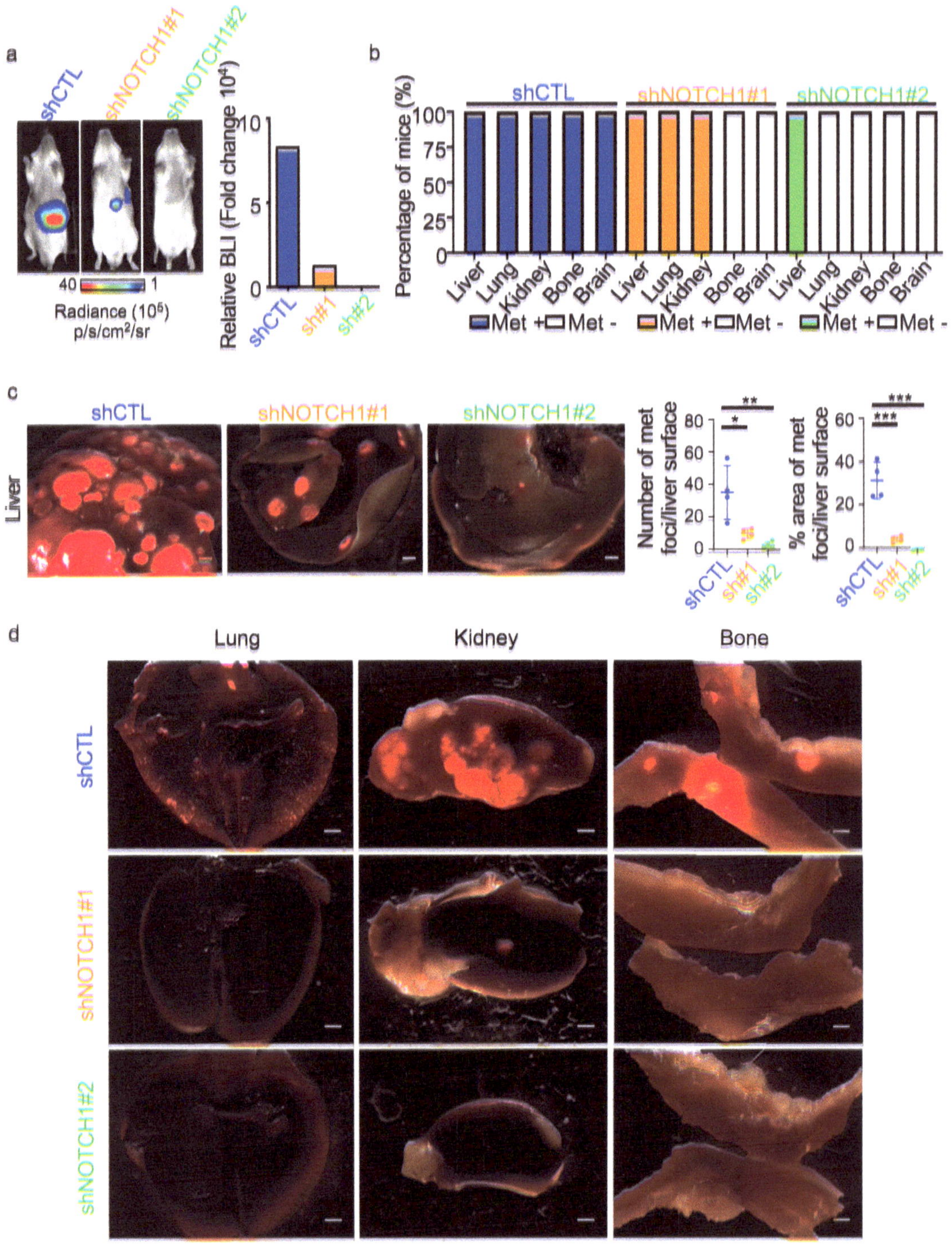

Fig. 2 NOTCH1 knockdown reduces metastatic colonization in vivo. (**a**) TD-NEPC-shCTL and the two independent clones of NOTCH1 knockdown were injected into male mice via intracardiac injection. At Day 21, whole BLI intensity (photons/s/cm^2/surface radiance) is captured and presented. (**b**) Percentage of mice with metastases at various organ sites in each of the control and knockdown groups. (**c**) Representative images of the liver from TD-NEPC-shCTL, shNOTCH1#1, and shNOTCH1#2 are presented. The plots represent the quantification of the number of metastatic foci and percentage of area of metastatic foci per liver surface. Scale bar is 1 mm. *$P < 0.05$, **$P < 0.01$, ***$P < 0.005$, and the P values are determined by two-tailed t-test. (**d**) Representative images of various organs from sh-CTL and both NOTCH1 knockdown groups (scale bar for images = 1 mm)

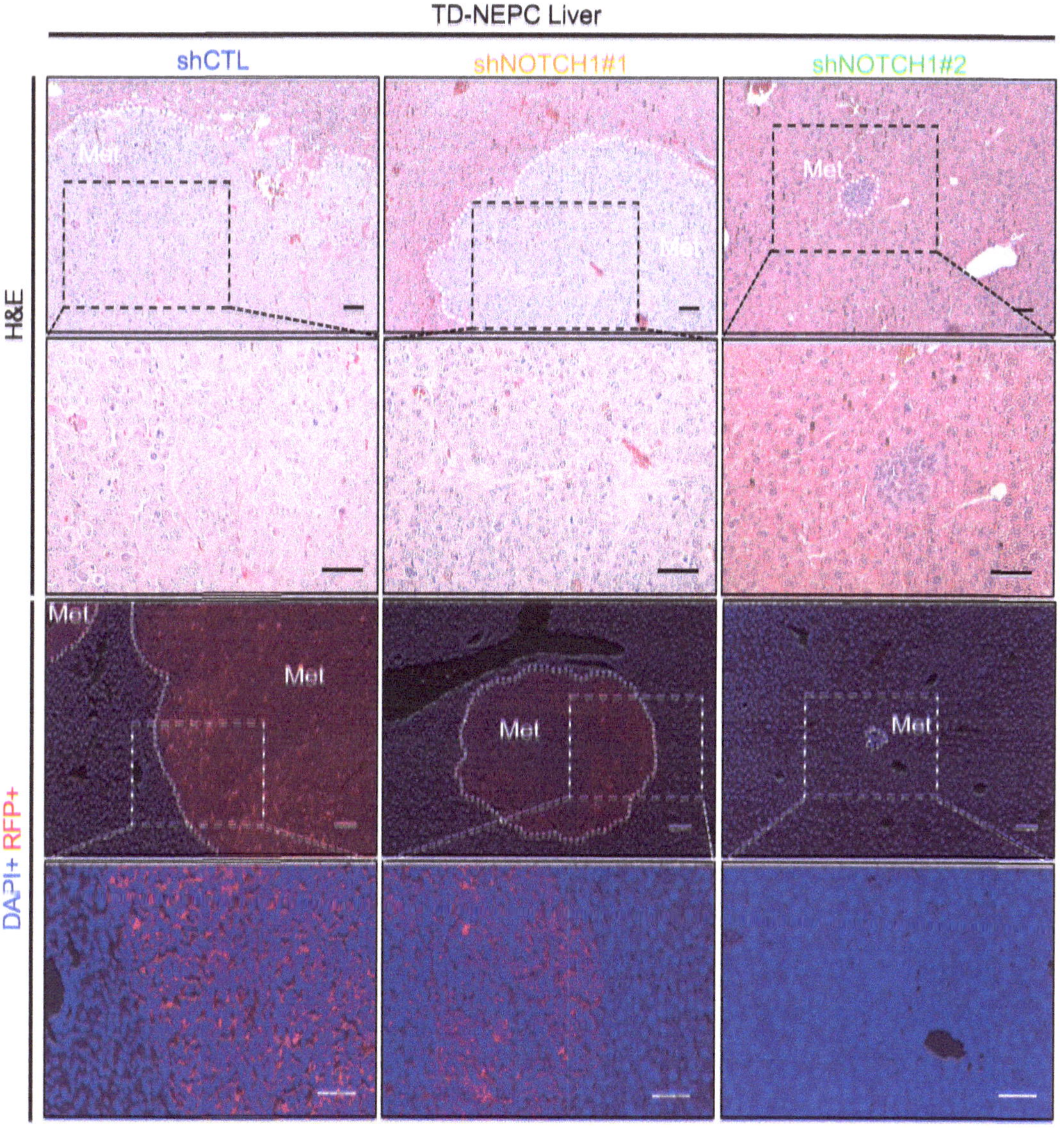

Fig. 3 Histopathology assessment of liver metastasis. Representative H&E and immunofluorescence staining images of livers are shown, and the metastatic regions are encircled by dotted white lines. The scale bars are 20 μm for H&E staining and 50 μm for immunofluorescence staining

3. Take high-resolution representative images of the metastatic sites under different magnifications and circle out the metastatic nodules (Fig. 3) (*see* **Note 7**).
4. After heating for 1 h at 64 °C, the slides with the tissues are rehydrated in 100%, 95%, and 70% ethyl alcohol for 5 min each. Wash the slides in the water for 10 min.
5. Mount the slices with DAPI Fluoromount solution.

6. Image the slides by DAPI and RFP filters. Take representative images of the metastatic nodules in the different organs (*see* **Note 8**).

4 Notes

1. Prostate cancer cell lines such as LNCaP, 22Rv1, C4-2, DU145, PC-3, and TD-NEPC cells express different levels of NOTCH1 as previously described [21, 30, 32]. In this protocol, we used a highly aggressive model of neuroendocrine prostate cancer (NEPC) to test the functional role of NOTCH1 on prostate cancer metastatic colonization.
2. Intracardiac injection model is a common tool to test the cancer cell metastatic colonization and homing to distant sites. By using this model, cells are directly injected into the left ventricle of the mouse heart which allows the cancer cells to disseminate to various organs including liver, lung, kidney, GU system, bone, and brain. Metastatic burden can be measured and quantified by whole body bioluminescence (BLI) imaging. Additionally, nodules in diverse organs can be accessed by histopathologic analysis 3 weeks post injection. The time of harvesting the organs varies between 2 and 4 weeks based on the cell lines and mouse strain that are used. Select the mice that are more than 6-week-old with at least 25 g weight to increase the survival rate post injection. All animal studies and procedures in the protocol have been approved by Stanford Administrative Panel on Laboratory Animal Care (APLAC), IACUC.
3. In order to confirm the syringe is still located in the left ventricle, monitor the continuously blood spurt flushes back into the syringe during the injection. Monitoring the blood spurt from the heart during injection is essential to ensure that the syringe is kept in the left ventricle at all times.
4. Image the mice on Day 0 to confirm the successful injection of the cancer cells into the left ventricle of the mouse heart. Whole body BLI can be performed on Days 14, 21, and 28 after the intracardiac injection to monitor the intensity of the metastatic burden.
5. Determining the time of harvesting the organs is important for the evaluation of metastatic nodules by histopathology. The nodules are supposed to be big enough for detection by fluorescent microscope, and in proper sizes for quantifying the metastatic burden by counting the numbers and measuring the size of the metastatic nodules.

6. In this protocol, the cells are engineered to express RFP and luciferase for quantification of metastatic nodules. The fluorescence imaging is a precise and quantitative tool to investigate the metastatic progression and develop novel therapeutic strategies for prostate cancer. The high-resolution fluorescence imaging can be used to quantify the incidence of metastasis in diverse organs, and utilized to visualize the metastatic sites and quantify the size and numbers of the metastatic nodules. Consistent with previous results [30], knockdown of NOTCH1 decreases the metastatic incidences in diverse organs, including liver, lung, kidney, bone, and brain. Moreover, there is a significant decrease in the numbers and size of metastatic nodules in the liver upon NOTCH1 knockdown when compared to the control.
7. The metastatic nodules can be accessed by H&E staining and immunofluorescence imaging. The representative H&E staining and immunofluorescence images of the liver from shCTL, shNOTCH1#1, and shNOTCH1#2 mouse are shown in Fig. 3.
8. In conclusion, this model facilitates the investigation of the mechanisms underlying the function of NOTCH receptors in metastatic colonization in prostate cancer and can be further applied to other types of cancer. We further demonstrate that loss of NOTCH1 significantly decreases metastatic colonization in prostate cancer utilizing this model.

Acknowledgments

T.S. is supported by the Canary Foundation, the National Institutes of Health/National Cancer Institute (NCI) R37CA240822 and R01CA244281. Shiqin Liu and En-chi Hsu contributed equally to this work.

References

1. Siegel RL, Miller KD, Fuchs HE, Jemal A (2021) Cancer statistics, 2021. CA Cancer J Clin 71:7–33
2. Sharifi N, Dahut WL, Steinberg SM et al (2005) A retrospective study of the time to clinical endpoints for advanced prostate cancer. BJU Int 96:985–989
3. Fizazi K, Shore N, Tammela TL et al (2019) Darolutamide in nonmetastatic, castration-resistant prostate cancer. N Engl J Med 380: 1235–1246
4. Scher HI, Fizazi K, Saad F et al (2012) Increased survival with enzalutamide in prostate cancer after chemotherapy. N Engl J Med 367:1187–1197
5. Beer TM, Armstrong AJ, Rathkopf DE et al (2014) Enzalutamide in metastatic prostate cancer before chemotherapy. N Engl J Med 371:424–433
6. de Bono JS, Logothetis CJ, Molina A et al (2011) Abiraterone and increased survival in metastatic prostate cancer. N Engl J Med 364: 1995–2005
7. Rathkopf DE, Antonarakis ES, Shore ND et al (2017) Safety and antitumor activity of apalutamide (ARN-509) in metastatic castration-

resistant prostate cancer with and without prior abiraterone acetate and prednisone. Clin Cancer Res 23:3544–3551
8. Berthold DR, Pond GR, Soban F, de Wit R, Eisenberger M, Tannock IF (2008) Docetaxel plus prednisone or mitoxantrone plus prednisone for advanced prostate cancer: updated survival in the TAX 327 study. J Clin Oncol 26:242–245
9. de Bono JS, Oudard S, Ozguroglu M et al (2010) Prednisone plus cabazitaxel or mitoxantrone for metastatic castration-resistant prostate cancer progressing after docetaxel treatment: a randomised open-label trial. Lancet 376:1147–1154
10. Mateo J, Carreira S, Sandhu S et al (2015) DNA-repair defects and olaparib in metastatic prostate cancer. N Engl J Med 373:1697–1708
11. Kantoff PW, Higano CS, Shore ND et al (2010) Sipuleucel-T immunotherapy for castration-resistant prostate cancer. N Engl J Med 363:411–422
12. Antonarakis ES, Piulats JM, Gross-Goupil M et al (2020) Pembrolizumab for treatment-refractory metastatic castration-resistant prostate cancer: multicohort, open-label phase II KEYNOTE-199 study. J Clin Oncol 38: 395–405
13. Litwin MS, Tan HJ (2017) The diagnosis and treatment of prostate cancer: a review. JAMA 317:2532–2542
14. Matulewicz RS, Weiner AB, Schaeffer EM (2017) Active surveillance for prostate cancer. JAMA 318:2152
15. Liu S, Shen M, Hsu EC et al (2021) Discovery of PTN as a serum-based biomarker of pro-metastatic prostate cancer. Br J Cancer 124:896–900
16. Liu S, Garcia-Marques F, Zhang CA et al (2021) Discovery of CASP8 as a potential biomarker for high-risk prostate cancer through a high-multiplex immunoassay. Sci Rep 11:7612
17. Radtke F, Raj K (2003) The role of Notch in tumorigenesis: oncogene or tumour suppressor? Nat Rev Cancer 3:756–767
18. Kwon OJ, Zhang L, Wang J et al (2016) Notch promotes tumor metastasis in a prostate-specific Pten-null mouse model. J Clin Invest 126:2626–2641
19. Reedijk M, Odorcic S, Chang L et al (2005) High-level coexpression of JAG1 and NOTCH1 is observed in human breast cancer and is associated with poor overall survival. Cancer Res 65:8530–8537
20. Valdez JM, Zhang L, Su Q et al (2012) Notch and TGFbeta form a reciprocal positive regulatory loop that suppresses murine prostate basal stem/progenitor cell activity. Cell Stem Cell 11:676–688
21. Stoyanova T, Riedinger M, Lin S et al (2016) Activation of Notch1 synergizes with multiple pathways in promoting castration-resistant prostate cancer. Proc Natl Acad Sci U S A 113:E6457–E6E66
22. Yang Y, Ahn YH, Gibbons DL et al (2011) The Notch ligand Jagged2 promotes lung adenocarcinoma metastasis through a miR-200-dependent pathway in mice. J Clin Invest 121:1373–1385
23. Ranganathan P, Weaver KL, Capobianco AJ (2011) Notch signalling in solid tumours: a little bit of everything but not all the time. Nat Rev Cancer 11:338–351
24. Wieland E, Rodriguez-Vita J, Liebler SS et al (2017) Endothelial Notch1 activity facilitates metastasis. Cancer Cell 31:355–367
25. Ciofani M, Zuniga-Pflucker JC (2005) Notch promotes survival of pre-T cells at the beta-selection checkpoint by regulating cellular metabolism. Nat Immunol 6:881–888
26. Sanchez-Martin M, Ferrando A (2017) The NOTCH1-MYC highway toward T-cell acute lymphoblastic leukemia. Blood 129: 1124–1133
27. Aster JC, Pear WS, Blacklow SC (2017) The varied roles of Notch in cancer. Annu Rev Pathol 12:245–275
28. Nowell CS, Radtke F (2017) Notch as a tumour suppressor. Nat Rev Cancer 17: 145–159
29. Koch U, Radtke F (2010) Notch signaling in solid tumors. Curr Top Dev Biol 92:411–455
30. Rice MA, Hsu EC, Aslan M, Ghoochani A, Su A, Stoyanova T (2019) Loss of Notch1 activity inhibits prostate cancer growth and metastasis and sensitizes prostate cancer cells to antiandrogen therapies. Mol Cancer Ther 18:1230–1242
31. Kwon OJ, Valdez JM, Zhang L et al (2014) Increased Notch signalling inhibits anoikis and stimulates proliferation of prostate luminal epithelial cells. Nat Commun 5:4416
32. Bin Hafeez B, Adhami VM, Asim M et al (2009) Targeted knockdown of Notch1 inhibits invasion of human prostate cancer cells concomitant with inhibition of matrix metalloproteinase-9 and urokinase plasminogen activator. Clin Cancer Res 15:452–459
33. Gingrich JR, Barrios RJ, Morton RA et al (1996) Metastatic prostate cancer in a transgenic mouse. Cancer Res 56:4096–4102
34. Cui D, Dai J, Keller JM, Mizokami A, Xia S, Keller ET (2015) Notch pathway inhibition using PF-03084014, a gamma-secretase

inhibitor (GSI), enhances the antitumor effect of docetaxel in prostate cancer. Clin Cancer Res 21:4619–4629

35. Domingo-Domenech J, Vidal SJ, Rodriguez-Bravo V et al (2012) Suppression of acquired docetaxel resistance in prostate cancer through depletion of notch- and hedgehog-dependent tumor-initiating cells. Cancer Cell 22:373–388
36. Hsu EC, Rice MA, Bermudez A et al (2020) Trop2 is a driver of metastatic prostate cancer with neuroendocrine phenotype via PARP1. Proc Natl Acad Sci U S A 117:2032–2042
37. Sarbassov DD, Guertin DA, Ali SM, Sabatini DM (2005) Phosphorylation and regulation of Akt/PKB by the rictor-mTOR complex. Science 307:1098–1101
38. Buckup M, Rice MA, Hsu EC et al (2021) Plectin is a regulator of prostate cancer growth and metastasis. Oncogene 40:663–676

Chapter 19

Functional Studies of Genetic Variants Associated with Human Diseases in Notch Signaling-Related Genes Using *Drosophila*

Sheng-An Yang, Jose L. Salazar, David Li-Kroeger, and Shinya Yamamoto

Abstract

Rare variants in the many genes related to Notch signaling cause diverse Mendelian diseases that affect myriad organ systems. In addition, genome- and exome-wide association studies have linked common and rare variants in Notch-related genes to common diseases and phenotypic traits. Moreover, somatic mutations in these genes have been observed in many types of cancer, some of which are classified as oncogenic and others as tumor suppressive. While functional characterization of some of these variants has been performed through experimental studies, the number of "variants of unknown significance" identified in patients with diverse conditions keeps increasing as high-throughput sequencing technologies become more commonly used in the clinic. Furthermore, as disease gene discovery efforts identify rare variants in human genes that have yet to be linked to a disease, the demand for functional characterization of variants in these "genes of unknown significance" continues to increase. In this chapter, we describe a workflow to functionally characterize a rare variant in a Notch signaling related gene that was found to be associated with late-onset Alzheimer's disease. This pipeline involves informatic analysis of the variant of interest using diverse human and model organism databases, followed by in vivo experiments in the fruit fly *Drosophila melanogaster*. The protocol described here can be used to study variants that affect amino acids that are not conserved between human and fly. By "humanizing" the *almondex* gene in *Drosophila* with mutant alleles and heterologous genomic rescue constructs, a missense variant in *TM2D3* (*TM2 Domain Containing 3*) was shown to be functionally damaging. This, and similar approaches, greatly facilitate functional interpretations of genetic variants in the human genome and propel personalized medicine.

Key words *almondex*, Alzheimer's disease, *Drosophila melanogaster*, Functional genomics, Genome-wide association studies (GWAS), Notch signaling, *TM2D3*, Variant of unknown significance (VUS), Whole-exome sequencing (WES), Whole-genome sequencing (WGS)

Dongyu Jia (ed.), *Notch Signaling Research: Methods and Protocols*,
Methods in Molecular Biology, vol. 2472, https://doi.org/10.1007/978-1-0716-2201-8_19,

1 Introduction

1.1 Genetic Variants Associated with Human Diseases in Notch Signaling Related Genes and the Increasing Demand for Experimental Studies to Characterize Their Functional Consequences

Notch signaling is required for the development of most, if not all, organ systems, and continuous to be required postdevelopmentally for tissue homeostasis and physiology [1–3]. Experiments in diverse model organisms including fruit flies, nematode worms, and mice have revealed the importance of Notch signaling in various cell and tissue types in vivo [4, 5]. Additionally, information gathered by physicians and clinical geneticists has elucidated the role of this evolutionarily conserved signaling pathway in human health [6, 7]. Through genetic and genomic studies of rare disease patients, a number of inherited or de novo pathogenic variants in genes that are part of the core Notch signaling pathway have been discovered as causes of various diseases [8, 9]. These include *NOTCH1*, *DLL4*, and *RBPJ* in Adams–Oliver syndrome [10]; *JAG1* and *NOTCH2* in Alagille syndrome [11]; *DLL3*, *LFNG*, and *HES7* in spondylocostal dysostosis [12]; *NOTCH3* in CADASIL (cerebral autosomal dominant arteriopathy with subcortical infarcts and leukoencephalopathy) [13]; and *PSEN1* and *PSEN2* in early-onset Alzheimer's disease (EOAD) [14]. Many of these clinical studies were supported by experimental data that showed disease associated variants have functional consequences on Notch signaling, which solidified the pathogenicity of the variants identified in patients.

Discovery of new human disease genes is happening at an unprecedented pace due to advancements in high-throughput sequencing technologies including whole-exome sequencing (WES) [15, 16]. For example, dominantly inherited and de novo heterozygous variants in *DLL1* were identified as a cause of a new syndrome named "Neurodevelopmental disorder with nonspecific brain abnormalities and with or without seizures" in a recent study using this technology [17]. For some genes that have already been linked to a known disease, variants with different functional consequences can be shown to cause a completely different disease with a distinct phenotype, a phenomenon that is referred to as "phenotypic expansion" [18]. For example, the skeletal disorder Hajdu–Cheney syndrome was found to be caused by late truncation variants in *NOTCH2* in 2011 [19, 20], a gene previously identified in 2006 as a cause of Alagille syndrome which is a multisystem disorder primarily affecting the liver [21]. In another example, lateral meningocele syndrome and infantile myofibromatosis, which are rare pediatric disorders affecting different organ systems, were found to be caused by rare variants in *NOTCH3* in 2015 [22] and 2013 [23], respectively. The same *NOTCH3* gene was previously linked to CADASIL which is an adult-onset brain vascular disorder first reported in 1977 [24] and molecularly mapped in 1996 [25]. Hence, in addition to new human disease gene discoveries that are likely to continue for many years to come [26], documentation of new phenotypic expansions for known disease genes will equally be important for us to understand how rare genetic variants cause diverse human disorders. Because many genetic variants identified through these discovery studies require experimental validation, there is an increasing demand for collaborative studies that involve scientists who can perform functional assays on the candidate genes of interest [27].

In addition to variants that cause rare diseases, large-scale genome wide-association studies (GWAS) have found associations between certain variants in or near Notch-related genes and common diseases or other human traits. This includes the association of variants in *NOTCH4* to schizophrenia [28], *DLL1* to type 1 diabetes [29], and *ADAM10* and *NOTCH3* to late-onset Alzheimer's disease (LOAD) [30, 31]. More recently, an intronic variant in *NOTCH4* has been associated with the risk of developing critical illness upon SARS-CoV-2 infection in COVID-19 patients [32], suggesting that this catalog of variant–phenotype association in Notch-related genes will continue to increase as human genomics advances. While most variants identified through GWAS lie in putative regulatory regions of genes, which are often difficult to functionally validate, some studies have focused on identified coding variants that can be experimentally tested to determine their functional consequences [31, 33, 34]. As more and more personal WES and whole-genome sequencing (WGS) data become available through large scale epidemiological studies such as the "All of Us" research program in the United States [35], demand for experimental functional studies of variants associated with common diseases or traits will likely increase [36].

While rare inherited variants and germline mutations have been studied in various diseases and phenotypic traits, somatic mutations in Notch signaling related genes have also been studied in various types of cancer including leukemia and solid tumors [37]. In some cancers, activation of Notch signaling is oncogenic, while in others it is tumor suppressive [38, 39]. For example, mutations that cause ligand-independent activation of *NOTCH1* have been found in majority of T cell acute lymphoblastic leukemia (T-ALL) [40], whereas mutations that inactivate the same gene have been found in multiple types of squamous cell carcinoma [41]. Therefore, similar to inherited variants and germline mutations, it is important to functionally categorize somatic mutations that are found in cancer samples to determine whether they are truly pathogenic or not [42].

Functional studies of genetic variants linked to Notch signaling related genes have been primarily performed through cell based assays [43–45]. Although reporter assays and expression analysis of downstream target genes in well-established cell lines such as 293T, HeLa and CHO (Chinese Hamster Ovary) cells are highly sensitive and effective, some functional deficits may be missed because these systems do not fully recapitulate the complex cellular interactions that are present in an in vivo setting. Since Notch signaling can involve complex cellular interactions between the signal sending and receiving cells and is a highly context dependent pathway, it is ideal to study the consequence of disease-linked/associated variants in a whole organism for a more comprehensive analysis. To date, most in vivo functional characterization of disease associated variants in Notch signaling related genes have been carried out in the mouse [46, 47]. While this mammalian model organism is an excellent model system that has many similarities with human and can be manipulated through diverse genome editing technologies [48, 49], other popular genetic model organisms such as fruit fly, nematode worm and zebrafish have unique characteristics that become advantages when studying variants that have been associated with various diseases [50, 51].

The fruit fly, *Drosophila melanogaster*, is a genetic model organism that has been extensively used to perform functional characterization of disease-linked variants over the past two decades [52]. Drosophilists have access to numerous genetic technologies that allow them to manipulate the fly in almost any way one can imagine. These include sophisticated genome editing technology using CRISPR (Clustered Regularly Interspaced Short Palindromic Repeats) and highly specific site-directed transgenic technology using the ΦC31 integrase [53–55]. A vast amount of knowledge on gene expression and function in *Drosophila* is being actively incorporated into FlyBase (http://flybase.org/) by skilled curators and is accessible to anyone interested in taking advantage of this information [56]. Furthermore, tens of thousands of useful genetic reagents are publicly available from stock centers including the Bloomington *Drosophila* Stock Center (BDSC, https://bdsc.indiana.edu/), the Kyoto Stock Center (https://kyotofly.kit.jp/), the Vienna *Drosophila* Stock Center (https://stockcenter.vdrc.at), and the National Institute of Genetics of Japan (https://shigen.nig.ac.jp/fly/nigfly/). In addition, many affordable cellular, molecular and antibody resources can be obtained through *Drosophila* Genomics Resource Center (https://dgrc.bio.indiana.edu/) and Developmental Studies Hybridoma Bank (https://dshb.biology.uiowa.edu/). Because Notch signaling research originated from studies of flies with notched wings, and most components of the core signaling pathway are conserved between *Drosophila* and human, fly geneticists are in an excellent position to functionally study the effect of variants of unknown significance (VUS) in human genes whose fly ortholog have been implicated in Notch signaling.

In this chapter, we describe a workflow to assess the function of a rare variant identified in a Notch signaling related gene using *Drosophila melanogaster*. This protocol is specifically intended for readers who have a certain degree of expertise in fly genetics who are contacted by physicians and clinical geneticists that have identified a potential disease causing human variant in a gene that has been linked to Notch signaling in the past. If one is new to fly research, we direct the readers to the following resources to first familiarize themselves with basic principles and knowledge related to *Drosophila* work [57–59]. Scientists, including fly biologists, can register in matchmaking registries including ModelMatcher (https://www.modelmatcher.net/) to connect with clinicians, patients and other scientists who wish to collaborate on a gene of interest. Upon being contacted by a potential clinical collaborator, a scientist should first evaluate the quality of the variant of interest from a human genetics/genomics perspective, collect information about the gene of interest in human, flies and other species, gather previously published regents to study the gene of interest, generate new genetic reagents that are tailored, and perform in vivo assays that will allow functional assessment of the variant of interest. First, we describe the general workflow related to the nonbenchwork tasks associated with this process. Next, we will describe a series of protocols related to the functional study of a missense variant in *TM2D3* (*TM2 Domain Containing 3*) [33] as an example of how such a functional study can be designed and executed at the bench using *Drosophila*.

1.2 Bioinformatic Analysis of a Genetic Variant of Interest, Selection of an Appropriate Model System, and Overall Experimental Design

Once a rare variant in a gene of interest is identified in an undiagnosed patient as a potential cause of disease through WES or WGS, or a coding variant in a gene is found to be statistically significantly associated with a common disease or a phenotypic trait, physicians and clinical geneticists will often wish to understand the functional consequence of such a variant. While there are many bioinformatic programs designed to predict the functional consequence of genetic variants, there are still limitations to in silico methods [60]. For example, while most nonsense and frameshift variants lead to loss-of-function (LOF) of the gene of interest, some truncating variants can be gain-of-function (GOF) alleles [61]. Furthermore, functional consequences of missense variants are even more difficult to predict because there are so many possible outcomes [62]. A missense variant can be a complete LOF (amorph), partial LOF (hypomorph), gain of endogenous function (hypermorph), gain of negative/toxic function (antimorph, also often referred to as “dominant negative”), gain of new function (neomorph) or have no functional consequences (isomorph) [63]. Because there are no computational methods that can accurately classify variants according to these functional categories, clinicians often seek support from bench scientists to conduct experimental studies [64]. If you are contacted by a clinician who is requesting experimental validation of a variant of interest, we recommend one to first perform literature and database searches to gather information about the gene of interest before hitting the bench.

1.2.1 Gather Information About the Disease of Interest, Patient or Patient Cohort, the Method That Was Used to Identify the Candidate Variant and the Rationale That Was Used to Rule Out Other Candidates

1. For rare disease cases in which WES or WGS was performed to identify the candidate variant, it is often important to know whether the WES/WGS was performed as a singleton (sequence of the patient is available but not for other relatives) or as a trio (sequence of the patient and his/her parents). Since each personal genome has many polymorphisms that are irrelevant to the disease of interest, singleton sequence data possesses more noise compared to trio sequence data [65, 66]. Whether the variant of interest has been validated by Sanger sequencing or not can also be important because next-generation sequencing technologies still have considerable error rates [67]. Moreover, a scientist should discuss with the clinician how other candidate variants identified in the patient were ruled out during the variant prioritization process to understand the logic behind the variant of interest being selected as the prime candidate. This process is crucial because it minimizes the possibility that one is working on a variant that is not the actual cause of the disease the patient has.
2. For genetic variants that are identified in large-scale association studies through epidemiological studies, it is important for the scientists to know the statistical power of the study, to determine whether the case and controls were properly matched, and to discuss the rationale behind why the clinicians and biostatisticians think the variant of interest is likely to be directly contributing to the disease phenotype rather than being just a marker that is cosegregating with the true culprit [68]. Such communication with clinical collaborators during the early phase of the project is important because it provides a better understanding of the larger project and the precise reason that the variant should be experimentally tested.

1.2.2 Assess the Likelihood of the Variant Being Pathogenic Using Public Human Genomic Databases and Variant Pathogenicity Prediction Tools

For candidate variants for undiagnosed rare diseases, it is often useful to know whether damaging variants in the gene of interest are likely to be under selective pressure in the general population. Moreover, it is important to search whether the variant of interest has been previously reported in other individuals with or without disease. In addition, especially for missense variants, it is often useful to know whether multiple in silico pathogenicity prediction programs predict that the variant of interest is likely to have functional consequences or not. One can gather such information using an online tool called MARRVEL (Model organism Aggregated Resources for Rare Variant ExpLoration, http://marrvel.org/), which is a publicly accessible search engine that integrates data from various public databases [69]. MARRVEL aggregates information from population genomic databases such as gnomAD [70] and disease cohort databases including ClinVar [71] and DECIPHER [72]. In addition, MARRVEL provides pathogenicity prediction scores from nine algorithms including CADD [73] and REVEL [74]. To allow users to determine whether the gene of interest has been previously linked to a genetic disorder in humans, MARRVEL displays information from OMIM [75] in a concise manner. For more information about MARRVEL and how to interpret the information provided by this tool, we recommend the readers to reference the following tutorial articles [76, 77].

1.2.3 Identify the Orthologous Genes in Key Model Organisms and Gather Biological Information from Different Species

If the human gene of interest harboring the candidate variant has not been well characterized, it is often important to gather information about its orthologous genes in multiple experimental model organisms to judge whether the variant of interest in the candidate gene is likely to explain the phenotype of the patient. For example, if manipulation of the orthologous gene of interest has been shown to cause similar or related phenotypes in model organisms, the chances of a rare variant in this gene being pathogenic can increase. On the other hand, it is important to keep in mind that the lack of phenotypic similarities between model organisms and patients is usually insufficient to rule out a candidate variant because some genes may have acquired human specific roles. In addition, if the variant identified in the patient causes different types of functional alterations to the encoded protein compared to the genetic manipulations that have been performed in model organisms (e.g., knockout, knockdown, knockin, overexpression), there could be some discrepancies between the patient and model organism phenotypes. MARRVEL can also be used to quickly identify the orthologous candidate genes in eight species (mouse, rat, frog, zebrafish, fruit fly, nematode worm, fission yeast, budding yeast) based on an integrated ortholog prediction tool called DIOPT [78]. Users can further obtain a summary of Gene Ontology (GO) terms and expression information in each species. MARRVEL further provides hyperlinks to each model organism's gene pages for in-depth analysis of the previous literature and large scale initiatives [69]. Other integrative tools provided by the Monarch Initiative [79], the Alliance of Genome Resources [80], and Gene2Function [81] are also useful when performing cross species searches. Such "vertical integration" of gene function information across multiple species often helps formulate a hypothesis that can be experimentally tested in a model organism with a conserved ortholog.

1.2.4 Design Specific Experiments to Test Variant Function in the Proper Organism and Experimental Settings

Upon gathering sufficient evidence that the variant of interest is a good candidate worth functionally testing, one should select a model system to pursue this experimentally. This choice is highly dependent on the gene of interest, patient's phenotypes and scientists' expertise. For genes that encode proteins with known biochemical or cell biological roles, quantitative functional studies can be designed and conducted in vitro. For genes that encode proteins that are difficult to study in vitro, or those that do not have a known molecular function, experiments in vivo can often provide hints that cannot be obtained through in vitro work. When performing experiments in vivo, previous knowledge regarding phenotypes that are seen upon knockout, knockdown, and overexpression of the gene of interest provide informative entry points to design specific experiments. When multiple organisms can be used, one should consider the pros and cons of each model. For example, while a mouse model may lead to establishment of a true "disease model," it can be time, labor and budget consuming. If the purpose of the initial experiment is to simply test whether a variant of interest has functional consequences, invertebrate model organisms such as *Drosophila* and *C. elegans* will likely allow one to assess this in a shorter time frame with a smaller budget. For a more global overview on ways to design variant function studies in *Drosophila*, we refer the readers to the following tutorial and review articles [55, 82].

2 Materials

2.1 Materials for Mutant and Transgenic Fly Generation (for Subheadings 3.2–3.5)

2.1.1 Reagents

10 mM ATP (NEB, P0756S).

10× T4 ligation buffer (NEB, B0202S).

Antarctic phosphatase (NEB, M0289S).

BbsI (NEB, R0539S).

BsaI-HFv2 (NEB, R3733S).

CutSmart buffer (NEB, B7204S).

ddH_2O (double distilled autoclaved water).

DH5α competent bacteria.

LB media.

Max Efficiency Stbl2 competent bacteria (Thermo Fisher # 10268019).

Miniprep or Midiprep kit (e.g., QIAprep Spin Miniprep Kit or Plasmid Midi Kit, QIAGEN, 28104 or 12143).

NheI (NEB, R0131).

P[acman] clone CH322-146A15 (BACPAC resources, CH322-146A15).

pattB (DGRC, 1420).

pBH vector [83].

PCR Purification Kit (e.g., QIAquick PCR Purification Kit, QIAGEN, 27104).

pCFD3-dU6:3gRNA plasmid (Addgene #49410).

Q5® Site-Directed Mutagenesis Kit (NEB, E0554S).

T4 DNA ligase (NEB, M0202S).

T4 Polynucleotide Kinase (NEB, M0201S).

TM2D3 (NM_078474) Human Tagged ORF Clone (Origene, RC203014).

XhoI (NEB, R0146S).

XbaI (NEB, R0145S).

yellow[wing2+] template vector [84].

2.1.2 Equipment and Supplies

0.2 ml PCR tubes (e.g., Eppendorf).

1.5 ml centrifuge tubes (e.g. Eppendorf)

15 ml centrifuge tubes (e.g., Corning).

Centrifuges (e.g., Centrifuge 5424 and 5810R, Eppendorf).

Microinjection needles (e.g., Femtotips® I capillaries, Eppendorf).

Microinjector (e.g., Femtojet® 4×, Eppendorf).

Micropipettes and pipette tips (e.g., Gilson, Eppendorf).

Stereomicroscope (STEMI2000, Zeiss).

2.1.3 Fly Stocks

y^1 M{vas-int.Dm}ZH-2A w; PBac{y+-attP-3B}VK00037* (BDSC #24872).

y^1 w^* (isoX); nos-Cas9 stock (derived from BDSC #78782).

2.2 Materials for Egg Hatching Assay and Neurogenic Assay (Subheadings 3.7–3.8)

2.2.1 Reagents

Active yeast.

Dechorionation solution: 50% bleach in H_2O.

Devitellinization solution: 100% methanol–100% *n*-heptane.

Fixation solution: 3.7% formaldehyde in PBS (pH 7.4)/100% *n*-heptane.

Phosphate buffered saline (PBS).

Mounting solution (e.g., VECTASHIELD® Antifade Mounting Medium with DAPI, H-1200, Vector Labs).

Normal donkey serum (for blocking solution).

Primary antibody: rat monoclonal anti-ELAV antibody (use at 1: 100 dilution, DSHB, 7E8A10).

Rehydration and Washing solution: 0.05% Triton X-100 in 1× PBS (PBST).

Secondary antibody: Alexa Fluor® 488 AffiniPure Donkey Anti-Rat IgG (use at 1:200 dilution, Jackson ImmunoResearch).

2.2.2 Equipment and Supplies

1.5 ml microcentrifuge tubes (e.g., Eppendorf).

20 ml liquid scintillation vials (e.g., Millipore Sigma).

24-well cell culture plate (e.g., Corning).

Dissection microscope with a digital camera [e.g. MZ16 microscope (Leica) with MicroFire microscope camera (Optronics)].

Egg laying bottles and grape juice plates.

Laboratory rocker (e.g., Bio-Rad).

Laser confocal microscope (e.g., LSM 710, Zeiss).

Mesh filter (~70 μm).

Nail polish.

Paint brush.

2.2.3 Fly Stocks

BDSC Stock #10: *amx*^{1} *lz*^{g} *v*^{1}*/C(1)DX, y*^{1} *f*^{1} (*see* **Note 1**).

BDSC Stock #7770: *Df(1)Exel9049, w*^{1118}*/Binsinscy* (*see* **Note 2**).

y w amx$^{\Delta}$ (generated in Subheading 3.2).

amx[+] (generated in Subheading 3.5).

TM2D3[+] (generated in Subheading 3.5).

TM2D3[P155L] (generated in Subheading 3.5).

Detailed fly genotypes shown in Fig. 4*:*

- Experiments using the new *amx*$^{\Delta}$ allele.
  - *amx*^{1} *lz*^{g} *v*^{1}.
  - *amx*^{1} *lz*^{g} *v*^{1}*; amx[+]*.
  - *amx*^{1} *lz*^{g} *v*^{1}*; TM2D3[+]*.
  - *amx*^{1} *lz*^{g} *v*^{1}*; TM2D3[P155L]*.
  - *Df(1)Exel4049, w*^{1118}*/Binsincy*.
- Experiments using the new *amx*$^{\Delta}$ allele.
  - *y w amx*$^{\Delta}$.
  - *y w amx*$^{\Delta}$*; amx[+]*.
  - *y w amx*$^{\Delta}$*; TM2D3[+]*.
  - *y w amx*$^{\Delta}$*; TM2D3[P155L]*.

3 Methods

In this section, we describe the functional analysis of a rare missense variant in *TM2D3* that was identified as a novel risk allele of LOAD through a large epidemiological study by the CHARGE (Cohorts for Heart and Aging Research in Genomic Epidemiology) consortium. The CHARGE consortium is a collaborative study that integrates multiple large and well-phenotyped longitudinal cohort studies in the US and in Europe [85]. In one of their studies, the consortium performed an exome-wide association study using the Exome Chip technology [86] to identify coding single nucleotide polymorphisms (SNPs) associated with increased risk of LOAD [33]. Unlike EOAD which is caused by dominantly inherited pathogenic variants in a small number of genes (e.g., *PSEN1*, *PSEN2*, *APP*), LOAD is thought to be caused by multiple genetic and environmental factors that converge onto a common pathogenic pathway [87]. Through this work, a rare missense variant in *TM2D3* (rs139709573, NM_078474.3:c.464C > T, NP_510883.2:p.Pro155Leu) was identified as a risk allele for LOAD with an odds ratio of 7.5 [95% confidence interval (CI) = 3.5–15.9; $p = 6.6 \times 10^{-9}$] in an Icelandic cohort within the CHARGE consortium [33]. In addition to the increased risk, LOAD patients who carry this allele showed an earlier age-at-onset (hazard ratio = 5.3; 95% CI 2.7–10.5; $p = 1.1 \times 10^{-6}$) compared to LOAD patients who do not, indicating that this allele may also be somehow contributing to the acceleration of LOAD pathogenesis.

While epidemiological data suggested that *TM2D3* likely plays a role in LOAD, there were several issues that needed to be addressed before drawing a strong conclusion. First, the function of *TM2D3* was completely unknown at the time in vertebrate species including human, mouse, frog and zebrafish. Therefore, even if there was a strong associative signal between a rare variant in *TM2D3* with LOAD, it was difficult to develop a logical hypothesis to explain the molecular link between the gene and the disorder. Second, when the clinicians informatically assessed the probability of the variant of interest being pathogenic, multiple computational pathogenicity prediction programs including CADD and PolyPhen predicted that the p.P155L variant was likely benign. The p.P155 residue in human TM2D3 is not conserved in most species (F in mouse, N in *Xenopus* and zebrafish, R in Flies, V in Worm), even though the gene itself is highly conserved in metazoan species. Therefore, computational algorithms concluded that the Proline to Leucine change in this position is likely to be tolerated because this amino acid did not seem to be under selective evolutionary pressure. Finally, the p.P155L variant was identified as being significantly associated with LOAD in the Icelandic cohort within the CHARGE consortium but this signal was not seen in other cohorts. The reason for this was because this allele is about ten times more abundant in Iceland [major allele frequency (MAF): 0.45%] compared to other CHARGE cohorts in European countries and in the U.S. with European ancestry (MAF: 0–0.06%). This enrichment is likely due to a genetic drift that happened during the establishment of the Icelandic population

[88]. Therefore, to provide independent evidence that the p.P155L variant is associated with LOAD in other cohorts, ten times more patients and control individuals had to be recruited into these other cohorts, which was far beyond the scope of the original study. In summary, the lack of knowledge about the gene function, bioinformatic data predicting the variant likely has no functional consequence, and inability to obtain independent epidemiological and genomic evidence to support that p.P155L in *TM2D3* is a strong LOAD associated variant prohibited the clinicians to make strong statements regarding the validity of their interesting findings.

The counterargument that *TM2D3* is a strong LOAD candidate gene came from pioneering studies performed in *Drosophila* between 1930s and 2000s. The fruit fly ortholog of *TM2D3* is *almondex* (*amx*), which was identified in the mid-1930s through an *X*-linked mutant allele that caused slightly reduced and narrowed eyes, reminiscent of a shape of an almond seed [89]. *amx* mutants also showed female sterility, but the eye morphological phenotype was eventually lost in the *amx* mutant stocks that had been maintained [90]. Hence, *amx* became the official name of the gene that is responsible for female sterility phenotype in the original amx^1 allele. In a seminal paper in 1983, Lehman et al. showed that *amx* mutant females are sterile because all embryos laid by homozygous *amx* mutant mothers shows a strong neurogenic phenotype, similar to embryos defective in genes such as *Notch*, *Delta*, *neuralized*, and *Enhancer of split* [91]. Since these genes are core components of the Notch signaling pathway, this finding suggested that *amx* plays an important role in embryonic Notch signaling as a maternal effect gene [92, 93]. However, its molecular function remains unknown. In 2008, through genetic epistasis experiments, Michellod and Randsholt proposed that *amx* functions at the γ-secretase mediated cleavage step of Notch activation [94]. Because γ-secretase is linked to Alzheimer's disease through *PSEN1* and *PSEN2*, which encode the core catalytic subunit of this protein complex [95, 96], a potential functional link between *TM2D3* and LOAD emerged.

To further provide support that *TM2D3* and LOAD are indeed functionally linked, it became essential to provide functional evidence that the p.P155L variant in this gene alters the function of the encoded protein in some way. Because there were no cellular or biochemical assays that can report the function of TM2D3, we decided to test this in a heterologous system in *Drosophila* by first determining whether the function of *amx* and *TM2D3* in Notch signaling is conserved, and further assessing how the p.P155L variant impacts its function. To achieve this, we used the female sterility and embryonic neurogenic phenotype as functional readouts for *amx*/*TM2D3* function in vivo, and either obtained or generated genetic reagents to test our hypothesis. Below, we describe our experiments in detail in hopes that this protocol will stimulate ideas on performing functional studies of other variants in this, or other, genes whose *Drosophila* orthologs have been linked to Notch signaling.

3.1 Identification and Acquisition of Preexisting Genetic Reagents in Drosophila *That Are Useful for Functional Studies of the Variant of Interest*

Prior to performing the actual experiments in *Drosophila*, it is often important to gather information on what kind of genetic reagents (e.g., mutant strains, transgenic strains, reporter strains) have been generated for the gene of interest through previous studies, and which stocks are available from public stock centers.

1. Go to FlyBase (http://flybase.org/) and visit the gene page of interest. For *amx*, go to http://flybase.org/reports/FBgn0000077. Navigate the page to obtain information on mutant alleles and transgenes that have been reported in previous studies and corresponding phenotypes associated with each reagent. Although similar information can be obtained from general literature searches utilizing resources such as PubMed (https://pubmed.ncbi.nlm.nih.gov/), FlyBase presents manually curated information in a highly organized manner that greatly facilitates literature mining. For a general tutorial on how to navigate through FlyBase, we recommend the readers to the following articles [56, 97].
2. Identify the reagents useful for your study, and attempt to obtain them. If a genetic reagent is available from a public stock center, it will usually be listed under the "Stocks and Reagents" box in the FlyBase gene page. For *amx*, go to http://flybase.org/reports/FBgn0000077#stocks_reagents.
   If the genetic reagent of interest cannot be found in public stock centers, you can try to reach out to the corresponding author of a recent publication that used that allele/transgenic line to request the necessary reagent. For this study, we identified a classic mutant allele of *amx* from the BDSC that carries the amx^1 allele (FlyBase ID: FBal0000504) which has been used in most previous studies. Because we realized that this strain carries other mutations in nearby genes (which could be the case for many mutations that have been generated more than several decades ago), we also identified a small deficiency, *Df(1)Exel9049* (FlyBase ID: FBab0047358), that removes the coding region of *amx* as well as several neighboring genes. The two *X*-chromosomes from the following two strains from BDSC can be combined *in trans* to generate clean hemizygous *amx* female flies [amx^1/*Df(1)Exel9049*] that are female sterile and show maternal effect neurogenic phenotypes that can be used for functional studies (*see* Subheadings 3.7 and 3.8).

   BDSC Stock #10: amx^1 lz^g v^1/*C(1)DX*, y^1 f^1 (*see* **Note 1**).

   BDSC Stock #7770: *Df(1)Exel9049*, w^{1118}/*Binsinscy* (*see* **Note 2**).

3.2 Generation of New LOF Reagents for the Fly Gene of Interest

While previously reported mutant alleles are often very useful and can help kickstart a project, it can eventually become necessary to generate new mutant lines for your gene of interest. In the case of *amx*, the publicly available stock from the BDSC discussed above (*amx*^{1}) carries a very closely linked secondary mutation (*lz*^{g}) that causes a number of phenotypes independent of *amx* that can interfere with the functional interpretation when homozygous (hence, we can only use the *amx*^{1} chromosome *in trans* with a deficiency that covers *amx* but not *lz* to perform clean experiments). Below, we describe a protocol to generate a new clean null allele of *amx* using a CRISPR-based method.

3.2.1 Select an Effective Method to Knock Out the Gene of Interest

The CRISPR-based *yellow* wing dominant marker insertion cassette technique offers a straightforward approach to generate molecularly defined amorphic (null) alleles that can be generated using standard equipment in a *Drosophila* lab [84]. While it is possible to generate LOF alleles by CRISPR via designing a single guide RNA (sgRNA) to make a double stranded break within the coding region of a gene of interest and screen for small indel events that causes a frameshift through nonhomologous end joining (NHEJ) [98–101], this process can be tedious since it typically requires a labor intensive molecular screening step to identify the desired gene editing event. In addition, a frameshift allele may not always be a null allele because it has the potential to generate a partial LOF, a dominant negative or a GOF allele depending on the gene and the nature of the frameshift. A method based on homology directed repair (HDR) allows full replacement of the gene of interest, allowing the researcher to be certain that the allele they are generating will be a null allele. Moreover, by replacing the gene of interest with a knockin cassette that carries a dominant visible marker to replace the region of interest (ROI), it becomes possible to trace the desired gene editing event by visual inspection of the fly rather than relying on molecular techniques, greatly simplifying the mutant identification. Furthermore, the same visible marker can be used in downstream applications to follow the mutant allele, simplifying genetic crosses.

The technique described here uses CRISPR/Cas9 to induce double strand breaks flanking the ROI while providing a template to replace the open reading frame of the GOI with a construct that expresses the *yellow* gene driven by cis-regulatory elements that direct its expression in the fly wing (*yellow*[*wing2+*]) (Fig. 1) [84]. Specifically, we document the utilization of this approach to generate a molecularly defined null allele for *amx* by replacing the coding region of this gene with the *yellow*[*wing2+*] marker. This marker can be easily visualized in a *yellow*$^{-}$ background, thus allowing robust identification of the flies that underwent gene editing (*see* **Note 3**). Other dominant markers such as *yellow[body+]* [84], *Kozak-GAL4* [102] and *3xP3-DsRed* [103] can also be utilized depending on the application and specific needs.

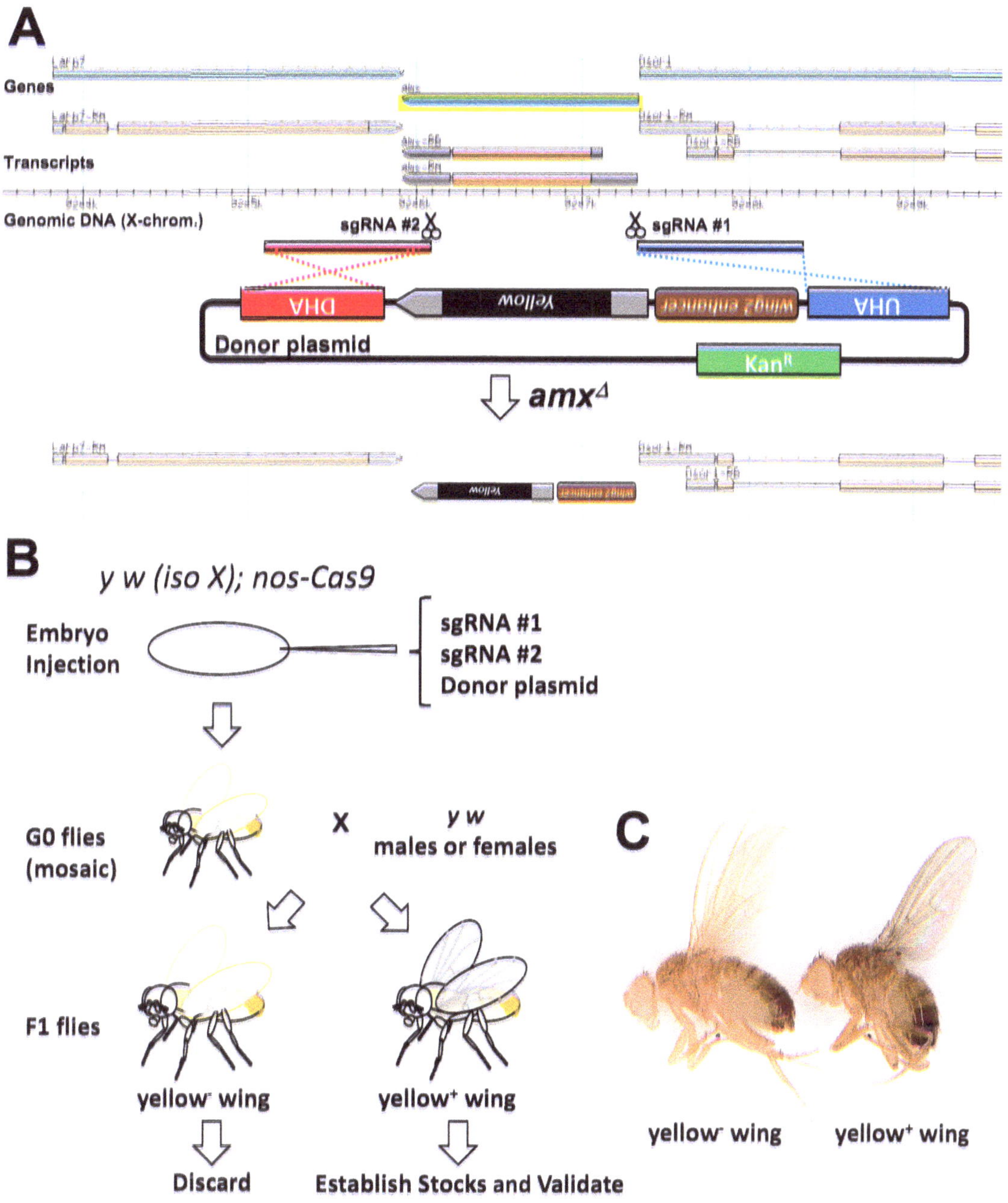

Fig. 1 Generation of a null allele of *amx* using the CRISPR/Cas9-based *yellow* wing dominant marker insertion strategy. (**a**) Schematic diagram of the *amx* locus on the *Drosophila X*-chromosome. The coding region of genes are shown as orange boxes and the untranslated regions are shown in gray. CRISPR-mediated homology directed repair (HDR) was used to knock out the *amx* gene by replacing it with the *yellow [wing2+]* cassette. The position of the sgRNA cleavage sites are shown using scissors. The upstream (UHA) and downstream homology arms (DHA) are shown using blue and red boxes, respectively. This donor plasmid also has a Kanamycin resistance (Kan^R) cassette shown using a green box, which will not be integrated into the locus of interest. (**b**) Schematic diagram of the microinjection and mutant selection process in vivo in flies. A cocktail of sgRNAs and donor plasmid is injected into the posterior pole of fly embryos that express Cas9 in

3.2.2 Define the ROI and Surrounding Sequences

Visualize the ROI using genome browsers (GBrowse or JBrowse) and download the FASTA format sequence of the gene of interest and surrounding regions from FlyBase (https://flybase.org/). Using these tools, one can identify the 5′UTR, coding sequence, 3′UTR of the gene of interest to determine the approximate ROI to delete (Fig. 1a). For *amx*, this information can be obtained through the following links.

GBrowse Link:
http://flybase.org/cgi-bin/gbrowse2/dmel?Search=1;name=FBgn0000077

JBrowse Link:
http://flybase.org/jbrowse/?data=data/json/dmel&loc=FBgn0000077

FASTA file of the extended gene region (*amx* ± 2 kb upstream and downstream)':
http://flybase.org/download/sequence/FBgn0000077/gene_extended

3.2.3 Select Appropriate sgRNA Sequences

To delete the ROI, one must identify two Cas9/sgRNA target sites that flank the ROI. The Cas9 endonuclease is directed to its target site by a combination of a PAM (protospacer adjacent motifs) sequence "NGG (N is any DNA base)" and a stretch of 20 nucleotides preceding the PAM complementary to the sgRNA sequence [104]. To identify Cas9 sites that flank the ROI, one can use the CRISPR Optimal Target Finder (http://targetfinder.flycrispr.neuro.brown.edu/) [103] as follows. Alternatively, DRSC/TRiP Functional Genomics Resources (https://www.flyrnai.org/crispr3/web/) also offers a similar platform with a genome viewer that can facilitate the process of sgRNA selection [105]. Throughout this protocol, we refer to the genomic region 5′ to the translational start site as "upstream" and the genomic region 3′ to the translational stop site as "downstream." Because *amx* is encoded on the antisense strand of the reference genome, the upstream region of *amx* is in the 3′ direction of the reference genome, and the downstream region of *amx* is in the 5′ direction (Fig. 1a).

Fig. 1 (continued) their germline [y^1 w^* *(iso X); nos-Cas9*]. After these embryos reach adulthood, they are crossed to a *yellow white* (*y w*) mutant strain and their progenies are screened for positive gene editing events. The flies with successful integration of the *yellow[wing2+]* cassette will have a darker wing color compared to flies that do not carry this cassette in a *yellow* mutant background. (**c**) Photograph of *y w* mutant male flies with (right) or without (left) the *yellow[wing2+]* cassette. The fly on the right is a null allele of *amx* generated through this strategy (*y w* amx^{Δ}), which has a darker wing color compared to a *y w* mutant with an intact *amx* gene

1. Copy and paste the *amx* 5′UTR sequence into CRISPR Optimal Target finder and select find CRISPR targets. This returns a list of potential sgRNA sequences in the 5′ UTR.
2. Evaluate the options and select the most appropriate sgRNAs. Select a target site near the coding sequence with minimal predicted off-targets, preferably with no off-targets on the same chromosome as the gene of interest. For *amx*, we selected 5′-TCCATTTAAGTTGTGACCATTGG-3′ (PAM sequences are underlined here) for the upstream cut site (sgRNA #1) and 5′-GAAGATCTTGCTATTCCTAATGG-3′ for the downstream cut site (sgRNA #2) as two gRNA sequences that flank the open reading frame of *amx* (Fig. 1a).
3. (Optional) Determine the predicted efficiency of CRISPR-mediated cleavage. Enter the sequence of the gRNAs into the DRSC/TRiP functional Genomics resources CRISPR Efficiency Prediction tool (https://www.flyrnai.org/evaluateCrispr/) [105]. This tool displays the predicted efficiency of each potential sgRNA in a genome viewer. In our experience, a sgRNA with a score of 5.5 or higher produces consistent results. If the score is low, select another sgRNA sequence and repeat this process until one identifies a good sgRNA that can be used.
4. Verify the sgRNA target sequences of the *Drosophila* strain that one will be editing to avoid polymorphisms at the sgRNA and PAM sites. This can be done by Sanger sequencing the region of interest using genomic DNA and appropriate primers, or through whole-genome sequencing using next-generation sequencing technologies [106]. If there is a polymorphism, redesign the sgRNA so the sequence matches with the genomic DNA of the strain of interest, or select another sgRNA target site.

3.2.4 Design and Subclone the sgRNA Expression Constructs

After selecting the two sgRNAs to cut the ROI, one must develop a construct that allows the expression of these sgRNAs in vivo. The following protocol follows the procedure developed by Port et al., [107] which is outlined in detail in http://www.crisprflydesign.org/wp-content/uploads/2014/05/Cloning-with-pCFD3.pdf.

1. Obtain the pCFD3-dU6:3gRNA plasmid (Addgene #49410). This plasmid allows the efficient expression of the sgRNA of interest upon injection into *Drosophila* based on the U6:3 promoter [107]. There are a number of other sgRNA expressing plasmids that can be used for this purpose, which are listed in https://www.crisprflydesign.org/plasmids/.
2. Order oligoDNAs to subclone the sgRNA sequences.

For sgRNA #1:

5′-gtcgGAAGATCTTGCTATTCCTAA-3′ (sense strand).

5′-aaacTTAGGAATAGCAAGATCTTC-3′ (antisense strand).

For sgRNA #2:

5′-gtcgTCCATTTAAGTTGTGACCAT-3′ (sense strand).

5′-aaacATGGTCACAACTTAAATGGA-3′ (antisense strand).

Here, "gtcg" and "aaac" sequences are added to the sense and antisense strands to generate an overhang to ligate the oligoDNA into the pCDF3-dU6:3gRNA plasmid cleaved with BbsI.

3. Set up the following reaction to anneal and phosphorylate the sense and antisense oligoDNA in PCR tubes to generate double stranded DNA with 5′-phosphorylation:

   1 μl of sense strand oligo (100 μM).

   1 μl of antisense strand oligo (100 μM).

   1 μl 10× T4 ligation buffer (NEB, B0202S).

   0.5 μl T4 Polynucleotide Kinase (NEB, M0201S)

   6.5 μl ddH_2O (double distilled autoclaved water)

   (10 μl in total)

   Place the PCR tubes in a thermocycler and set the following program.

   37 °C for 30 min.

   95 °C for 5 min.

   Ramp down to 25 °C at the slowest setting.

   End reaction.

4. Digest the pCFD3-dU6:3gRNA plasmid with BbsI (NEB, R0539S) and purify the digested product using a DNA purification kit according to the manufacturer's protocol (e.g., QIAGEN, QIAquick PCR Purification Kit, 28104).

5. Ligate, transform and grow the *E coli* colony carrying the sgRNA expression constructs. For each annealed oligo pair, set up the following reactions in PCR tubes.

   50 ng BbsI-digested pCFD3-dU6:3gRNA plasmid from Subheading 3.2.4, **step 4**.

   1 μl of 1:200 dilution of annealed oligos from Subheading 3.2.4, **step 3**.

   (sgRNA #1 and sgRNA #2 in separate tubes).

   1.5 μl 10× T4 ligation buffer (NEB, B0202S).

   1 μl T4 DNA ligase (NEB, M0202S).

ddH_2O to total volume of 15 μl.

(15 μl in total)

Incubate the tubes at room temperature for 10–30 min or overnight (o/n) at 16 °C. Transform the ligated DNA into a competent bacteria (e.g., DH5α) using standard transformation protocols. pCFD3-dU6:3gRNA carries an ampicillin resistance marker. Pick up single colonies (we recommend selecting at least 4 colonies from each reaction) and grow them on liquid LB media (~3 ml). Miniprep the culture and verify the sequence of the insert using Sanger sequencing using the standard M13 reverse primer (5′-CAGGAAACAGCTATGAC-3′). Select the colony that carries the correct insert and generate glycerol stocks for long-term storage of each sgRNA plasmid. Perform miniprep or midiprep to obtain sufficient amounts of sgRNA plasmids for injection (*see* Subheading 3.2.6).

3.2.5 Design and Assemble the HDR Donor Plasmid

Using Golden Gate cloning [108], assemble the *yellow[wing2+]* donor plasmid to replace the ROI. The *yellow[wing2+]* donor plasmid can be assembled from four DNA fragments: destination vector pBH (available from Drs. Benjamin Housden and Norbert Perrimon) [83], *yellow[wing2+]* template vector [84], and two homology arms corresponding to the upstream and downstream regions flanking the deleted segment (Fig. 2). Golden Gate cloning utilizes type IIS restriction enzymes to make unique pairs of 5-′-overhangs between each fragment to its neighboring fragment in the correct order for assembly [109]. By alternating digestion and ligation cycles in a thermal cycler multiple fragments can be assembled together. The specific protocol applied to construct the targeting plasmid to generate the *amx* null allele is described below. For general tutorials on Golden Gate cloning, we refer the readers to the following articles [110–112]. Alternatively, one can use other molecular strategies such as Gibson cloning [113] to assemble a donor plasmid for a specific gene of interest, which will not be described here.

1. Choose an appropriate type IIs restriction enzyme to be used for Golden Gate cloning. To generate the *yellow[wing2+]* donor plasmid to replace the coding region of *amx*, we chose BsaI because the recognition sequence of this enzyme is not present in the homology arms that we designed.
2. Design and generate the upstream and downstream homology arms. Design PCR primers to amplify the upstream and downstream homology arms. For details on how to design primers and amplicons for Golden Gate cloning, *see* Marillonnet and Werner [110].
   (a) Design the primers to amplify the upstream homology arm (UHA). For the UHA, the end of the annealing

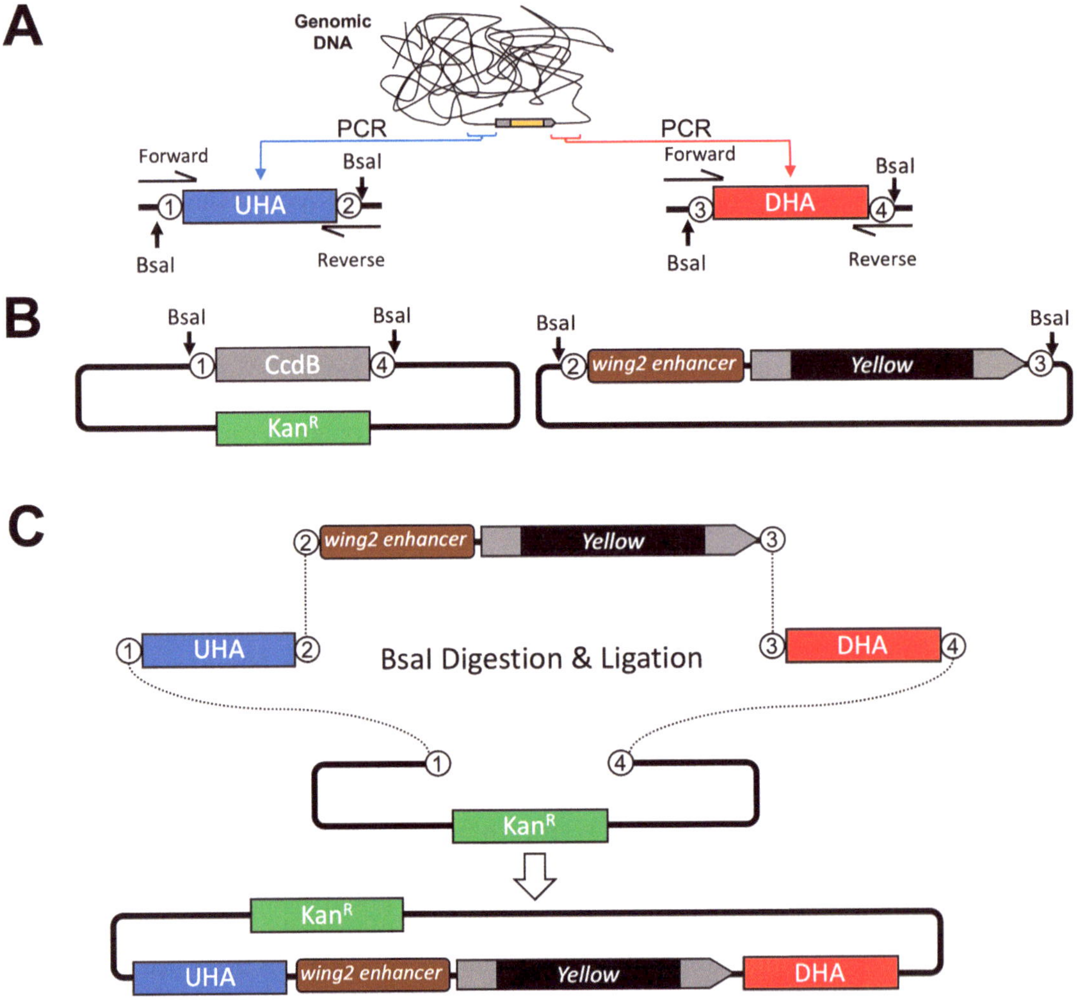

Fig. 2 Generation of an *amx-yellow[wing2+]* homology directed repair donor construct using the Golden Gate cloning strategy. (**a**) In order to clone the upstream (UHA) and downstream homology arms (DHA) of the homology directed repair (HDR) donor construct, perform PCR using specific primers and genomic fly DNA. In addition to the segments that anneal with the genomic DNA, the primers designed here have features that facilitate the subcloning of these fragments using the Golden Gate strategy. (**b**) In addition to the two homology arms generated by PCR, this protocol requires two plasmids, one that provides the vector backbone of the final product (pBH vector shown on the left) and another that provides the *yellow[wing2+]* cassette. (**c**) Assembly of the *amx-yellow[wing2+]* HDR plasmid through the Golden Gate reaction. By mixing the UHA and DHA from (**a**), the two plasmids from (**b**), a type IIs restriction enzyme BsaI and a DNA ligase, the four segments will be assembled into one plasmid through repetitive digestion and ligation reactions based on the specific overhangs created by the BsaI digestion (shown as overhangs (1) to (4))

portion of the reverse primer is defined by the Cas9 cut site of dsRNA #1. For the UHA of *amx*, we designed and synthesized the following two oligo DNA to amplify a ~1 kb region upstream of the sgRNA #1 cut site.

Forward Primer: 5′-agagagGGTCTCgTATAgagagagtacct gctctttcactcc-3′.

Reverse Primer: 5′-ctctctGGTCTCtTTCCcattggccgc tttagtcgtaggag-3′.

Here, "ctctct" and "agagag" are random nucleotides 5′ to the BsaI binding site that were added to facilitate the digestion reaction, "GGTCTC" is the recognition sequence for BsaI, the following "t" or "g" is a random single base spacer sequence required to align the BsaI cut at the overhang, "TATA" on the forward primer is the overhang to anneal with the pBH vector (overhang (1) in Fig. 2a), "TTCC" is the overhang to anneal with the *yellow[wing2+]* cassette (overhang (2) in Fig. 2a) and the underlined sequences are the annealing portions of the primer.

(b) Design the primers to amplify the downstream homology arm (DHA), similar to Subheading 3.2.5, **step 2a**. For the DHA, the beginning of the annealing portion of the forward primer is defined by the Cas9 cut site of dsRNA #2. For the DHA of *amx*, we designed and synthesized the following two oligo DNA to amplify a ~1 kb region downstream of the sgRNA #2 cut site.

Forward Primer:

5′-agagagGGTCTCgATCCggaatagcaagatcttctcaaaaacgtg tac-3′.

Reverse Primer: 5′-agagagGGTCTCgGACCgagtgctccct gctaaaaccatgc-3′.

Here, "agagag" are random nucleotides 5′ to the BsaI binding site that were added to facilitate the digestion reaction, "GGTCTC" is the recognition sequence for BsaI, the following "g" is a random single base spacer sequence required for BsaI, "GACC" on the forward primer is the overhang to anneal with the pBH vector (overhang (4) in Fig. 2a), "ATCC" is the overhang to anneal with the yellow[wing2+] cassette (overhang (3) in Fig. 2a) and the underlined sequence is the annealing portions of the primers for PCR amplification of genomic DNA.

(c) Amplify the UHA and DHA by PCR based amplification of genomic DNA extracted from the flies that will be used for gene targeting (Fig. 2a). To knock out *amx*, we utilized the y^1 *w** *(isoX); nos-Cas9* stock (nos-Cas9 Derived from BDSC #78782). This strain expresses Cas9 in the germline and its *X*-chromosome has been isogenized and fully sequenced [84].

3. Perform the Golden Gate assembly reaction (Fig. 2b, c).

   Set up the Golden Gate reaction by preparing the following solution. In a PCR tube, add the following.

   20 ng pBH-donor plasmid.

   Equimolar amounts of upstream homology arm.

   Equimolar amounts downstream homology arm.

   Equimolar amounts yellow wing insert plasmid.

   1 μl CutSmart buffer (NEB, B7204S).

   1 μl 10 mM ATP.

   0.5 μl T4 ligase (NEB M0202S).

   0.5 μl restriction enzyme BsaI-HFv2 (NEB, R3733S)

   ddH_2O to total volume of 10 μl.

   (10 μl in total)

   Perform the Golden Gate reaction in a thermal cycler using the following program:

   Step 1: 37 °C for 3 min.

   Step 2: 20 °C for 2 min.

   Step 3: Repeat "**Steps 1 and 2**" nine times.

   Step 4: 37 °C for 5 min.

   Step 5: End program.

4. Transform the product into competent bacteria (Max Efficiency Stbl2, Thermo Fisher) and grow these cells on an LB plate appropriate selection media (pBH is Kanamycin resistant). Pick up single colonies (we recommend selecting 4–8 colonies) and grow them on liquid LB media (~3 ml). Miniprep the culture and verify the sequence through DNA fingerprinting and/or Sanger sequencing. Select the colony that carries the correct insert and generate glycerol stocks for long-term storage of the HDR donor template plasmid. Perform miniprep or midiprep to obtain a sufficient amount of sgRNA plasmids for injection (*see* Subheading 3.2.6).

3.2.6 Microinject the sgRNA and HDR Donor Plasmids and Establish the Genome Edited Stocks

1. Make a sgRNA/Donor vector injection solution containing the following.

   Homology Directed Repair Donor vector (200–300 ng/μl) from subheading 3.2.5.

   Upstream sgRNA expression construct (25 ng/μl) from Subheading 3.2.4 (sgRNA #1).

Downstream sgRNA expression construct (25 ng/μl) from Subheading 3.2.4 (sgRNA #2).

(30 μl in total)

2. Microinject the injection solution into flies that express Cas9 in the germline. To generate the *amx* null allele, we microinjected the injection solution into the posterior region of y^1 w^* *(iso X); nos-Cas9* embryos that were within 1 h after egg laying using standard transgenic techniques [114] (Fig. 1b). Injected embryos are designated G0, which are mosaic animals.
3. Raise G0 to adults and set crosses to obtain F1 flies. Collect males and virgin females and cross single males to 3–5 virgin *yellow white* (*y w*) mutant females and 2–3 virgin females to 1–2 *y w* males in individual vials. These crosses produce F1 offspring with potential gene knockout and yellow wing cassette insertion.
4. Screen the F1 flies for insertion of cassette and establish stable stocks. Raise F1 offspring to adults. Anesthetize F1 adult flies using CO_2 and a diffuser pad. Under a stereomicroscope (0.63–5×) at low power, examine wings of each F1 fly for dark pigment resulting from the yellow wing dominant marker construct (Fig. 1c). Pick male flies with dark colored wings and individually cross them to appropriate balancer stocks (e.g., *FM7c*).
5. Molecularly validate the targeting event. Collect 6–8 adult flies, isolate Genomic DNA from each balanced line and perform PCR across each homology arm to the insert followed by Sanger sequencing to confirm the editing event. The PCR primers must start from the genomic DNA flanking the homology arms to ensure correct insertion. We recommend designing primers ~50 base pairs outside of the homology arm ends and the primers within the yellow[wing2] insert. Primers 5-′-GCACTGCGCTCAAACTATTAGTACCCC-3′ and 5-′-CGGAGCCAGACTTCAGACGGG-3′ can be used to sequence the upstream and downstream homology arms, respectively.

3.3 Generation of a Genomic Rescue Construct for the Fly Gene of Interest

In addition to mutant flies, it is important to generate flies that express a reference and variant protein of interest to assess the functional consequence of the genetic variant of interest. This can be accomplished using the UAS/GAL4 system [115] by cloning the human or fly cDNA under the control of the upstream activation sequence (UAS) and expressing the reference and variant forms with ubiquitous or tissue/gene specific GAL4 drivers [82]. Although many GAL4s, including gene specific T2A-GAL4 [116] drivers, are available to express the proteins of interest in any tissue or developmental time point, the protein of interest expressed through this system tends to be overexpressed compared to endogenous levels. Alternatively, one can engineer the variant of interest in the endogenous gene using scarless genome editing [84, 117] or perform rescue experiments of a LOF allele using a genomic rescue construct that contains the variant of interest [118]. However, this strategy does not allow testing of variants that affect nonconserved amino acids. As a hybrid approach, one can first humanize the fly genomic rescue construct and then introduce the variant of interest. Here, we describe the methodology we used to humanize the fly *amx* genomic rescue construct with human *TM2D3*, and to further introduce the p.P155L variant linked to AD that affects a residue that is not conserved between *Drosophila* Amx and human TM2D3 [33] (Fig. 3).

3.3.1 Design of a Genomic Rescue Fragment for the Gene of Interest

In order to generate a genomic rescue construct that correctly expresses a specific gene of interest, it is often necessary to capture not only the coding sequence but also its accompanying upstream and downstream untranslated regions (UTRs), the promoter sequence, and the cis-regulatory elements (CRE) necessary for the correct temporal and spatial expression pattern. Ideally, this genomic rescue fragment (GRF) should not include the complete sequence of neighboring genes, so that it is only capable of expressing the specific gene of interest, and can therefore show specificity of any given phenotype it rescues. For most genes, the CRE that drive gene expression are not well mapped. Therefore, the boundaries of the GRF are generally determined by the surrounding genomic structure. The GRF should be limited in size by the constraints of the plasmid that are being used. Here, we used pattB (DGRC, 1420) as the plasmid backbone, which is suited to integrate DNA fragments that are <6 kb into specific attP docking sites using the ΦC31 integrase-mediated transgenesis system [119]. This construct carries a mini-*white*$^{+}$ marker [120], which allows users to screen for transgenic flies by examining their eye color in a *white*$^{-}$ background. For larger GFRs, one can utilize fosmid or BAC (bacterial artificial chromosome) based backbones [121, 122], which will not be discussed here.

1. Examine the sequences flanking the gene of interest by using the GBrowse or JBrowse feature on FlyBase as in Subheading 3.2.2.
2. Select a genomic fragment that likely contains all the CRE of the gene of interest. For *amx*, we selected an approximately 3.3 kb fragment that includes a short segment of the 3′ end of the upstream gene *Larp7*, the coding sequence and associated UTRs of *amx*, and a short segment of the downstream gene *Dsor1* including most of *Dsor1* intron 1 (Fig. 3a). Partial sequences in the neighboring genes were included in case they contain distant CRE that regulate the expression of *amx*,

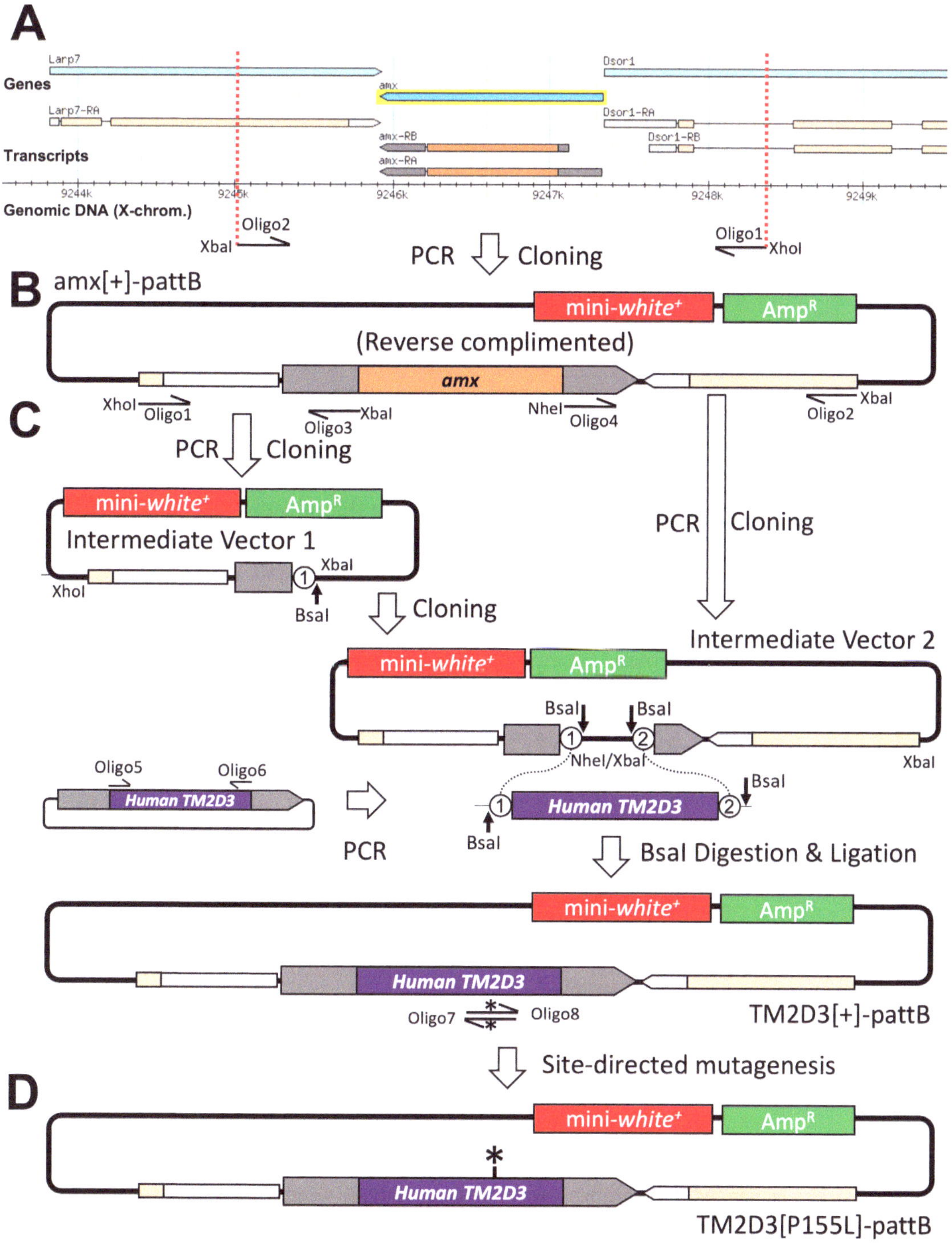

Fig. 3 Generation of a series of *amx* genomic rescue constructs to test a disease-associated variant in *TM2D3*. (**a**) Schematic diagram of the *amx* locus on the *Drosophila X*-chromosome. The coding region of genes are shown as orange boxes and the untranslated regions (UTR) are shown in gray. Two primers (Oligo 1 and 2) were designed to amplify *a* ~ 3.3 kb region that contains the *amx* gene and neighboring sequences. (**b**) Construction of the amx[+]-attB genomic rescue construct by subcloning the PCR amplicon from (**a**) into pattB using XhoI and XbaI digestion and ligation. pattB contains a mini-white^{+} cassette (red box) as a dominant

but functional *Larp7* and *Dsor1* will not be generated from this GRF due to lack of intact coding regions.

3. Design PCR primers flanking the chosen sequence with appropriate binding sites for restriction endonucleases to subclone this fragment into pattB. We selected the following two oligo DNAs and Xho I and Xba I as upstream and downstream restriction enzymes, respectively.

   Forward Primer (Oligo 1): 5′-ctctctCTCGAGgttatgttgcctacattttggtgctcac-3′.

   Reverse Primer (Oligo 2): 5′-cttctTCTAGAgcgtcgcatcgtcagtgaggc-3′.

   Here, "ctctct" and "cttct" are random nucleotides 5′ to the restriction enzyme binding site that were added to facilitate the digestion reaction, "CTCGAG" and "TCTAGA" are recognition sequences for XhoI and XbaI, respectively. The underlined sequences are the annealing portions of the primers.

3.3.2 Perform a PCR Reaction to Amplify the GFR

Different sources of genomic DNA can be used as a template, including genomic DNA isolated from a wild-type fly strain or BAC clones that contain the GFR. We performed PCR from a P [acman] clone CH322-146A15 (BACPAC resources, CH322-146A15), which is a clone that was generated from genomic DNA extracted from the *Drosophila* strain used to build the reference genome [123, 124].

3.3.3 Subclone the GFR into the pattB Plasmid

Digest the PCR product and pattB plasmid with appropriate restriction endonucleases. Here, we used XhoI and XbaI. After dephosphorylation of the pattB plasmid using a DNA phosphatase (e.g., Antarctic Phosphatase, NEB, M0289S), perform a ligation reaction using a DNA ligase (e.g., T4 DNA ligase, NEB, M0202S), according to the manufacturer's protocol to generate the genomic rescue plasmid.

Fig. 3 (continued) visual marker that can be used to track the transgenesis event in a *white*⁻ mutant background. It also carries an Ampicillin resistance (Amp^R) cassette (green box) for selection in bacteria. (**c**) Construction of a humanized *amx* genomic rescue construct (TM2D3[+]-attB) based on amx[+]-attB and two intermediate vectors. Human reference *TM2D3* coding sequence was PCR amplified from a cDNA plasmid and integrated into a plasmid in which the *amx* coding sequence was replaced by a spacer sequence flanking two BsaI recognition sites (Intermediate Vector 2). A construct in which the upstream region of the *amx* genomic rescue amplicon with a BsaI site was cloned into pattB (Intermediate Vector 1) as a precursor of Intermediate Vector 2. Golden Gate cloning with BsaI and overhangs (1)/(2) allows swapping of the spacer sequence in Intermediate Vector 2 with human *TM2D3* coding sequence to generate the humanized genomic rescue construct. (**d**) Generation of the humanized *amx* genomic rescue construct carrying the variant of interest (TM2D3[P155L]-attB) through site-directed mutagenesis. The variant of interest is shown using an asterisk

3.3.4 Validate the GFR-pattB Genomic Rescue Construct Using Sanger Sequencing

First perform end sequencing using sequencing primers that anneal to the vector backbone. For pattB, one can use the following primers.

pattB insert sequencing primer 1:

5′-CATTTAATTTAGAAAATGCTTGGATTTCACTG-3′.

pattB insert sequencing primer 2:

5′-CCGCACCGCGGCTTCGAGACC-3′.

To verify that the entire GFR is correct, design gene specific sequencing primers to cover the whole GFR. Identify a colony with the proper sequence and keep a glycerol stock for long term storage. We refer to this construct as amx-pattB (Fig. 3b). Perform miniprep or midiprep to obtain a sufficient amount of this genomic rescue construct for injection (*see* Subheading 3.5).

3.4 Design and Construct a Humanized Genomic Rescue Construct

Since the *amx* gene contains only a single exon, the coding region of the fly *amx* genomic rescue construct can be relatively easily replaced with a human *TM2D3* coding sequence to generate a humanization construct. From this construct, human TM2D3 protein will be expressed in the same spatial and temporal pattern of the fly Amx protein, likely at comparable levels (although there may be some differences in translational efficiency and protein stability due to sequence differences). The following protocol describes the precise method used to construct the humanized genomic rescue construct in Jakobsdottir et al. [33]. Although such cloning can be performed through simple restriction enzyme digestion and ligation reaction if appropriate restriction enzymes cut sites can be found, absence of convenient cut sites forced us to perform this in a multistep procedure, involving the generation of intermediate plasmids that are suitable for Golden Gate cloning (Fig. 3c). It is worth noting that recent advances in multifragment assembly have greatly simplified cloning, including the development of commercial kits such as the Gibson Assembly® Master Mix (NEB, E2611S) and NEBuilder® HiFi DNA Assembly Master Mix (NEB, E2621S). These and other techniques can now be used instead of the method we used, but the method described here can be more cost-effective since it does not require researchers to purchase customized commercial kits.

1. Obtain a human cDNA plasmid that contains the full-length open reading frame. *TM2D3* has two splicing isoforms and the difference between the two is the presence or absence of an alternatively spliced exon. We obtained the longer isoform of *TM2D3* (NM_078474) from a commercial source (Origene, RC203014) for this project.
2. Clone the upstream region of the GRF with a BsaI restriction site at the 3′ end into the pattB vector to make the first

intermediate vector. The forward primer for the upstream region is the same as the forward oligo used for the cloning of the amx GRF (Oligo 1, *see* Subheading 3.3.1, **step 3**). The reverse primer will provide a reverse facing BsaI site to leave an upstream 5′ overhang to insert the human *TM2D3* coding sequence and an XbaI site to insert into the pattB plasmid.

Reverse Primer (Oligo 3).

5′-agagagTCTAGAagagaGGTCTCg<u>tttggtcagatagagcgggg aatacgc</u>-3′.

Here, "agagag" is a series of random nucleotides 5′ to the restriction enzyme binding site that was added to facilitate the digestion reaction, "TCTAGA" is a recognition sequence for XbaI, "agaga" is a spacer sequence, "GGTCTC" is the recognition sequence for BsaI, the following "g" is a random single base spacer sequence required for BsaI, and the underlined sequence is the annealing portions of the primers.

Perform PCR using pattB-amx plasmid as a template and subclone the amplicon using restriction endonucleases XhoI and XbaI into pattB. Confirm the plasmid using Sanger sequencing and generate a glycerol stock. We refer to this vector as "Intermediate Vector 1."

3. Clone the downstream region of the GRF with a BsaI restriction site at its 5′ end into the Intermediate Vector 1 to make the second intermediate vector. The forward primer will provide a forward facing BsaI site to leave a downstream 5′ overhang to insert the human *TM2D3* coding sequence, and is flanked by NheI upstream of the BsaI site that can be used to clone the insert into Intermediate Vector 1. The reverse primer for the downstream region is the same as the reverse primer used for the cloning of the amx GRF (Oligo 2, *see* Subheading 3.3.1).

   Forward Primer (Oligo 4):

   5′-gcggcGCTAGCGGTCTCt<u>tttagttttacaggggttaataggtcataag ctaac</u>-3′.

   Here, "gcggc" is a series of random nucleotides 5′ to the restriction enzyme binding site that was added to facilitate the digestion reaction, "GCTAGC" is a recognition sequence for NheI, "GGTCTC" is the recognition sequence for BsaI, the following "t" is a random single base pair spacer sequence required for BsaI, and the underlined sequence is the annealing portions of the primer.

   Perform PCR using pattB-amx plasmid as a template and Oligo 4 and Oligo 2 as forward and reverse primers, respectively, to subclone the amplicon using NheI and XbaI into Intermediate Vector 1 (from Subheading 3.4, **step 2**). It is important to keep in mind that this insert can ligate in either the forward (correct) or reverse (incorrect) orientation. In the correct orientation, an NheI/XbaI hybrid site internal to the BsaI flanked spacer sequence is formed, providing a way to distinguish the two products. Confirm the correct plasmid using Sanger sequencing and generate a glycerol stock. We refer to this vector as "Intermediate Vector 2" (*see* **Note 4**). The region between the two BsaI sites of this plasmid is a spacer sequence that will be removed and replaced with the human *TM2D3* coding sequence in subsequent steps.

4. Insert the *TM2D3* coding sequence into the *amx* genomic locus using Golden Gate assembly. The Intermediate Vector 2 from Subheading 3.4, **step 3** now has the upstream and downstream components flanking an ends-in BsaI removable cassette that will leave two different 5′ overhangs when digested with the BsaI restriction enzyme. The *TM2D3* coding sequence can be swapped with this cassette to make the final humanized rescue construct. Although the pattB vector has two additional BsaI sites within its sequences, Golden Gate assembly can be used to insert the *TM2D3* coding sequence while preserving the integrity of the pattB backbone using the following procedure:

   (a) PCR amplify the *TM2D3* coding sequence from a human cDNA plasmid (Origene, RC203014) and flank the amplicon with BsaI restriction sites using the following primers. In addition to sequences that anneal with the *TM2D3* open reading frame, additional sequences have been added to generate 5′ overhangs compatible with the upstream and downstream cohesive ends generated when the spacer cassette is removed from Intermediate Vector 2 upon BsaI digestion.

   Forward Primer (Oligo 5).

   5′-gcggcGCTAGCGGTCTCtCAAA<u>atggcgggaggggtgct ccc</u>-3′.

   Reverse Primer (Oligo 6).

   5-

   ′-agagagTCTAGAagagaGGTCTCg<u>TAAActaaatgta-caaagagccatctgctgg</u>-3′.

   Here, "agagag" and "gcggc" are series of random nucleotides 5′ to the restriction enzyme binding site that were added to facilitate the digestion reaction, "GCTAGC" and "TCTAGA" are recognition sequences for NheI and XbaI which were inserted for validation purposes but are not strictly necessary (will be cleaved off upon BsaI digestion), "GGTCTC" is the recognition sequence for BsaI, the following "t" or "g" are random single base spacer sequences required for BsaI, "CAAA" and "TAAA" are overhangs (overhangs (1) and (2), respectively, in Fig. 3c) for ligating the *TM2D3* coding sequence in place of the spacer sequence that is removed upon BsaI digestion, and the underlined sequences are the annealing portions of the primers.

(b) Set up a Golden Gate reaction in a PCR tube including the following:

20 ng Intermediate Vector 2 from Subheading 3.4, **step 3**.

Equimolar amounts of the *TM2D3* amplicon from Subheading 3.4, **step 4a**.

1 μl CutSmart buffer (NEB, B7204S).

1 μl 10 mM ATP.

0.5 μl T4 ligase (NEB M0202S).

0.5 μl restriction enzyme BsaI-HFv2 (NEB, R3733S)

ddH_2O to total volume of 10 μl.

(10 μl in total)

(c) Run the Golden Gate program in a thermal Cycler.

Step 1: 37 °C for 3 min.

Step 2: 20 °C for 2 min.

Step 3: Repeat **Steps 1** and **2** ten times.

Step 4: 20 °C for 15 min.

(This final ligation step is necessary because pattB has additional BsaI sites)

(d) Transform the product into chemically competent bacteria (e.g., DH5α or equivalent strain) and grow these cells on an LB plate with appropriate selection media (pattB is Ampicillin resistant) o/n. Pick up single colonies (we recommend selecting 8–10 colonies) and grow them on liquid LB media (~3 ml) o/n. Miniprep the culture and verify the construct via DNA fingerprinting and Sanger sequencing. We refer to this plasmid as TM2D3[+]-attB. Select the colony that carries the correct insert and generate glycerol stocks for long-term storage of this humanized genomic rescue plasmid.

5. Introduce the variant of interest into the humanized genomic rescue construct. Perform site directed mutagenesis of the humanized genomic rescue construct to introduce the variant of interest (Fig. 3d). Here, we describe a protocol using the Q5® Site-Directed Mutagenesis system (NEB) to introduce the p.P155L into the humanized *amx* genomic rescue construct (TM2D3[P155L]-pattB). Additional methods such as QuikChange II Site-directed mutagenesis system (Agilent) and

Phusion Site-Directed Mutagenesis system (ThermoFisher) can also be employed, which will not be discussed here.

(a) Set up a mutagenesis reaction using high-fidelity DNA polymerase (Q5 Hot Start High-Fidelity DNA Polymerase included in the Q5® Site-Directed Mutagenesis kit) and appropriate primer pairs in a PCR tube. Perform a mutagenic PCR reaction according to the manufacturer's protocol. To introduce the p.P155L variant (NM_078474.3:c.464C>T), we designed and used the following mutagenesis primers and followed the protocol for Q5® Site-Directed Mutagenesis.

Forward mutagenesis primer.

5′-CCTGTCCTCGGCAGCGCTACCTTGCCAAC TGCACGGTGCGGG-3′ (forward).

Reverse mutagenesis primer.

5′-CCCGCACCGTGCAGTTGGCAAGGTAGCGCTG CCGAGGACAGG-3′ (reverse).

Here, the underlined base corresponds to the mutation that will be introduced.

(b) Ligate the amplified product and digest the template plasmid using the KLD enzyme mix included in the Q5® Site-Directed Mutagenesis kit based on the manufacturer's protocol.

(c) Transform the product into chemically competent bacteria (e.g., DH5α or equivalent strain) and grow these cells on an LB plate with appropriate selection media (pattB is Ampicillin resistant) o/n. Pick up single colonies (we recommend selecting 12 colonies) and grow them on liquid LB media (~3 ml) o/n. Miniprep the culture and verify the construct via DNA fingerprinting and Sanger sequencing (*see* **Note 5**). Select the colony that carries the correct insertion and generate glycerol stocks for long-term storage. We refer to this plasmid as TM2D3 [P155L]-pattB. Perform miniprep or midiprep to obtain sufficient amounts of this mutagenized humanized genomic rescue plasmid for injection (*see* Subheading 3.5).

3.5 Generation of Transgenic Flies That Carry the Original and Humanized Genomic Rescue Constructs

1. Make a plasmid injection solution. Adjust the genomic rescue constructs from Subheadings 3.3 and 3.4 to 200–300 ng/μl with ddH_2O.

2. Microinject the injection solution into flies that express ΦC31 integrase in the germline. A number of stocks are available from BDSC (https://bdsc.indiana.edu/stocks/phic31/phic31_int.html). For this project, we selected a line which expresses ΦC31 integrase in the germline using a *vasa* promoter and

carries an *attB* docking site on the second chromosome (y^1 *M{vas-int.Dm}ZH-2A w**; *PBac{y+-attP-3B}VK00037*, BDSC: 24872). Inject the solution prepared in Subheading 3.5, **step 1** into the posterior region of y^1 *M{vas-int.Dm}ZH-2A w**; *PBac{y+-attP-3B}VK00037* embryos that were within 1 h after egg laying using standard transgenic techniques [114] (Fig. 1b). Injected embryos are designated G0 and are mosaic.

3. Raise G0 to adults and set crosses to obtain F1 flies. Collect males and virgin females and cross single males to 3–5 virgin *yellow*$^-$ and 2–3 virgin females to 1–2 *yellow*$^-$ males in individual vials. These crosses produce F1 offspring with potential transgene insertion.
4. Screen the F1 flies for insertion of transgene and establish stable stocks. Raise F1 offspring to adults. Anesthetize F1 adult flies using CO_2 and a diffuser pad. Under a stereomicroscope (0.63–5×) at low power, examine wings of each F1 fly for red eye color (Fig. 1c). Pick male flies with dark colored wings and individually cross them to appropriate balancer stocks (e.g., *CyO*). Crossing scheme to establish stable stocks using the ΦC31 transgenesis system can be found in [82].
5. Molecularly validate the targeting event. Collect 6–8 adult flies, isolate Genomic DNA from each balanced line and perform PCR with appropriate primer pairs to confirm the successful integration of the transgenes [119, 121].

3.6 Establishment of Fly Stocks To Be Used for Functional Assessment of the Variant of Interest

To perform rescue experiments, one must combine the genomic rescue transgenes generated in Subheading 3.5 with mutant alleles obtained or generated in Subheadings 3.1 or 3.2. The *amx* gene is located on the *X*-chromosome, and the genomic rescue constructs have been inserted onto the second chromosome. Using standard genetic techniques in *Drosophila*, construct the following fly strains that have an *amx* mutant allele and a genomic rescue transgene. For tutorials on designing fly crossing schemes, we refer the readers to the "Fly Pushing" book by Greenspan [57].

amx^1 *lz*^{g} *v*^{1}; *amx[+]*.

amx^1 *lz*^{g} *v*^{1}; *TM2D3[+]*.

amx^1 *lz*^{g} *v*^{1}; *TM2D3[P155L]*.

y w amx$^\Delta$; *amx[+]*.

y w amx$^\Delta$; *TM2D3[+]*.

y w amx$^\Delta$; *TM2D3[P155L]*.

3.7 Functional Study of the Variant of Interest Based on Female Fertility

Studies have reported that maternal loss of *amx* in *Drosophila* leads to production of embryos that fail to hatch from eggs [89]. Therefore, egg hatching rate can be used as a functional measure of *amx* activity in vivo. Furthermore, if the humanized genomic rescue transgene also rescues this defect, egg hatching rate can be used as a quantitative measure to assess the functional difference between the reference and variant human *TM2D3* transgenes. Here, we present a detailed protocol of egg hatching assay and analysis.

3.7.1 Generate or Collect the amx Mutant Flies for the Experiment

1. For the experiment using the classic *amx*^{1} allele from Subheading 3.1, cross the *amx*^{1} *lz*^{g} *v*^{1} males with or without the rescue transgene on the second chromosome (*−; amx[+]; TM2D3[+]; TM2D3[P155L]*) to *Df(1)Exel4049, w*^{1118}*/Binsinscy* virgin females (P0 generation).
2. In the following generation (F1), collect virgin female flies that are *amx*^{1} *lz*^{g} *v*^{1} */Df(1)Exel4049, w*^{1118}*; (with or without rescue transgene)/+*.
3. For the experiment using the new *amx*$^{\Delta}$ allele from Subheading 3.2, collect virgin females from respective stocks (*see* **Note 6**).

3.7.2 Embryo Collection, Quantification and Analysis

1. Cross the *amx* mutant females to *amx* mutant males (*amx*^{1} *lz*^{g} *v*^{1}/Y or *y w amx*$^{\Delta}$/Y) and place them in a plastic bottle with a grape juice plate supplemented with active yeast paste. Keep the cross at 25 °C for 2 days.
2. Replace with a new grape plate and allow flies to lay eggs (F2 generation) on the plate with some yeast paste at 25 °C for ~6 h.
3. Place the grape plates with embryos in a humid-controlled incubator for 24 h at 25 °C.
4. Use a fine paint brush washed with water to collect embryos on a mesh filter.
5. Dechlorinate embryos with 50% bleach for 2 min. Check under a dissection microscope to confirm that the outer egg shell (chorion) has been successfully removed.
6. Rinse the dechorionated embryos with running water to completely remove the bleach.
7. Move the embryos into a 24-well cell culture plate well that contains ~500 μl of PBS. To facilitate the qualification process, do not overcrowd the embryos (<100 embryos per well) so that the embryos do not overlap.
8. Let the embryos float to the top of the PBS and take a photograph of each well.
9. Incubate the plate at 25 °C for 24 h. During this time period, all normal embryos should have already hatched from their egg case.
10. Take a photograph of each well after the 24-h incubation and count the number of larvae that have successfully hatched. Embryos that look completely white should be excluded from

the calculation because they are most likely to be unfertilized eggs.

11. Calculate the egg hatching rate by dividing the number of larvae by the number of eggs being laid (% = hatched eggs/ total eggs × 100). Repeat the process at least three times to obtain biological replicates (Fig. 4a, b) [33, 125].

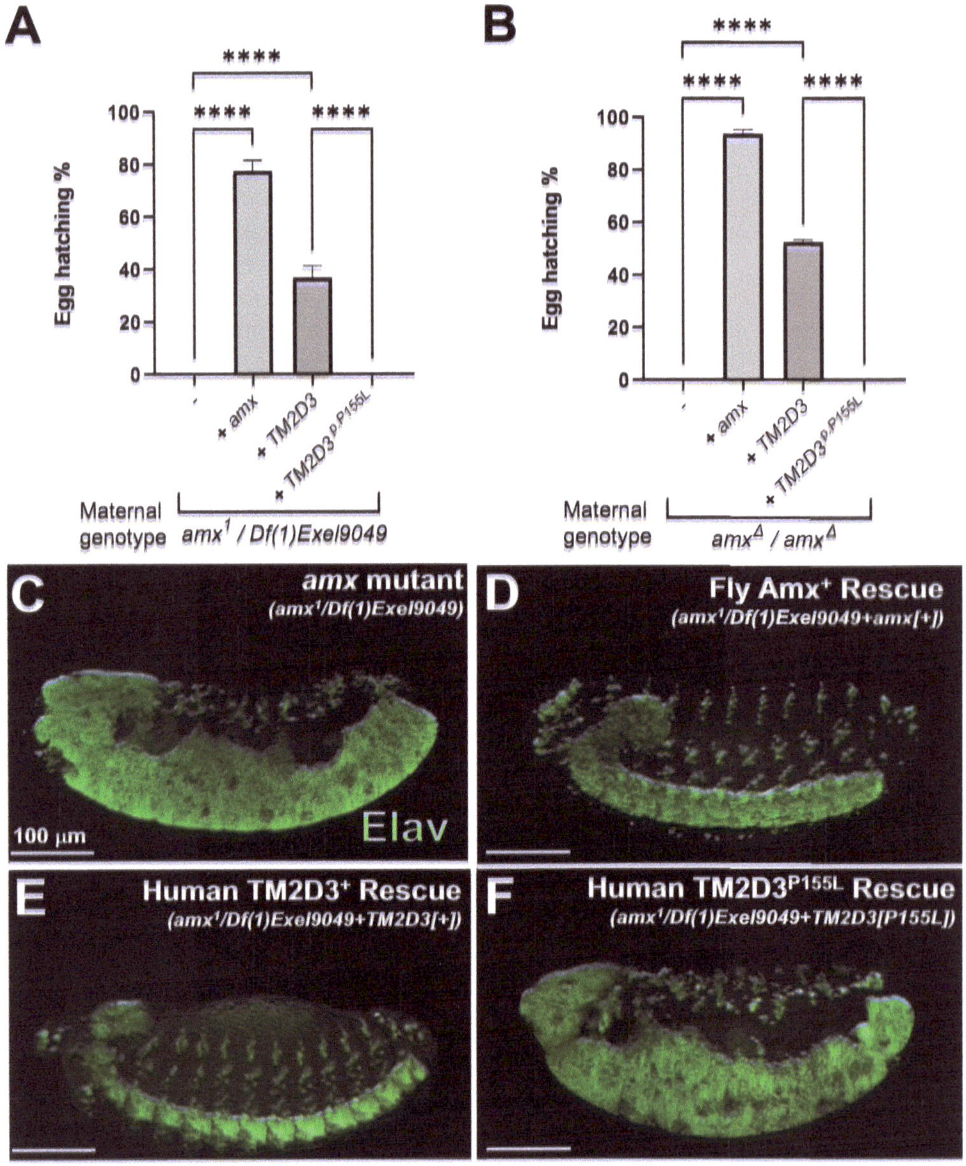

Fig. 4 Functional studies of a late-onset Alzheimer's disease-associated rare missense variant in *TM2D3* based on Notch signaling related phenotypes in vivo in *Drosophila*. (**a**) Results of an egg hatching assay using the classical *amx* allele over a molecularly defined deficiency obtained from the Bloomington *Drosophila* Stock Center (BDSC). None of the eggs laid by the *amx*^{1}/*Df(1)Exel9094* mothers hatch, while a significant portion of

3.8 Functional Study of the Variant of Interest Based on the Embryonic Neurogenic Phenotype

In addition to failure of hatching, embryos laid by *amx* defective mothers also show a robust embryonic neurogenic phenotype, which is a sign of Notch signaling mediated lateral inhibition [126, 127]. Therefore, the embryonic neurogenic phenotype can be used as an indirect measure of Notch signaling activity in developing embryos to assess the functional impact of the disease-associated variant in vivo.

3.8.1 Generate or Collect the amx Mutant Flies for the Experiment

Follow the crossing scheme outlined in Subheading 3.7.1 to obtain the necessary flies.

3.8.2 Embryo Collection and Fixation

1. Cross the *amx* mutant females with or without genomic rescue transgenes to *amx* mutant males (amx^1 lz^g v^1/Y or *y w* amx^Δ/Y) and collect/process the embryos as in Subheading 3.7.2, **steps 1–6**.
2. Use a fine brush to transfer the embryos to 3 ml of fixation solution in a 20 ml scintillation vial.
3. Add 3 ml of 100% heptane (1:1 to formaldehyde/PBS) to the scintillation vial. *n*-heptane and formaldehyde/PBS form two phase layers and dechorionated embryos remain at the interface.
4. Incubate the embryos for 20 min at room temperature on a laboratory rocker to fix the embryos while constantly shaking the vials.
5. Remove the lower formaldehyde/PBS layer using a micropipette.

Fig. 4 (continued) this can be rescued by introduction of a wild-type *amx* genomic rescue construct (+*amx*). The maximum rescue expected in this experiment is 75% because *Df(1)Exel9094* removes a nearby essential gene [*Df(1)9094/Y* males are lethal regardless of *amx* function]. Humanized *amx* genomic rescue transgene (+*TM2D3*) has about 50% activity of the fly *amx* genomic rescue transgene and can rescue the egg hatching defect of amx^1/*Df(1)Exel9094* progenies. In contrast, the humanized *amx* genomic rescue construct with the late-onset Alzheimer's disease (LOAD) associated p.P155L variant (+$TM2D3^{p.P155L}$) lacks this activity. (**b**) Results of an egg hatching assay using the newly generated *amx* null allele via CRISPR. The results obtained from amx^Δ are similar to the results using amx^1/*Df(1)Exel9094*, further confirming the results using the classic allele. (**c–f**) Representative images of the nervous system of embryos that are mutant for *amx* rescued by different genomic rescue constructs. Neurons are visualized by immunofluorescence staining against Elav, a panneuronal nuclear marker. (**c**) Embryos from amx^1/*Df(1)Exel9094* mothers show a strong neurogenic phenotype, which is a classic sign of Notch signaling defect during embryonic neurogenesis. (**d**) A wild-type *amx* genomic rescue transgene can fully rescue this defect. (**e**) A humanized *amx* genomic transgene expressing the reference human *TM2D3* can also rescue this defect in a subset of animals. (**f**) A humanized *amx* genomic transgene expressing human *TM2D3* with the p.P155L variant cannot rescue the neurogenic defect, indicating this missense variant is deleterious to a Notch signaling related function of this gene in vivo. Similar results have been obtained using the amx^Δ allele (data not shown). Scale bars = 100 μm. Images were adopted from [33]

6. Add 3 ml of 100% methanol (methanol will sink below the *n*-heptane layer) and then shake vigorously by hand for 30 s.
7. Place the scintillation vial on the bench until two phase layers of heptane and methanol are formed. Fixed and devitellinated embryos will settle to the bottom in the lower (methanol) layer.
8. Remove the upper *n*-heptane layer, the debris at the interface and most of the methanol until embryos are only covered in a small amount of methanol.
9. Wash the embryos with fresh 100% methanol to remove any remaining fixative and *n*-heptane. Transfer the embryos in methanol to 1.5 ml centrifuge tubes. Embryos in this condition can be preserved long-term at −20 °C until use.

3.8.3 Staining and Mounting

1. Rehydrate and permeabilize the embryos from Subheading 3.8.4 by incubating them in 0.05% Triton X in PBS (PBST) three times for 20 min each at room temperature.
2. Incubate the embryos with a primary antibody against a neuronal nuclear marker (rat monoclonal anti-ELAV antibody, 1: 100, DSHB, 7E8A10) in a blocking solution (5% normal donkey serum in PBST) at 4 °C o/n.
3. Wash the embryos three times with PBST for 20 min each at room temperature.
4. Incubate the embryos with a secondary antibody (Alexa Fluor® 488 AffiniPure Donkey Anti-Rat IgG, 1:200, Jackson ImmunoResearch) in blocking solution (5% normal donkey serum in PBST) for 2 h at room temperature.
5. Wash the embryos three times with PBST for 20 min each at room temperature.
6. Remove as much PBST as possible and add sufficient amount of mounting media (VECTASHIELD® Antifade Mounting Medium with DAPI, H-1200, Vector Labs) into the tube to cover sample. Transfer the embryos in the mounting media on to a glass slide with a trimmed pipette tip and cover with a coverslip. Secure the coverslip with nail polish.

3.8.4 Imaging and Analysis

Image the samples with a fluorescence microscope with an appropriate laser and filter. For best results, use a laser confocal microscope and make a *z*-stack out of multiple optical sections (Fig. 4c–f) [33, 125].

4 Notes

1. *C(1)DX*, y^1 f^1 (FlyBase ID: FBab0000080) that is used to maintain the amx^1 lz^g v^1 chromosome is a compound *X*-chromosome. All males in this stock are amx^1 lz^g v^1/Y and all females in this stock are *C(1)DX*, y^1 f^1/Y.

2. *Binsinscy* (FlyBase ID: FBba0000019) that is used to maintain the *Df(1)Exel9049, w^{1118}* chromosome is a balancer chromosome for the *X*-chromosome. This chromosome carries a recessive female sterile mutation in addition to several dominant and recessive markers. Although there are homozygous *Binsinscy* females present in this stock, they can be distinguished from the *Df(1)Exel9049, w^{1118}/Binsinscy* flies based on the semidominant *Bar* mutation (B^1) present on *Binsincy.*
3. Another advantage of using this system is that the *yellow [wing2+]* knockin mutant flies can be further used to perform scarless gene editing of the gene of interest. For more information on this application, *see* Li-Kroeger et al. [84].
4. One benefit of generating this Intermediate Vector 2 is that it can be used as a template to make additional constructs that utilize the regulatory region of *amx*. For example, one can insert components such as GAL4, EGFP, or Luciferase instead of human *TM2D3* to generate additional reagents to study this *Drosophila* gene in more detail.
5. To avoid the risk of introducing second site mutations during the mutagenic PCR step, one can subclone the humanized genomic fragment into a new attB plasmid. This will eliminate the possibility that there is a second site mutation on the vector backbone that may interfere with the transgenesis process (e.g., mutation in the *attB* site or *mini-white*$^{+}$ marker).
6. One advantage of using a clean null allele is that it reduces the amount of fly work required to perform the egg hatching and embryonic neurogenic assays. Because the *amx^{Δ}* chromosome does not carry any other mutations that affect this assay, one can use this chromosome in a homozygous state by collecting virgin females directly from each stock. This is in contrast to the *amx^1* chromosome which carries a closely linked *lz^g* mutation that needs to be neutralized by placing this chromosome over a deficiency that does not affect *lz*.

Acknowledgments

J.L.S. received support from the National Institutes of Health (NIH) training grant T32GM008307. S.Y. received support from the NIH (R01AG071557), Alzheimer's Association (NIRH-15-364099), and Nancy Chang PhD, Award for Research Excellence from Baylor College of Medicine related to the project discussed in this article. We thank J. Michael Harnish for providing useful feedback.

References

1. Kopan R, Ilagan MXG (2009) The canonical Notch signaling pathway: unfolding the activation mechanism. Cell 137:216–233
2. Ables JL, Breunig JJ, Eisch AJ et al (2011) Not(ch) just development: Notch signalling in the adult brain. Nat Rev Neurosci 12: 269–283
3. Yamamoto S, Schulze KL, Bellen HJ (2014) Introduction to Notch signaling. Methods Mol Biol 1187:1–14
4. Artavanis-Tsakonas S, Muskavitch MAT (2010) Notch: the past, the present, and the future, Notch signaling. Elsevier, Amsterdam, pp 1–29
5. Greenwald I (2012) Notch and the awesome power of genetics. Genetics 191:655–669
6. Louvi A, Artavanis-Tsakonas S (2012) Notch and disease: a growing field. Semin Cell Dev Biol 23:473–480
7. Ho DM, Artavanis-Tsakonas S, Louvi A (2020) The Notch pathway in CNS homeostasis and neurodegeneration. Wiley interdisciplinary reviews. Dev Biol 9:e358
8. Mašek J, Andersson ER (2017) The developmental biology of genetic Notch disorders. Development 144:1743–1763
9. Salazar JL, Yamamoto S (2018) Integration of drosophila and human genetics to understand notch signaling related diseases. Adv Exp Med Biol 1066:141–185
10. Hassed S, Li S, Mulvihill J et al (2017) Adams-Oliver syndrome review of the literature: refining the diagnostic phenotype. Am J Med Genet A 173:790–800
11. Gilbert MA, Spinner NB (2017) Alagille syndrome: genetics and functional models. Curr Pathobiol Rep 5:233–241
12. Nóbrega A, Maia-Fernandes AC, Andrade RP (2021) Altered cogs of the clock: insights into the embryonic etiology of spondylocostal dysostosis. J Dev Biol 9
13. Mizuno T, Mizuta I, Watanabe-Hosomi A et al (2020) Clinical and genetic aspects of CADASIL. Front Aging Neurosci 12:91
14. Selkoe DJ, Hardy J (2016) The amyloid hypothesis of Alzheimer's disease at 25 years. EMBO Mol Med 8:595–608
15. Splinter K, Adams DR, Bacino CA et al (2018) Effect of genetic diagnosis on patients with previously undiagnosed disease. N Engl J Med 379:2131–2139
16. Posey JE, O'Donnell-Luria AH, Chong JX et al (2019) Insights into genetics, human biology and disease gleaned from family based genomic studies. Genet Med 21: 798–812
17. Fischer-Zirnsak B, Segebrecht L, Schubach M et al (2019) Haploinsufficiency of the notch ligand DLL1 causes variable neurodevelopmental disorders. Am J Hum Genet 105: 631–639
18. Chong JX, Buckingham KJ, Jhangiani SN et al (2015) The genetic basis of mendelian phenotypes: discoveries, challenges, and opportunities. Am J Hum Genet 97:199–215
19. Simpson MA, Irving MD, Asilmaz E et al (2011) Mutations in NOTCH2 cause Hajdu-Cheney syndrome, a disorder of severe and progressive bone loss. Nat Genet 43: 303–305
20. Isidor B, Lindenbaum P, Pichon O et al (2011) Truncating mutations in the last exon of NOTCH2 cause a rare skeletal disorder with osteoporosis. Nat Genet 43: 306–308
21. McDaniell R, Warthen DM, Sanchez-Lara PA et al (2006) NOTCH2 mutations cause Alagille syndrome, a heterogeneous disorder of the NOTCH signaling pathway. Am J Hum Genet 79:169–173
22. Gripp KW, Robbins KM, Sobreira NL et al (2015) Truncating mutations in the last exon of NOTCH3 cause lateral meningocele syndrome. Am J Med Genet A 167A:271–281
23. Martignetti JA, Tian L, Li D et al (2013) Mutations in PDGFRB cause autosomal-dominant infantile myofibromatosis. Am J Hum Genet 92:1001–1007
24. Stevens DL, Hewlett RH, Brownell B (1977) Chronic familial vascular encephalopathy. Lancet 1:1364–1365
25. Joutel A, Corpechot C, Ducros A et al (1996) Notch3 mutations in CADASIL, a hereditary adult-onset condition causing stroke and dementia. Nature 383:707–710
26. Bamshad MJ, Nickerson DA, Chong JX (2019) Mendelian gene discovery: fast and furious with no end in sight. Am J Hum Genet 105:448–455
27. Baldridge D, Wangler MF, Bowman AN et al (2021) Model organisms contribute to diagnosis and discovery in the undiagnosed diseases network: current state and a future vision. Orphanet J Rare Dis 16:206
28. Wei J, Hemmings GP (2000) The NOTCH4 locus is associated with susceptibility to schizophrenia. Nat Genet 25:376–377
29. Qu J, Qu H-Q, Bradfield JP et al (2021) Association of DLL1 with type 1 diabetes in

patients characterized by low polygenic risk score. Metab Clin Exp 114:154418

30. Kunkle BW, Grenier-Boley B, Sims R et al (2019) Genetic meta-analysis of diagnosed Alzheimer's disease identifies new risk loci and implicates Aβ, tau, immunity and lipid processing. Nat Genet 51:414–430
31. Patel D, Mez J, Vardarajan BN et al (2019) Association of rare coding mutations with alzheimer disease and other dementias among adults of European Ancestry. JAMA Netw Open 2:e191350
32. Pairo-Castineira E, Clohisey S, Klaric L et al (2021) Genetic mechanisms of critical illness in COVID-19. Nature 591:92–98
33. Jakobsdottir J, van der Lee SJ, Bis JC et al (2016) Rare functional variant in TM2D3 is associated with late-onset Alzheimer's disease. PLoS Genet 12:e1006327
34. Zhou X, Li H, Guo S et al (2019) Associations of multiple NOTCH4 exonic variants with systemic sclerosis. J Rheumatol 46:184–189
35. All of Us Research Program Investigators, Denny JC, Rutter JL et al (2019) The "All of Us" Research Program. N Engl J Med 381: 668–676
36. Lichou F, Trynka G (2020) Functional studies of GWAS variants are gaining momentum. Nat Commun 11:6283
37. Aster JC, Pear WS, Blacklow SC (2017) The varied roles of notch in cancer. Annu Rev Pathol 12:245–275
38. Lobry C, Oh P, Mansour MR et al (2014) Notch signaling: switching an oncogene to a tumor suppressor. Blood 123:2451–2459
39. Yamamoto S (2020) Making sense out of missense mutations: mechanistic dissection of Notch receptors through structure-function studies in Drosophila. Develop Growth Differ 62:15–34
40. Weng AP, Ferrando AA, Lee W et al (2004) Activating mutations of NOTCH1 in human T cell acute lymphoblastic leukemia. Science 306:269–271
41. Wang NJ, Sanborn Z, Arnett KL et al (2011) Loss-of-function mutations in Notch receptors in cutaneous and lung squamous cell carcinoma. Proc Natl Acad Sci U S A 108: 17761–17766
42. Katoh M, Katoh M (2020) Precision medicine for human cancers with Notch signaling dysregulation (review). Int J Mol Med 45: 279–297
43. Ilagan MXG, Kopan R (2014) Monitoring Notch activation in cultured mammalian cells: transcriptional reporter assays. Methods Mol Biol 1187:143–154
44. Ilagan MXG, Kopan R (2014) Monitoring Notch activation in cultured mammalian cells: luciferase complementation imaging assays. Methods Mol Biol 1187:155–168
45. Serneels L, Tesseur I, De Strooper B (2014) Assay to probe proteolytic processing of Notch by γ-secretase. Methods Mol Biol 1187:223–229
46. Gridley T, Groves AK (2014) Overview of genetic tools and techniques to study Notch signaling in mice. Methods Mol Biol 1187: 47–61
47. Marathe S, Alberi L (2014) Monitoring Notch activity in the mouse. Methods Mol Biol 1187:115–129
48. Justice MJ, Dhillon P (2016) Using the mouse to model human disease: increasing validity and reproducibility. Dis Model Mech 9:101–103
49. Clark JF, Dinsmore CJ, Soriano P (2020) A most formidable arsenal: genetic technologies for building a better mouse. Genes Dev 34: 1256–1286
50. Wangler MF, Yamamoto S, Chao H-T et al (2017) Model organisms facilitate rare disease diagnosis and therapeutic research. Genetics 207:9–27
51. Boycott KM, Campeau PM, Howley HE et al (2020) The Canadian rare diseases models and mechanisms (RDMM) network: connecting understudied genes to model organisms. Am J Hum Genet 106:143–152
52. Bellen HJ, Yamamoto S (2015) Morgan's legacy: fruit flies and the functional annotation of conserved genes. Cell 163:12–14
53. Venken KJT, Sarrion-Perdigones A, Vandeventer PJ et al (2016) Genome engineering: Drosophila melanogaster and beyond. Wiley interdisciplinary reviews. Dev Biol 5:233–267
54. Kanca O, Bellen HJ, Schnorrer F (2017) Gene tagging strategies to assess protein expression, localization, and function in drosophila. Genetics 207:389–412
55. Bellen HJ, Wangler MF, Yamamoto S (2019) The fruit fly at the interface of diagnosis and pathogenic mechanisms of rare and common human diseases. Hum Mol Genet 28: R207–R214
56. Larkin A, Marygold SJ, Antonazzo G et al (2021) FlyBase: updates to the Drosophila melanogaster knowledge base. Nucleic Acids Res 49:D899–D907
57. Greenspan RJ (2004) Fly pushing: the theory and practice of drosophila genetics. Cold Spring Harbor Laboratory Press, Cold Spring Harbor

58. Dahmann C (2018) Drosophila: methods and protocols. Springer, New York
59. Ashburn MD, Golic KG, Hawley RS (2005) Drosophila: a laboratory handbook. Cold Spring Harbor Laboratory Press, Cold Spring Harbor
60. Ghosh R, Oak N, Plon SE (2017) Evaluation of in silico algorithms for use with ACMG/AMP clinical variant interpretation guidelines. Genome Biol 18:225
61. Coban-Akdemir Z, White JJ, Song X et al (2018) Identifying genes whose mutant transcripts cause dominant disease traits by potential gain-of-function alleles. Am J Hum Genet 103:171–187
62. Katsonis P, Koire A, Wilson SJ et al (2014) Single nucleotide variations: biological impact and theoretical interpretation. Protein Sci 23: 1650–1666
63. Muller HJ (1932) Further studies on the nature and causes of gene mutations. In: International congress on genetics, vol 6.1, pp 213–255
64. Manolio TA, Fowler DM, Starita LM et al (2017) Bedside back to bench: building bridges between basic and clinical genomic research. Cell 169:6–12
65. Sawyer SL, Hartley T, Dyment DA et al (2016) Utility of whole-exome sequencing for those near the end of the diagnostic odyssey: time to address gaps in care. Clin Genet 89:275–284
66. Eldomery MK, Coban-Akdemir Z, Harel T et al (2017) Lessons learned from additional research analyses of unsolved clinical exome cases. Genome Med 9:26
67. Ma X, Shao Y, Tian L et al (2019) Analysis of error profiles in deep next-generation sequencing data. Genome Biol 20:50
68. Tam V, Patel N, Turcotte M et al (2019) Benefits and limitations of genome-wide association studies. Nat Rev Genet 20:467–484
69. Wang J, Al-Ouran R, Hu Y et al (2017) MARRVEL: integration of human and model organism genetic resources to facilitate functional annotation of the human genome. Am J Hum Genet 100:843–853
70. Karczewski KJ, Francioli LC, Tiao G et al (2020) The mutational constraint spectrum quantified from variation in 141,456 humans. Nature 581:434–443
71. Landrum MJ, Chitipiralla S, Brown GR et al (2020) ClinVar: improvements to accessing data. Nucleic Acids Res 48:D835–D844
72. Bragin E, Chatzimichali EA, Wright CF et al (2014) DECIPHER: database for the interpretation of phenotype-linked plausibly pathogenic sequence and copy-number variation. Nucleic Acids Res 42:D993–D1000
73. Rentzsch P, Witten D, Cooper GM et al (2019) CADD: predicting the deleteriousness of variants throughout the human genome. Nucleic Acids Res 47:D886–D894
74. Ioannidis NM, Rothstein JH, Pejaver V et al (2016) REVEL: an ensemble method for predicting the pathogenicity of rare missense variants. Am J Hum Genet 99:877–885
75. Amberger JS, Bocchini CA, Scott AF et al (2019) OMIM.org: leveraging knowledge across phenotype-gene relationships. Nucleic Acids Res 47:D1038–D1043
76. Wang J, Liu Z, Bellen HJ et al (2019) Navigating MARRVEL, a web-based tool that integrates human genomics and model organism genetics information. J Vis Exp (150). https://doi.org/10.3791/59542
77. Wang J, Mao D, Fazal F et al (2019) Using MARRVEL v1.2 for bioinformatics analysis of human genes and variant pathogenicity. Curr Protoc Bioinformatics 67:e85
78. Hu Y, Flockhart I, Vinayagam A et al (2011) An integrative approach to ortholog prediction for disease-focused and other functional studies. BMC Bioinformatics 12:357
79. Shefchek KA, Harris NL, Gargano M et al (2020) The Monarch Initiative in 2019: an integrative data and analytic platform connecting phenotypes to genotypes across species. Nucleic Acids Res 48:D704–D715
80. Alliance of Genome Resources Consortium (2020) Alliance of Genome Resources Portal: unified model organism research platform. Nucleic Acids Res 48:D650–D658
81. Hu Y, Comjean A, Mohr SE et al (2017) Gene2function: an integrated online resource for gene function discovery. G3 (Bethesda MD) 7:2855–2858
82. Harnish JM, Deal SL, Chao H-T et al (2019) In vivo functional study of disease-associated rare human variants using Drosophila. J Vis Exp (150). https://doi.org/10.3791/59658
83. Housden BE, Perrimon N (2016) Design and generation of donor constructs for genome engineering in Drosophila. Cold Spring Harbor Protoc (9). https://doi.org/10.1101/pdb.prot090787. https://pubmed.ncbi.nlm.nih.gov/27587780/
84. Li-Kroeger D, Kanca O, Lee P-T et al (2018) An expanded toolkit for gene tagging based on MiMIC and scarless CRISPR tagging in Drosophila. eLife 7:e38709
85. Psaty BM, O'Donnell CJ, Gudnason V et al (2009) Cohorts for Heart and Aging Research in Genomic Epidemiology

(CHARGE) Consortium: design of prospective meta-analyses of genome-wide association studies from 5 cohorts. Circ Cardiovasc Genet 2:73–80
86. Grove ML, Yu B, Cochran BJ et al (2013) Best practices and joint calling of the HumanExome BeadChip: the CHARGE Consortium. PLoS One 8:e68095
87. Sims R, Hill M, Williams J (2020) The multiplex model of the genetics of Alzheimer's disease. Nat Neurosci 23:311–322
88. Ebenesersdóttir SS, Sandoval-Velasco M, Gunnarsdóttir ED et al (2018) Ancient genomes from Iceland reveal the making of a human population. Science 360:1028–1032
89. Shannon MP (1972) Characterization of the female-sterile mutant almondex of Drosophila melanogaster. Genetica 43:244–256
90. Green MM, Green KC (1956) A cytogenetic analysis of the lozenge pseudoalleles in Drosophila. Zeitschrift fur induktive Abstammungs- und Vererbungslehre 87:708–721
91. Lehmann R, Jiménez F, Dietrich U et al (1983) On the phenotype and development of mutants of early neurogenesis inDrosophila melanogaster. Wilhelm Rouxs Arch Dev Biol 192:62–74
92. de-la-Concha A, Dietrich U, Weigel D et al (1988) Functional interactions of neurogenic genes of Drosophila melanogaster. Genetics 118:499–508
93. Michellod M-A, Forquignon F, Santamaria P et al (2003) Differential requirements for the neurogenic gene almondex during Drosophila melanogaster development. Genesis 37: 113–122
94. Michellod M-A, Randsholt NB (2008) Implication of the Drosophila beta-amyloid peptide binding-like protein AMX in Notch signaling during early neurogenesis. Brain Res Bull 75: 305–309
95. Sherrington R, Rogaev EI, Liang Y et al (1995) Cloning of a gene bearing missense mutations in early-onset familial Alzheimer's disease. Nature 375:754–760
96. Rogaev EI, Sherrington R, Rogaeva EA et al (1995) Familial Alzheimer's disease in kindreds with missense mutations in a gene on chromosome 1 related to the Alzheimer's disease type 3 gene. Nature 376:775–778
97. McQuilton P, St Pierre SE, Thurmond J et al (2012) FlyBase 101--the basics of navigating FlyBase. Nucleic Acids Res 40:D706–D714
98. Gratz SJ, Cummings AM, Nguyen JN et al (2013) Genome engineering of Drosophila with the CRISPR RNA-guided Cas9 nuclease. Genetics 194:1029–1035
99. Bassett AR, Tibbit C, Ponting CP et al (2013) Highly efficient targeted mutagenesis of Drosophila with the CRISPR/Cas9 system. Cell Rep 4:220–228
100. Yu Z, Ren M, Wang Z et al (2013) Highly efficient genome modifications mediated by CRISPR/Cas9 in Drosophila. Genetics 195: 289–291
101. Kondo S, Ueda R (2013) Highly improved gene targeting by germline-specific Cas9 expression in Drosophila. Genetics 195: 715–721
102. Kanca O, Andrews JC, Lee P-T et al (2019) De novo variants in WDR37 are associated with epilepsy, Colobomas, dysmorphism, developmental delay, intellectual disability, and cerebellar hypoplasia. Am J Hum Genet 105:413–424
103. Gratz SJ, Ukken FP, Rubinstein CD et al (2014) Highly specific and efficient CRISPR/Cas9-catalyzed homology-directed repair in Drosophila. Genetics 196:961–971
104. Wiedenheft B, Sternberg SH, Doudna JA (2012) RNA-guided genetic silencing systems in bacteria and archaea. Nature 482: 331–338
105. Housden BE, Valvezan AJ, Kelley C et al (2015) Identification of potential drug targets for tuberous sclerosis complex by synthetic screens combining CRISPR-based knockouts with RNAi. Sci Signal 8:rs9
106. Haelterman NA, Jiang L, Li Y et al (2014) Large-scale identification of chemically induced mutations in Drosophila melanogaster. Genome Res 24:1707–1718
107. Port F, Chen H-M, Lee T et al (2014) Optimized CRISPR/Cas tools for efficient germline and somatic genome engineering in Drosophila. Proc Natl Acad Sci U S A 111: E2967–E2976
108. Engler C, Kandzia R, Marillonnet S (2008) A one pot, one step, precision cloning method with high throughput capability. PLoS One 3: e3647
109. Casini A, Storch M, Baldwin GS et al (2015) Bricks and blueprints: methods and standards for DNA assembly. Nat Rev Mol Cell Biol 16: 568–576
110. Marillonnet S, Werner S (2015) Assembly of multigene constructs using golden gate cloning. Methods Mol Biol 1321:269–284
111. Marillonnet S, Grützner R (2020) Synthetic DNA assembly using golden gate cloning and the hierarchical modular cloning pipeline. Curr Protoc Mol Biol 130:e115
112. Mukherjee M, Caroll E, Wang ZQ (2021) Rapid assembly of multi-gene constructs

using modular golden gate cloning. J Vis Exp (168). https://doi.org/10.3791/61993. https://pubmed.ncbi.nlm.nih.gov/33616121/
113. Gibson DG, Young L, Chuang R-Y et al (2009) Enzymatic assembly of DNA molecules up to several hundred kilobases. Nat Methods 6:343–345
114. Bachmann A, Knust E (2008) The use of P-element transposons to generate transgenic flies. Methods Mol Biol 420:61–77
115. Brand AH, Perrimon N (1993) Targeted gene expression as a means of altering cell fates and generating dominant phenotypes. Development 118:401–415
116. Lee P-T, Zirin J, Kanca O et al (2018) A gene-specific T2A-GAL4 library for Drosophila. eLife 7:e35574
117. Feng S, Lu S, Grueber WB et al (2021) Scarless engineering of the Drosophila genome near any site-specific integration site. Genetics 217:iyab012
118. Venken KJT, Bellen HJ (2014) Chemical mutagens, transposons, and transgenes to interrogate gene function in Drosophila melanogaster. Methods 68:15–28
119. Bischof J, Björklund M, Furger E et al (2013) A versatile platform for creating a comprehensive UAS-ORFeome library in Drosophila. Development 140:2434–2442
120. Pirrotta V (1988) Vectors for P-mediated transformation in Drosophila. Biotechnology 10:437–456
121. Venken KJT, He Y, Hoskins RA et al (2006) P[acman]: a BAC transgenic platform for targeted insertion of large DNA fragments in D. melanogaster. Science 314:1747–1751
122. Sarov M, Barz C, Jambor H et al (2016) A genome-wide resource for the analysis of protein localisation in Drosophila. eLife 5:e12068
123. Adams MD, Celniker SE, Holt RA et al (2000) The genome sequence of Drosophila melanogaster. Science 287:2185–2195
124. Venken KJT, Carlson JW, Schulze KL et al (2009) Versatile P[acman] BAC libraries for transgenesis studies in Drosophila melanogaster. Nat Methods 6:431–434
125. Salazar JL, Yang S.-A, Lin YQ et al (2021) Alzheimer's disease-associated TM2D genes regulate Notch signaling and neuronal function in Drosophila. PLoS Genet 17:e1009962. https://doi.org/10.1371/journal.pgen.1009962. https://pubmed.ncbi.nlm.nih.gov/34905536/
126. Muskavitch MA (1994) Delta-notch signaling and Drosophila cell fate choice. Dev Biol 166:415–430
127. Lehmann R, Dietrich U, Jiménez F et al (1981) Mutations of early neurogenesis inDrosophila. Wilhelm Rouxs Arch Dev Biol 190:226–229

Chapter 20

Bioinformatics Tools to Understand Notch

Ashley Avila, Roxana Gonzalez Tascon, and Dongyu Jia

Abstract

As a result of the culmination of data, and the fast-paced advancement of new research, all the biological information collected can make it difficult to sort data. This is oftentimes experienced when learning about the human genome. Fortunately, with the advancement of technology, the field of bioinformatics has emerged which has allowed for the creation of a variety of biological databases. These biological databases provide a condensed reservoir of organized information that is easy to use and topic-specific. Here, we provide a list of 39 biological databases that help break down the fundamental details of a gene. This chapter uses the NOTCH1 gene as an example to demonstrate how biological databases can be used to extract gene information. Five sections were created to highlight the major areas needed to build a comprehensive foundation of NOTCH1. The first section lists databases containing basic gene and protein product information. The next section consists of protein interactions and signaling pathway databases which are essential in understanding the biological processes a gene product is involved in. Gene expression and disease databases are the next two sections which are connected since disease results from the aberrant expression of a gene product. The last database section examines model organisms which serve a key role in the study of human genetic diseases. Using these databases, we can elucidate NOTCH1's gene/protein structure, expression, and vital physiological function through the Notch signaling pathway.

Key words Notch signaling, Bioinformatics tools

1 Introduction

The Notch signaling pathway is well known for being a highly conserved pathway that regulates cell–cell interactions in which a ligand (Delta-like or Jagged) of a signaling cell binds to a receptor (NOTCH1-4) of a neighboring cell. The composition of the receptors found in this signaling pathway consists of three domains: extracellular, transmembrane, and intracellular. Among them, the extracellular and intracellular domains are the most studied ones. The Notch extracellular domain (NECD) of the receptor contains

Supplementary Information The online version contains supplementary material available at [https://doi.org/10.1007/978-1-0716-2201-8_20].

Dongyu Jia (ed.), *Notch Signaling Research: Methods and Protocols*,
Methods in Molecular Biology, vol. 2472, https://doi.org/10.1007/978-1-0716-2201-8_20,

N-terminal epidermal growth factor-like (EGF) repeats, Lin12/Notch repeats (LNRs), and a heterodimerization domain. The Lin12/Notch repeats (LNRs), and heterodimerization domain are referred to as the negative regulatory region (NRR). Meanwhile, a protein-binding RAM region can be found in the Notch intracellular domain (NICD) of the receptor alongside ankyrin repeats, and a PEST domain [1].

These transmembrane receptors are each encoded by different genes. Notch signaling genes (NOTCH1-4) provide the instructions needed for the making of their corresponding receptor proteins. The pathway becomes activated when the ligand binds to its specific binding site on the Notch receptor. Once activated, a series of proteolytic cleavages occur, one of them resulting in the cleavage of the NICD of the receptor inside the cell. The cleaved section translocates to the nucleus where it can activate the transcription of downstream target genes [2].

Notch signaling plays important roles in many biological processes. This signaling pathway has been observed to play a factor in cell differentiation involved in processes such as embryonic development, T-cell commitment, and cancer development. It has also been associated with controlling cell homeostasis, survival, and proliferation [3]. However, Notch signaling can sometimes go wrong and lead to different types of cancers and diseases. Dysfunction in the pathway can be caused due to mutations in the signaling genes. Cell differentiation can be affected and can cause undifferentiated cells to become more susceptible to malignant transformations [4].

Ongoing research of the Notch signaling pathway is very important. More research is needed as there is still information about this pathway that is not yet well understood. Being able to have a clearer understanding of this pathway can be very beneficial to the field of medicine. Some studies have shown that Notch signaling might cross-talk with other oncogenic signaling pathways, and this could be associated with tumor aggressiveness. Learning more information on how these interactions occur and what can result from them is important as it can provide more insight into the origins of some aggressive tumors [4].

High expression levels of some Notch receptors and ligands have been associated with low cancer patient survival. Continued research of this pathway can allow these Notch genes to be associated with multiple types of cancers and be used as prognostic markers. It could greatly help in combating drug resistance in cancer therapies and recurrence of tumors. This could allow for an increase in cancer patient survival. There is also some complexity associated with Notch signaling such as how it can perform as a tumor suppressor or promoter depending on the type of cell and tissue context it is found in. This is yet another factor in which more research is greatly beneficial for its understanding [4].

The human genome has been studied throughout the years, and new advances continue to emerge as more studies are conducted and more information is obtained. Bioinformatics has been used for over 50 years and continues to grow in the world of science. It has been presented to be very useful in genomics. It allows for the biological data collected throughout the years to be compared and analyzed to aid in the understanding of many diseases. Here, we provide a list of 39 biological databases that are helpful in breaking down the fundamental details of genes by using the Notch gene NOTCH1 as an example.

2 Materials

2.1 Basic Gene and Protein Information

2.1.1 Basic Gene Information

1. GeneCards https://www.genecards.org/
2. Gene-NCBI https://www.ncbi.nlm.nih.gov/gene/
3. UCSC Genome Browser http://genome.ucsc.edu/

2.1.2 Basic Protein Information

1. Uniprot https://www.uniprot.org/
2. NCBI Protein Blast https://blast.ncbi.nlm.nih.gov/Blast.cgi/
3. PhosphoSitePlus https://www.phosphosite.org/homeAction
4. ProteomicsDB https://www.proteomicsdb.org/
5. RCSB Protein Data Bank https://www.rcsb.org/
6. AlphaFold Protein Structure Database https://alphafold.ebi.ac.uk/

2.2 Gene Expression Patterns in Cells, Tissues, and Cancers

2.2.1 Gene Expression in Cells and Tissues

1. The Human Protein Atlas https://www.proteinatlas.org/
2. Genotype-Tissue Expression (GTEx) Portal https://gtexportal.org/home/
3. Gene Expression Omnibus (GEO) https://www.ncbi.nlm.nih.gov/geo/

2.2.2 Gene Expression in Cancers

1. Gene Expression Profiling Interactive Analysis 2 (GEPIA2) http://gepia2.cancer-pku.cn/
2. Gene Expression database of Normal and Tumor tissues 2 (GENT2) http://gent2.appex.kr/gent2/
3. Kaplan-Meier Plotter http://kmplot.com/analysis/
4. Dependency Map (DepMap) Portal https://depmap.org/portal/
5. Prediction of Clinical Outcomes from Genomic Profiles (PRECOG) https://precog.stanford.edu/

2.3 Protein Interactions and Signaling Pathways

2.3.1 Protein Interactions

1. STRING https://string-db.org/
2. BioGRID https://thebiogrid.org/
3. The SIGnaling Network Open Resource (SIGNOR) https://signor.uniroma2.it/
4. InnateDB https://www.innatedb.com/.

2.3.2 Signaling Pathways

1. Reactome https://reactome.org/
2. NDEx Integrated Query http://iquery.ndexbio.org/
3. Kyoto Encyclopedia of Genes and Genomes (KEGG) https://www.genome.jp/kegg/
4. SABiosciences pathway analysis (Supplementary Table 1).

2.4 Gene-Disease Associations

2.4.1 Gene-Common Human Diseases Associations

1. Online Mendelian Inheritance in Man (OMIM) https://www.omim.org/
2. DisGeNET https://www.disgenet.org/search
3. ClinVar https://www.ncbi.nlm.nih.gov/clinvar
4. Model organism Aggregated Resources for Rare Variant ExpLoration (MARRVEL) http://marrvel.org/

2.4.2 Gene-Cancer Associations

1. cBioPortal https://www.cbioportal.org/
2. ImmunoBase https://genetics.opentargets.org/immunobase

2.5 Model Organism Databases

2.5.1 Collection of Model Organisms

1. Alliance of Genome Resources https://www.alliancegenome.org/
2. Ensembl https://useast.ensembl.org/index.html

2.5.2 Organism Specific

1. FlyBase https://flybase.org/
2. Rat Genome Database (RGD) https://rgd.mcw.edu/rgdweb/homepage/
3. The Zebrafish Information Network (ZFIN) https://zfin.org/
4. Mouse Genome Informatics (MGI) http://www.informatics.jax.org/
5. WormBase https://wormbase.org/
6. Saccharomyces Genome Database (SGD) https://www.yeastgenome.org/

3 Methods

3.1 Basic Information over NOTCH1 Gene and Protein

3.1.1 NOTCH1 Gene

1. GeneCards https://www.genecards.org/

   Description: A database that provides a variety of detailed genomics data regarding human genes. Important information noted from this database is DNA regulatory sequences such as promoters and enhancers of a gene. GeneCards also contains a section related to gene products for research experiments.

   Step 1. Enter **NOTCH1** into the **Explore a Gene** search bar and click **GO**.

   Step 2. Select the **Genomics** section.

   Step 3. In this section you will find information regarding **promoters** and **enhancers** for NOTCH1.

   Step 4. Click on the **Products** section to view a wide assortment of **NOTCH1 research products** from 20 different companies.

2. Gene-NCBI https://www.ncbi.nlm.nih.gov/gene/

   Description: A database with in-depth gene information from a variety of species. Notable information from this database includes gene locus, gene type (ncRNA or protein-coding), exon count, gene isoforms, accession numbers, and FASTA sequences.

   Step 1. Input **NOTCH1** into the search bar and click **Search**.

   Step 2. Select **NOTCH1—notch receptor 1** and the species selected is *Homo sapiens.*

   Step 3. The **Table of contents** is located on the right side of the page. Go to **Summary** to find information about the **Gene type** of NOTCH1.

   Step 4. Now go to **Genomic context** using the Table of contents to view NOTCH1 chromosomal **location** and **Exon count** for the gene.

   Step 5. In the Table of contents click on **General protein information** to see information regarding **mRNA** and **protein isoforms** produced by a gene. NOTCH1 only has one curated mRNA and protein isoform listed.

   Step 6. Click on the protein reference sequence, **NP_060087.3,** for NOTCH1 to view the **accession** number and **FASTA** sequence which can be used for **NCBI Protein BLAST**.

3. UCSC Genome Browser http://genome.ucsc.edu/

   Description: An extremely popular browser tool for quickly viewing, analyzing, and downloading a target portion of a genome with aligned annotation available. The step-by-step instruction of utilizing this browser has been previously published [5, 6].

3.1.2 NOTCH1 Protein

1. Uniprot https://www.uniprot.org/
   Description: A database with proteomic data from protein coding genes. Highlighted features of this database include protein topology and 3D structure visualization.

   Step 1. Enter **NOTCH1** into the search bar and click on **Search**.

   Step 2. Select **P46531** which is human NOTCH1 protein.

   Step 3. On the right side of the page click, on the **Subcellular location** tab.

   Step 4. Scroll down to the **Topology** section to view information regarding the Notch1 receptor amino acid length for its extracellular, transmembrane and cytoplasmic sections.

   Step 5. To view **3D structures** of the Notch1 receptor go to **Structure**.

   Step 6. Click on the row of interest. Each row provides information about the **method** used to obtain the structure. **Amino acid positions**, and **resolution**.

2. NCBI Protein Blast https://blast.ncbi.nlm.nih.gov/Blast.cgi/
   Description: A database used to compare the alignment of proteins to find areas of similarity.

   Step 1. Click on **Protein BLAST**.

   Step 2. Select **Align two or more sequences**.

   Step 3. In this example *Homo sapiens* NOTCH1 and *Drosophila melanogaster* Notch isoform A proteins will be compared. Using **Gene-NCBI, find** the **accession number,** or **FASTA sequence** of these proteins.

   Step 4. Input **accession number**, or **FASTA sequence** of one organism into the **Query Sequence** and the other organism into the **Subject Sequence**.

   Step 5. Select **BLAST**.

   Step 6. In the **Descriptions** tab will be alignment results such as **Query Cover**, **E value**, and **Percent Identity**.

   Step 7. Select the **Graphic Summary** tab to see a visual alignment of the two proteins with **Alignment Scores**.

3. PhosphoSitePlus https://www.phosphosite.org/homeAction
   Description: A database with information about posttranslational modifications (PTMs) and the upstream and downstream regulation involved. This database also provides a detailed list of the biological, and molecular functions a protein is involved in.

Step 1. Enter **NOTCH1** into the search bar and click on **Search**.

Step 2. Select **Notch 1** under **PROTEIN**.

Step 3. On the **Protein Information** tab will be a detailed list of the **molecular functions** and **biological processes** that Notch1 receptor protein is involved in.

Step 4. Select the **Site Table** tab to see a detailed list of the **PTMs** that are also displayed on the **diagram**.

Step 5. On the sequence diagram, place the cursor over the colored **circles** to see **PTM class**, **amino acid position**, and **residue**.

Step 6. To learn about **kinases** involved in Notch1 receptor protein PTMs, and **regulatory proteins** affected, select the **Upstream** tab. **Treatment** information is also available for certain PTMs.

Step 7. To learn about the **downstream effects** as a result of PTMs on Notch1 receptor protein select the **Downstream** tab.

4. ProteomicsDB https://www.proteomicsdb.org/

Description: A database that displays protein posttranslational modifications (PTMs) such as acetylation, phosphorylation, and ubiquitination. Protein domains are displayed with start and end positions.

Step 1. Under **Tools** select the **image** above Proteins.

Step 2. In the search bar enter **NOTCH1** and click on the search button.

Step 3. Click on **NOTCH1** under **Protein**.

Step 4. In **Sequence Coverage and Domains** will be the Notch1 receptor protein PTMs. Place the cursor over the PTMs colored **squares** to see amino acid **position** and **residue** affected.

Step 5. To learn about **Notch1 receptor protein domains**, place the cursor over the sequence diagram which will display information about the **name** of the domain and the amino acid **start/end** position.

5. RCSB Protein Data Bank https://www.rcsb.org/

Description: A database that provides 3D protein structures for protein-coding genes.

Step 1. Enter **NOTCH1** in the search bar and select **NOTCH1** under **in Gene Name**.

Step 2. Now under **Refinements** on the right side of the page select **Homo sapiens**, and submit by pressing on the **green**

triangle found on the top or bottom of the Refinements section. You can now view Notch1 receptor protein structures from humans (*see* **Note 1**).

6. AlphaFold Protein Structure Database https://alphafold.ebi.ac.uk/

   Description: An AI database recently developed by Deepmind that can accurately predict protein structures down to atomic accuracy [7]. The website provides its open-source code.

   Step 1. Enter **NOTCH1** into the search bar, and select **Search**.

   Step 2. Click on **Neurogenic locus notch homolog protein 1** with the **Source Organism** *Homo sapiens.*

   Step 3. A **3D predicted model** of the Notch1 receptor protein will be displayed.

   Step 4. Protein is able to be viewed from all angles by **clicking** on the image and **moving** the cursor in the desired direction.

   Step 5. Select an area on the protein to see a detailed breakdown to the amino acid level. A **Confidence score** is assigned to each amino acid in the structure.

3.2 NOTCH1 Expression in Cells, Tissues, and Cancer

3.2.1 NOTCH1 Expression in Cells and Tissue

1. The Human Protein Atlas https://www.proteinatlas.org/

   Description: A database with information on gene expression levels in tissue type, cell type, pathology, brain, blood, and cell line.

   Step 1. Enter **NOTCH1** into the search bar and click on **Search**.

   Step 2. Select the row with **NOTCH1** gene.

   Step 3. On the **Summary** page will be a condensed overview of information from the **tissue, cell type, pathology, brain, blood, and cell atlas**.

   Step 4. To view more detailed information about NOTCH1 expression select a **specific atlas** which are all located in the **top right corner**.

2. Genotype-Tissue Expression (GTEx) Portal https://gtexportal.org/home/

   Description: A database with tissue and single cell gene expression levels.

   Step 1. Enter **NOTCH1** into the search bar with the words **By gene ID** and submit search.

   Step 2. On the left side of the page select **Bulk Tissue Expression**.

Step 3. A graph with different types of **tissues** and NOTCH1 expression levels measured by **transcripts per million (TPM)** will be displayed.

Step 4. At the top of the graph in **SUBSET**, select **Sex** to see tissue expression in **males vs. females**.

Step 5. Now go to **Single Cell Expression** to see percent of cells that express NOTCH1 in a **specific cell type** sample size.

3. Gene Expression Omnibus (GEO) https://www.ncbi.nlm.nih.gov/geo/

   Description: A database with gene expression data from curated microarray experiments.

   Step 1. In the search bar use **keywords** to describe the type of experiment you are looking for that you would like to know the NOTCH1 expression levels.

   Step 2. In this example, **Thymocyte** was entered into the search bar, and the smaller **number** result was selected.

   Step 3. On the left side of the page under **Entry type** select **Datasets**.

   Step 4. For this example, **E2A transcription factor deficiency effect on DN2 thymocyte** was selected.

   Step 5. Enter **NOTCH1** into the **Find gene name or symbol** search bar and click on **Go**.

   Step 6. The results page will show **bar graphs** of the NOTCH1 gene expression from the experiments. Click on graph to see values and more details.

3.2.2 NOTCH1 Expression in Cancer

1. Gene Expression Profiling Interactive Analysis 2 (GEPIA2) http://gepia2.cancer-pku.cn/

   Description: A database that displays gene expression levels from human tumor and normal tissue samples. Data is presented using an interactive body map, dot plot, and box plot.

   Step 1. In the **Single Gene Analysis** tab enter **NOTCH1** into the search bar and then click on **GoPIA!**

   Step 2. Place the cursor over the tumor and normal tissue **body maps** to see the **median expression** of NOTCH1.

   Step 3. Scroll down to see **dot plot** and **box plot** which displays NOTCH1 expression by **transcripts per million (TPM)**.

2. Gene Expression database of Normal and Tumor Tissues 2 (GENT2) http://gent2.appex.kr/gent2/

   Description: A newly updated database that displays gene expression levels in normal and cancer tissues using box plots. Survival analysis for gene expression in cancer patients is also available that is displayed using Kaplan-Meier plots.

Step 1. To look at NOTCH1 gene expression **across tissues**, select **Search-Gene profile**.

Step 2. Enter **NOTCH1** into the **Keyword** search bar, and click Search.

Step 3. **Box plots** of NOTCH1 gene expression levels in **normal** and **cancer tissues** will be displayed.

Step 4. To view NOTCH1 expression levels by **cancer tissue** and **subtype**, go to **Search-Subtype profile**.

Step 5. Select **cancer tissue**, **subtype**, and **prognosis**. Enter **NOTCH1** into the **Gene Symbol** search bar and click Search.

Step 6. **Box plots** of NOTCH1 gene expression levels of **cancer tissues** subtypes will be displayed as well as **Kaplan-Meier plots**.

Step 7. To view **NOTCH1 expression levels** in **cancer tissues** and cancer **prognosis**, go to **Meta-survival analysis**.

Step 8. Select **cancer tissue** and type of **prognosis** which are **Progression-free survival**, **Recurrence-free survival**, and **Overall survival**. Enter **NOTCH1** into the **Gene Symbol** search bar and click Search.

Step 9. A **forest plot** is displayed for the survival analysis. Statistical values such as **hazard ratio**, ***p*-value**, and ***z*-value** are found under the plot.

3. Kaplan–Meier Plotter http://kmplot.com/analysis/

   Description: A database with gene expression data on 21 cancer types. Kaplan–Meier plots are available for mRNA, and protein expression data (*see* **Note 2**).

Step 1. If using **KM Plotter for pan-cancer,** enter NOTCH1 into the **Gene symbol** text box. Select a cancer type or select all. Click on **Draw Kaplan-Meier plot**.

Step 2. If using a **KM Plotter for breast/ovarian/lung cancer,** enter NOTCH1 into the **Affy id/Gene symbol** text box. Click on **Draw Kaplan-Meier plot**.

Step 3. To use **KM Plotter for breast protein** enter gene name into **Select gene symbol/Uniprot ID** textbox and select type of **dataset**. Click on **Draw Kaplan-Meier plot**.

4. Dependency Map (DepMap) Portal https://depmap.org/portal/

   Description: A database with information on gene expression levels in cancer cell lines.

Step 1. Input **NOTCH1** into the search bar and click on **NOTCH1 notch receptor 1**.

Step 2. On the **Overview** tab, go to the **Expression & CN** section and look under **Expression 21Q2 Public** to see a graph displaying the median expression of NOTCH1 mRNA from ~1400 cell lines.

Step 3. In the same section click on **View details**.

Step 4. This page displays NOTCH1 mRNA expression from the cancer cell lines using **box plots**, and categorized by cell lineage.

5. Prediction of Clinical Outcomes from Genomic Profiles (PRECOG) https://precog.stanford.edu/

   Description: A database containing survival *Z* scores for genes based on survival outcome and gene expression in cancer patients. Survival curves are also available using Kaplan-Meier plots. A video instruction has been previously provided by our lab [8].

Step 1. Click on **Register** to create an account. Requires an academic email address.

Step 2. **Sign in** to PRECOG.

Step 3. Select **View details** under **Meta-*Z* analysis**.

Step 4. Enter **NOTCH1** into the search bar to see **survival *Z* score** values for 39 different types of cancer and an unweighted meta-*Z* score of all cancers.

Step 5. Select ***Z* score** value under cancer type to see **Kaplan-Meier plots**.

3.3 NOTCH1 Protein Interactions and Signaling Pathways

3.3.1 NOTCH1 Protein Interactions

1. STRING https://string-db.org/

   Description: A database with protein interactions displayed in a string diagram.

Step 1. Click on **Search**.

Step 2. Enter **NOTCH1** in the **Protein Name** search bar, and for **Organism** select ***Homo sapiens***. Click the **Search** button.

Step 3. Press **Continue** if the box **Homo sapiens NOTCH1** is selected.

Step 4. A **string diagram** showing Notch1 receptor protein interactions will be displayed. Select a **node** to learn more about a protein.

2. BioGRID https://thebiogrid.org/

   Description: A database with protein interactions that are displayed with high and low throughput evidence. A string diagram of all of a protein's interactors is also available.

Step 1. Enter **NOTCH1** into the search bar and click **Submit Identifier Search**.

Step 2. On the **Interactors** tab will be displayed the Notch1 receptor protein interactors and amount of **evidence** that supports the association. Evidence is separated by **high and low throughput** studies.

Step 3. Click on the **Network** tab for a **string diagram** of all the Notch1 receptor protein interactions.

3. The SIGnaling Network Open Resource (SIGNOR) https://signor.uniroma2.it/

   Description: A database of protein interactions with an interactive map and evidence scores.

   Step 1. Enter **NOTCH1** in the ***Homo sapiens*** tab and click Search.

   Step 2. Click on **GET RELATIONS** for NOTCH1 result.

   Step 3. A **table** and **interactive map** of Notch1 receptor protein interactions will be displayed.

   Step 4. Go to the table to see the **mechanism of action** through which the Notch1 receptor protein interacts with other proteins. An **evidence score** is assigned to each interaction.

   Step 5. Click the **plus** sign on the table for additional information about the interaction.

4. InnateDB https://www.innatedb.com/

   Description: A database with protein interaction data displaying the interaction type and biomolecule interactors involved.

   Step 1. Under **Search InnateDB** select **Homo sapiens,** and enter **NOTCH1**. Click the search button.

   Step 2. Click on **number** under **Experimentally validated interactions**.

   Step 3. The results page will display NOTCH1 receptor protein **interactors** along with information about the **interaction type**, biomolecule **interactor types**, and in some cases the **tissue** where the interaction was found.

   Step 4. Select **Details** to learn more about a specific interaction.

3.3.2 NOTCH1 Signaling Pathways

1. Reactome https://reactome.org/

   Description: A database with normal and disease gene pathways that can be viewed by cell location. Disease pathways display gene mutants.

   Step 1. Click on **Pathway Browser**.

   Step 2. Enter **NOTCH1** into the search bar and click on the search button.

   Step 3. Results will display a variety of **NOTCH1 variants** with **compartment locations**.

Step 4. Select the result of interest and then go to the **Details** box.

Step 5. Select **GO** next to the desired pathway to see a detailed visual diagram.

2. NDEx Integrated Query http://iquery.ndexbio.org/

   Description: A database with compiled normal and disease gene pathways from other online sources and research papers.

   Step 1. Enter **NOTCH1** into the search bar and click on the search button.

   Step 2. Go to **RELEVANT PATHWAYS** and select a pathway. NOTCH1 will be highlighted **red** in the pathway.

   Step 3. Go to **PATHWAY FIGURES** and select a pathway.

   Step 4. Look under **NETWORK** to find the **Paper Link** to learn more about the pathway in the figure.

3. Kyoto Encyclopedia of Genes and Genomes (KEGG) https://www.genome.jp/kegg/

   Description: A database with normal function and disease gene pathways.

   Step 1. Select **KEGG GENES** under **Data-oriented entry points**.

   Step 2. Enter **NOTCH1** in the **Search hsa for** search bar and click Go.

   Step 3. Click on the number **4851** above **K02599 Notch 1**.

   Step 4. Go to **Pathway** and select **hsa(#)** next to the pathway you would like to see visualized.

   Step 5. **NOTCH1** gene will be highlighted **red** in pathway and use the **Scale** button to zoom in or out.

4. SABiosciences pathway Analysis (Supplementary Table 1)

   Description: QIAGEN previously generated its SABiosciences array gene tables, which included 84 distinct literature-based curated genes that represent each molecular pathway. Many researchers, including us, analyzed the overlap between the list of interested genes and the curated genes that represent 67 different pathways to understand the involved molecular pathways [9, 10]. Here, we provided the complete list of the 67 pathways and their associated 84 genes (Supplementary Table 1).

3.4 Investigating the Gene-Disease Associations in Relation to NOTCH1

3.4.1 NOTCH1 and Common Human Disease Associations

1. OMIM https://www.omim.org/

   Description: Database used to find diseases a gene has been linked to in humans.

   Step 1. Enter **NOTCH1** into the search bar and select the **search** button.

   Step 2. Select **NOTCH1** in the results page.

Step 3. Go to the **Table of Contents** on the left side of the page and select **Gene Phenotype-Relationships**, **Molecular Genetics**, and **Allelic Variants** for information regarding the different types of diseases associated with NOTCH1.

2. DisGeNET https://www.disgenet.org/search

   Description: Database that contains disease-gene associations with assigned statistical scores. Each disease has displayed the number of variants from the gene and the types of mutations with nucleotide change displayed.

   Step 1. On the home page, select **genes**, and then enter **NOTCH1** into the search bar. Press submit.

   Step 2. Scroll down and **four options** will appear: Summary of GDAs, Evidences for GDAs, Summary of VDAs, and Evidences for VDAs.

   Step 3. Select **Summary of GDAs**, and under **Score gda** you will find the **Gene Disease-Association Score** (*see* **Note 3**).

   Step 4. On the same page look under **N. SNPs gda** for the number of variants associated with that disease. Click on the **number** to see specific details about the variants.

   Step 5. Select **Summary of VDAs**, and under **Score vda** you will find the **Variant Disease-Association Score** (*see* **Note 3**).

   Step 6. On the same page look under **DSI v** for the disease specificity index which is a score based on if a NOTCH1 variant is associated with few or many diseases (*see* **Note 4**).

   Step 7. Select **Evidences for GDAs or VDAs** to see PubMed reference number (**PMID**) for the stated NOTCH1 disease-associations.

3. ClinVar https://www.ncbi.nlm.nih.gov/clinvar

   Description: Database with information over different gene variants and their association to disease.

   Step 1. Enter **NOTCH1** into the search bar and click **Search**. On the left side of the page you will find filters that can narrow your results (*see* **Note 5**).

   Step 2. To narrow results to NOTCH1, select **Single gene** under **Variant-gene relationship**.

   Step 3. Then select **Pathogenic** under **Clinical significance** for the strongest disease associations.

4. MARRVEL (Model Organism Aggregated Resources for Rare Variant ExpLoration) http://marrvel.org/

Description: A database that provides a compiled view of the information from four disease databases. A detailed video instruction has been provided by the developers [11].

Step 1. Enter **NOTCH1** into the search bar and click on the gene when it appears as an option. Click on Search.

Step 2. On the results page to the right will be listed **OMIM, ClinVar, Geno2MP, and DECIPHER**.

Step 3. Click on a database to learn about **NOTCH1 variants** and the **diseases/phenotypes** associated.

Step 4. To learn additional information about a database click on **More on (database)**.

3.4.2 NOTCH1 and Cancer

1. cBioPortal https://www.cbioportal.org/

Description: A database that provides a gene's alteration frequency from 32 cancer studies totaling 10,953 patients. Kaplan-Meier survival analysis is available to compare patients in a cancer study that do or do not have a gene alteration. A video instruction has been previously provided by our lab [8].

Step 1. Go to the **Quick Search** tab and enter **NOTCH1** into the search bar. Select **GENE NOTCH1**.

Step 2. On the **Cancer Types Summary** page hover the cursor over a specific bar graph to see NOTCH1 percentage alteration frequencies and types of alterations observed.

Step 3. To view NOTCH1 survival analysis using Kaplan-Meier method hover over the bar graph of the cancer study of interest and select **Query this study for NOTCH1**.

Step 4 Click on **Comparison/Survival** and then select **Survival.**

Step 5. On the left side of the page you have the option to view graphs of **Disease free, Overall, Disease-specific, and Progression Free** survival by clicking on the desired option.

2. ImmunoBase https://genetics.opentargets.org/immunobase

A database that provides information on the roles a gene can play in promoting cancer. Another feature shows different types of cancers with mutations in the selected gene and drug treatment used.

Step 1. In the search bar enter **NOTCH1** and select the gene in the drop down list.

Step 2. Now click on **Target profile overview**.

Step 3. Select **Cancer Biomarkers** on the profile page for a list of cancers and mutations found in NOTCH1. Use the horizontal scroll bar to view the drug treatment used for the type of cancer and mutation.

Step 4. Scroll down or click on **Cancer Hallmarks** on the profile page to view NOTCH1 roles in cancer when normal activity is altered.

3.5 NOTCH1 Human Homologs in Different Organisms

3.5.1 Human NOTCH1 Homolog in Several Organisms

1. Alliance of Genome Resources https://www.alliancegenome.org/

   Description: A database with human gene homolog information on six of the most popular model organisms in genetics.

   Step 1. Type in **NOTCH1** into the search bar and click on the search tab.

   Step 2. Select **Gene** on the results page on the left side (*see* **Note 6**).

2. Ensembl https://useast.ensembl.org/index.html

   Description: A database that provides gene information from 280 species. Displays species with and without human gene homologs.

   Step 1. Select **Human** and then enter **NOTCH1**, press **Go**.

   Step 2. Select **NOTCH1 (Human Gene)**.

   Step 3. Click **Orthologues** on the left side of the page under **Comparative Genomics**.

   Step 4. Scroll down and information will appear under the **Orthologues** section.

3.5.2 Human NOTCH1 Homolog in One Organism

1. FlyBase https://flybase.org/

   Description: A comprehensive genomic database of the model organism *Drosophila melanogaster*.

   Step 1. On the home page, scroll down to **QuickSearch,** and select **Homologs**.

   Step 2. In the Input section select **H. Sapiens**, and enter **NOTCH1**.

   Step 3. In the Output section unselect all boxes by clicking on un/check all, and select only D. melanogaster (Fruit fly), press **Search.**

   Step 4. The results page will display orthologues of NOTCH1 in Drosophila rated with a score based on how similar. Select **FlyBase** for gene **N** (NOTCH) which has the highest score.

   Step 5. On the Drosophila Notch gene information page go to the **Report Sections** on the right and select **Orthologs**.

   Step 6. The **Human Orthologs** section will display the human homologs with the option to do a protein alignment and display if a transgene fly stock has been created.

2. Rat Genome Database (RGD) https://rgd.mcw.edu/rgdweb/homepage/

   Description: A database with extensive biological data over the model organism *Rattus norvegicus.*

   Step 1. Select **Orthologs** on the home page.

   Step 2. In the **Enter Input Details** section select **Human** as the species, and type **NOTCH1** in the **Enter Gene Symbols** text box.

   Step 3. In the **Enter Output details** section select **Rat** as the species and click **submit**.

   Step 4. In the **Ortholog Report** the Rat ortholog of the NOTCH1 gene will be listed.

   Step 5. Click on **NOTCH1** to learn more about the **human gene**, or select **Notch1** to learn more about the **rat gene**.

3. The Zebrafish Information Network (ZFIN) https://zfin.org/

   Description: A database of genomic information regarding the model organism *Danio rerio.*

   Step 1. Use the **Alliance of Genome Resources** database to discover if NOTCH1 has a homolog in *Danio rerio.*

   Step 2. There are two zebrafish homologs in NOTCH1 so enter either **notch1a** or **notch1b** into the **search bar** of ZFIN and click on Search.

   Step 3. Click on **notch1a** or **notch1b** in the results page.

   Step 4. On the right side of the page click on **Orthology** to see information about **homologs** in humans and other species with cited literature evidence.

4. Mouse Genome Informatics (MGI) http://www.informatics.jax.org/

   Description: A database with genomic information from the model organism *Mus musculus.*

   Step 1. Use the **Alliance of Genome Resources** database to discover if NOTCH1 has a homolog in *Mus musculus.*

   Step 2. There is a NOTCH1 human homolog in mice called Notch1. Enter **Notch1** into the search bar on MGI and click on **QuickSearch**.

   Step 3. Select **Notch1** under **Symbol**.

   Step 4. Go to **Homology** and click on **more**.

   Step 5. Click on **Notch1 stringent orthology** to see human and other species gene **homologs.**

5. WormBase https://wormbase.org/

   Description: A database containing genomic information over several species of nematode. The most studied species is *Caenorhabditis elegans.*

   Step 1. Enter **NOTCH1** into the **Search directory**, and select **Notch1** ***Homo sapiens.***

   Step 2. On the left side of the page under **Page Content** select **Homology**.

   Step 3. In the **Homology** section will be the species of nematode and NOTCH1 orthologs. **Click** on **lin-12** or **glp-1** to learn more about these genes.

6. Saccharomyces Genome Database (SGD) https://www.yeastgenome.org/

   Description: A database with gene information from the model organism *Saccharomyces cerevisiae.*

   Step 1. In the top right corner of the web page select **YeastMine**.

   Step 2. Select the **Homology** tab.

   Step 3. Click on **Human Gene** → **Yeast Homolog(s)** → **OMIM Disease Phenotype**.

   Step 4. Enter **NOTCH1** into the search bar, and select **Show Results**. NOTCH1 does not have a yeast homolog and displays **No Results**.

   Step 5. If the human gene did contain a yeast homolog then the gene in *Saccharomyces cerevisiae* would be found under **Homologue.**

4 Notes

1. This database does not contain a complete protein structure of NOTCH1. Structures available are fragments of Notch1 receptor which may also be bonded to other macromolecules or ligands. Mutated versions of Notch1 receptor fragments are also available.
2. The Kaplan-Meier Plotter database has a wide variety of parameters, but for this example only gene name and cancer type will be discussed.
3. A score closer to 1 indicates a higher level of evidence between NOTCH1 and the disease.
4. A score closer to 1 indicates the variant is more disease specific meaning it is associated with fewer diseases.

5. Type of filters include: origin of allele, clinical significance, mutation type, number of nucleotide base pairs, method data was collected, molecular consequence, and review status.
6. Database also contains an allele tab with known allele variants and a model tab containing phenotype descriptions of the effects seen from these allele variants.

Acknowledgments

We only listed 39 biological databases in our chapter, but there are far more databases available for researchers. For instance, Wikipedia has a long list of biological databases (https://en.wikipedia.org/wiki/List_of_biological_databases). We apologize for those not included. This work was supported by the Start-Up funding, Faculty Research Committee (FRC) Award, College of Science and Mathematics Research Grant Award, and Undergraduate Research Assistant Funding from Georgia Southern University and the Georgia Southern Biology Undergraduate Research Grant. AA was also supported by Chandler Scholarship and The College Office of Undergraduate Research at Georgia Southern University.

References

1. Aster JC, Pear WS, Blacklow SC (2017) The varied roles of Notch in cancer. Annu Rev Pathol 12:245–275. https://doi.org/10.1146/annurev-pathol-052016-100127
2. Hori K, Sen A, Artavanis-Tsakonas S (2013) Notch signaling at a glance. J Cell Sci 126 (pt 10):2135–2140. https://doi.org/10.1242/jcs.127308
3. Terragni J, Zhang G, Sun Z, Pradhan S, Song L, Crawford GE, Lacey M, Ehrlich M (2014) Notch signaling genes: myogenic DNA hypomethylation and 5-hydroxymethylcytosine. Epigenetics 9(6): 842–850. https://doi.org/10.4161/epi.28597
4. Wang Z, Li Y, Ahmad A, Azmi AS, Banerjee S, Kong D, Sarkar FH (2010) Targeting Notch signaling pathway to overcome drug resistance for cancer therapy. Biochim Biophys Acta 1806(2):258–267. https://doi.org/10.1016/j.bbcan.2010.06.001
5. Karolchik D, Hinrichs AS, Kent WJ (2007) The UCSC genome browser. In: Current protocols in bioinformatics, chapter 1:unit 1 4. https://doi.org/10.1002/0471250953.bi0104s17
6. Zweig AS, Karolchik D, Kuhn RM, Haussler D, Kent WJ (2008) UCSC genome browser tutorial. Genomics 92(2):75–84. https://doi.org/10.1016/j.ygeno.2008.02.003
7. Jumper J, Evans R, Pritzel A, Green T, Figurnov M, Ronneberger O, Tunyasuvunakool K, Bates R, Zidek A, Potapenko A, Bridgland A, Meyer C, Kohl SAA, Ballard AJ, Cowie A, Romera-Paredes B, Nikolov S, Jain R, Adler J, Back T, Petersen S, Reiman D, Clancy E, Zielinski M, Steinegger M, Pacholska M, Berghammer T, Bodenstein S, Silver D, Vinyals O, Senior AW, Kavukcuoglu K, Kohli P, Hassabis D (2021) Highly accurate protein structure prediction with AlphaFold. Nature 596(7873):583–589. https://doi.org/10.1038/s41586-021-03819-2
8. Defreitas S, Rowe M, Paculis L, Jia D (2020) Integration of bioinformatics approaches and experimental validations to understand the role of Notch signaling in ovarian cancer. J Vis Exp 155. https://doi.org/10.3791/60502
9. Jia D, Liu Z, Deng N, Tan TZ, Huang RY, Taylor-Harding B, Cheon DJ, Lawrenson K, Wiedemeyer WR, Walts AE, Karlan BY, Orsulic

S (2016) A COL11A1-correlated pan-cancer gene signature of activated fibroblasts for the prioritization of therapeutic targets. Cancer Lett 382(2):203–214. https://doi.org/10.1016/j.canlet.2016.09.001

10. Yu Z, Xu Q, Wang G, Rowe M, Driskell C, Xie Q, Wu M, Jia D (2019) DNA topoisomerase II alpha and RAD21 cohesin complex component are predicted as potential therapeutic targets in bladder cancer. Oncol Lett 18(1): 518–528. https://doi.org/10.3892/ol.2019.10365

11. Wang J, Liu Z, Bellen HJ, Yamamoto S (2019) Navigating MARRVEL, a web-based tool that integrates human genomics and model organism genetics information. J Vis Exp 150. https://doi.org/10.3791/59542

Index

Dongyu Jia (ed.), *Notch Signaling Research: Methods and Protocols*,
Methods in Molecular Biology, vol. 2472, https://doi.org/10.1007/978-1-0716-2201-8,

MIX
Papier aus verantwortungsvollen Quellen
Paper from responsible sources
FSC® C105338

If you have any concerns about our products,
you can contact us on
ProductSafety@springernature.com

In case Publisher is established outside the EU,
the EU authorized representative is:
Springer Nature Customer Service Center GmbH
Europaplatz 3, 69115 Heidelberg, Germany

Printed by Libri Plureos GmbH
in Hamburg, Germany